CLASSICAL AND QUANTUM BLACK HOLES

Series in High Energy Physics, Cosmology and Gravitation

CLASSICAL AND QUANTUM BLACK HOLES

Edited by

Pietro Fré

Department of Physics,
University of Turin

Vittorio Gorini

Department of Chemical, Mathematical and Physical Sciences,
University of Insubria at Como

Giulio Magli

Department of Mathematics,
Politecnico of Milan

and

Ugo Moschella

Department of Chemical, Mathematical and Physical Sciences,
University of Insubria at Como

CRC Press
Taylor & Francis Group
Boca Raton London New York

CRC Press is an imprint of the
Taylor & Francis Group, an **informa** business

CRC Press
Taylor & Francis Group
6000 Broken Sound Parkway NW, Suite 300
Boca Raton, FL 33487-2742

First issued in paperback 2019

© 1999 by Taylor & Francis Group, LLC
CRC Press is an imprint of Taylor & Francis Group, an Informa business

No claim to original U.S. Government works

ISBN-13: 978-0-7503-0627-0 (hbk)
ISBN-13: 978-0-367-39947-4 (pbk)

British Library Cataloguing-in-Publication Data

A catalogue record for this book is available from the British Library.

Library of Congress Cataloging-in-Publication Data are available

**Visit the Taylor & Francis Web site at
http://www.taylorandfrancis.com**

**and the CRC Press Web site at
http://www.crcpress.com**

Contents

PART 3

BPS black holes in supergravity:
Duality groups, p-branes, central charges and entropy
Riccardo D'Auria and Pietro Fré   **137**

Preface

Black holes are among the most fascinating objects the human mind has been capable of imagining, and as such represent one of the highest levels of human thought. They occupy a central place in contemporary research, and this is true for very different areas, ranging from observational astrophysics through theoretical physics to pure mathematics. During the spring of 1998, SIGRAV—Società Italiana di Relatività e Gravitazione (Italian Society of Relativity and Gravitation) promoted the organization of a doctoral school on *The Physics of Black Holes*, which took place at the Alessandro Volta Centre for Scientific Culture, located in the beautiful environment of Villa Olmo in Como, Italy. This book brings together the five courses which were given there by distinguished scientists, covering many different aspects of this rich and vast subject such as black hole thermodynamics and black holes in string theory, in supergravity, in astrophysics and on the computer. An introduction endeavours to uncover the *fil rouge* linking all these courses.

The school was made possible by the financial support of the Physics Department of the University of Milan and of the Institute of Chemical, Physical and Mathematical Sciences of the Como Section of the University of Milan, as well as by funds from an ASI research project coordinated by Professor I Ciufolini.

We are grateful to all the members of the scientific organizing committee and to the scientific coordinator at Centro Volta, Professor G Casati, for their invaluable help in the organization. We also acknowledge the essential organizational support of the secretarial conference staff of Centro Volta, Maria Giovanna Falasconi, Chiara Stefanetti and Nadia Tansini.

P Fré, V Gorini, G Magli and U Moschella
16 March 1999

Chapter 1

The physics of black holes (an overview)

Vittorio Gorini[1], Giulio Magli[2] and Ugo Moschella[1]
[1] Department of Chemical, Physical and Mathematical Sciences,
University of Insubria, Via Lucini 3, 22100 Como, Italy
[2] Department of Mathematics, Politecnico di Milano,
Piazza Leonardo da Vinci, 20133 Milan, Italy

The history of black holes goes back to far earlier than the development of general relativity. In the late 18th century, John Michell (1784) and Pierre Simon de Laplace (1796) observed independently that light (at that time, according to Newton's theory, thought to be made of corpuscles) emitted radially from a spherical body of mass M and radius R would eventually fall back toward the body if $r < 2GM/c^2$. Therefore, such an object would be invisible to a far away observer.

This idea was abandoned and left dormant after the taking over of the wave theory of light, following the two-slit interference experiment by Young (1801).

Its revival came when, shortly after the formulation by Einstein of the theory of general relativity in 1915, Karl Schwarzschild discovered the exact solution of the gravitational field equations outside a spherically symmetric body (star). This solution did indeed exhibit a 'singularity' at the critical radius of Michell and Laplace $r = R_0 = 2GM/c^2$. To wit, it was recognized that, due to the frequency shift in a gravitational field, any light emitted outward at the critical distance would undergo an infinite red shift and would thereby be unable to reach any outside observer: a 'star' with a radius smaller than R_0 would 'appear dark'.

However, in the years following the advent of general relativity, and until very much later, almost no one, including Einstein and Eddington in the first place, was willing to believe that anything so outrageous could exist anywhere in the universe. Moreover there were arguments for this which appeared to be convincing. To mention one: shortly after he had found his vacuum solution, Schwarzschild was able to match it to another exact solution describing the gravitational field inside the star, under the simplifying assumption of constant star density. The resulting pressure would depend on the radius R of the body and, for a given R, would grow as one approaches the star centre. The central pressure would remain finite

1

as long as $R > (9/8)R_0$, but it would become infinite for $R = (9/8)R_0$. This was taken as an indication that no celestial body with such a radius, or smaller, could exist, since it would require an infinite pressure of the matter inside the star to counterbalance the inward pull of gravity. This sounded plausible to most, since people were accustomed to think of a star as a body in equilibrium, its inward pull of gravity being balanced by its internal pressure. No truly dynamical situation such as contraction or collapse was considered.

On the other hand, the problem of what happened to a star after it had exhausted its internal supply of energy production was a central issue in the 1920s. This matter was linked to the issue of explaining the stability of white dwarfs, stars whose masses were of the order of the Sun's mass, but whose radii were more than an order of magnitude smaller (although still safely off Schwarzschild's gravitational radius), thus leading to densities as high as millions of grams per cubic centimetre. The answer to this problem was essentially provided by Fowler in 1926 who proved that in such objects gravity was balanced by the degeneracy pressure of free electrons. The issue was taken up again by Chandrasekhar who, in 1931, also showed that, due to the softening of the equation of state for electrons at relativistic energies, there was a limit of about 1.4 solar masses for the mass of a stable white dwarf (*the Chandrasekhar limit*), beyond which the degeneracy pressure would be unable to balance further contraction produced by the inward pull of gravity.

But what if a star, contracting after its nuclear fuel had exhausted, did not eject enough matter-radiation to reduce its mass below the Chandrasekhar limit? In this case, further contraction was expected and the problem arose of whether the body may settle into another stable equilibrium configuration with a smaller radius. This led eventually to the concept of the neutron star.

The possibility that normal stars would develop cores at their centres with nuclear densities, responsible for the production of the star's heat, was suggested by Landau as early as 1932, just before the discovery of the neutron. This suggestion can in a way be considered the precursor, though in a different setting and with different purposes, of the concept of the neutron star. This concept was pioneered by Fritz Zwicky in the first half of the 1930s to explain the origin of supernovae and cosmic rays. He suggested that the implosion of a massive star at the end of its thermonuclear evolution might give rise to a central core of extremely packed neutrons, the energy released in the core's implosion being responsible for the blowing off of the outer layers of the star and thus the supernova phenomenon.

Just as in white dwarfs gravity is balanced by the electron degeneracy pressure, a neutron star would be supported against its own gravity by the pressure of degenerate neutrons. The first systematic treatment of this issue, making full use of general relativity, was carried out by Oppenheimer and Volkoff, and by Tolman, in 1939, who also sought for an upper limit to the mass of a neutron star, analogous to the Chandrasekhar limit for white dwarfs. The computation of this limit would have required a knowledge of the equation of state at nuclear matter densities (of the order of 10^{15} grams per cubic centimetre), which in turn implied a detailed knowledge of the nuclear force. But even without such a detailed knowledge,

Oppenheimer, Volkoff and Tolman were able to conclude that there is a maximum limit to the mass of a neutron star, lying between a fraction of and several solar masses.

Thus a neutron star is a kind of giant nucleus, having a mass of the order of the mass of the sun and made almost entirely of neutrons, with a few residual protons and electrons: in the course of the core formation in a supernova explosion electrons are absorbed by protons via inverse beta decay, yielding neutrons and flying off neutrinos.

This was all hypothetical in the late 1930s, and neutron stars had to wait until 1967 to be discovered by chance as pulsars. These objects are indeed firmly interpreted today as rapidly spinning neutron stars. What is important to note here is that the radius of a typical neutron star is of the order of several kilometres, thus not much larger than the gravitational radius (recall that the gravitational radius of the sun is a little less than three kilometres). As to the upper limit for the mass of such objects, there is still some uncertainty today connected with the incomplete knowledge of the nuclear force, but a reasonable estimate is quite safely believed to be around three solar masses (for details see Treves and Haardt, this volume).

No physical process is known which can produce a pressure able to balance the gravitational pull in stars exceeding the neutron star mass limit. We are thus led to admit that in the process of implosion either the star is able to radiate away a sufficient amount of mass–energy to settle below the neutron star limit (as it is likely to happen in most supernova explosions) or that the contraction continues up to the formation of a status of infinite density, i.e., a singularity. It is at this point that black holes came into play as a possible solution of the puzzle, from the work by Oppenheimer and Snyder (1939). They were, indeed, able to construct a solution of the Einstein field equations composed by a collapsing homogeneous sphere of dust matched with the Schwarzschild vacuum. The continued gravitational contraction of this (very idealized) model of a 'star' led to the formation of a singularity causally disconnected from far away observers due to the formation of an event horizon. Indeed, as seen by an observer far away from the collapsing cloud, the contraction slows down and 'freezes' exactly at $r = R_0$. Due to gravitational time dilation, the process requires an infinite amount of time as measured by the far away observer. Correspondingly, any hypothetical electromagnetic information sent off from the surface of the collapsing object to the outside observer becomes increasingly redshifted, and the redshift tends to infinity as the surface approaches the horizon. Equivalently, this can in turn be reinterpreted as the indication that the escape velocity of a massive object fired off radially from the collapsing surface tends to the speed of light as the surface closes on the horizon (recall Michell and Laplace). On the other hand, for an observer riding on the surface of the star, the collapse goes right through the critical radius and reaches the singularity at $r = 0$ in a finite amount of proper time. Once the matter has crossed the critical radius there is no way for it to come out any more, since any time- or lightlike future oriented curve starting at any point with $r < R_0$ ends up at the (spacelike) singularity $r = 0$. In other words, any $r = $ const hypersurface is a *trapped surface*.

Further progress was halted because of World War II and the subsequent cold war until the late 1950s and early 1960s, when computer simulations of realistic spherical collapses taking into account pressure, nuclear reactions, radiation, mass ejection etc proved that implosion of a sufficiently large mass would produce a black hole, as in the Oppenheimer–Snyder model. This eventually convinced relativists and astrophysicists alike, and John Wheeler in the first place, that black holes could indeed be formed in the collapse of sufficiently massive stars at the end of their lives (the term 'black hole' was actually coined by Wheeler very late (1967)). Theoretical research on black holes flared up at an almost frenzied pace and a sequel of new discoveries took place: the 'golden age' of black hole classical physics, which would last until the mid-1970s, had started.

First of all, due to the work of Kruskal and, independently, of Szekeres, the mathematical structure of the Schwarzschild singularity was definitively understood through the discovery of the maximal analytical extension of the Schwarzschild spacetime. In the same years, Roy Kerr was investigating a class of algebraically special vacuum spacetimes which admitted a preferred null direction. He found out that, among the possible solutions, there was one representing the gravitational field of a stationary, rotating object (Kerr 1963). The multipole structure of this solution is very special since it is uniquely characterized by only two parameters, the mass M and angular momentum per unit mass a (if a vanishes one recovers the Schwarzschild solution).

Immediately thereafter the causal structure of the Kerr spacetime was discovered. For $a < M$, three 'subsequent' hypersurfaces contain the Kerr singularity. The external surface, called the static limit, can be crossed in both senses. However an observer who crosses the static limit cannot remain at rest with respect to a far away colleague since, due to the relativistic dragging of inertial frames, it is obliged to rotate in the same direction as the black hole. The second surface is an event horizon, i.e. a one-way membrane very much like the Schwarzschild one. The region in between the static limit and the horizon is called the ergosphere. As noted by Penrose (1969), the presence of this region allows for energy extraction from a spinning black hole (this is the reason for the name 'ergosphere'). In particular, for example, the Penrose process may be at least partly responsible for the relativistic jets seen in many galactic nuclei (see Kiefer, this volume). In contrast, no energy can be extracted from a Schwarzschild black hole by purely classical (i.e. non-quantum) processes. The third surface is called a Cauchy horizon. The reason for this name is that, due to the timelike nature of the Kerr singularity, beyond it predictability is lost. It has the peculiar property of being a surface of infinite blueshift: an observer crossing it would see the whole of the past of his universe in an infinitesimally small amount of time. The Kerr singularity is at best visualized as a 'ring' of radius a in the 'xy-plane' of a suitably chosen Cartesian system of coordinates.

It is worthwhile to recall here that the relativistic dragging of inertial frames in the vicinity of a rotating body (Lense–Thirring effect) becomes particularly prominent close to the horizon of a Kerr black hole. Indeed, as noted above, the extreme manifestation of this effect takes place within the ergosphere, where no test body can remain static, being inevitably dragged around by the spinning of the hole. In particular, an accretion disc around a spinning black hole (or a neutron star), tilted relative to the equatorial plane of the spinning body, has to precess in a prograde direction along the spin axis, due to the dragging effect (see Treves and Haardt, this volume, for a recent possible detection of this effect in neutron stars, and prospects of such a detection for black holes as well).

The Schwarzschild solution is the unique spherically symmetric vacuum solution of the Einstein field equations (Birkhoff theorem). In principle, one should expect the stationary axisymmetric black hole solutions to be a large family, since no uniqueness theorem such as Birkhoff's holds for the vacuum spacetime geometry outside an axisymmetric star. It is, therefore, really a remarkable fact that such a uniqueness theorem (conjectured by Carter and proved by Robinson) turns out to hold in the case of black hole solutions.

Let us state the Kerr uniqueness theorem in the traditional way (Chandrasekhar 1983): all stationary, vacuum solutions of the Einstein field equations which are asymptotically flat and contain a regular horizon with no singularities outside the horizon are given by a two-parameter family. Since we know one such family (the Kerr one) this family is unique ('a black hole has no hair', Wheeler). Therefore, by virtue of this 'no hair theorem', the only memory of the nature, structure and composition of any object which collapses to form a stationary black hole is embodied in the mass and the angular momentum of the hole, any residual hair being quickly radiated away in the collapse process. The very same effect of 'hair-cutting' has to be expected in dynamical processes involving black holes, like merging of two of them (see below).

The 'no hair' theorem can be extended to the case of an electrically charged black hole, meaning that a charged black hole is uniquely characterized by its mass, angular momentum per unit mass and charge. Historically, the spherically symmetric charged black hole was discovered in 1918 by Reissner and Nördstrom, while the charged generalization of the Kerr black hole was found by Newman shortly after Kerr's discovery. Of course, one does not expect charged black holes to have any astrophysical significance, since any residual charge resulting from the collapse would be very quickly neutralized by the surrounding medium. However, charged spherically symmetric black holes play a relevant role at the theoretical level since their causal structure is very similar to that of the uncharged Kerr holes, while their mathematical structure is much simpler. This is particularly relevant in the studies on the stability of the Cauchy horizon (see below).

Recently, the no hair theorem has been extended to the presence of non-Abelian charges, namely to the black hole solutions of the Einstein–Yang–Mills equations. However, these solutions seem to be unstable (see e.g the relevant contributions in Burko and Ori 1997).

The causal structure of the Kerr black hole changes completely, turning into a naked singularity (i.e. a singularity visible to far away observers), if the parameter a is greater than M (in the very special case in which $a = M$ only one horizon survives; this case is usually referred to as the *extremal* Kerr solution).

The fact that the Kerr family contains naked singularities leads us to the issue of *cosmic censorship*. This terminology was initiated by R Penrose who wrote: *Does there exist a cosmic censor who forbids the occurrence of naked singularities, clothing each one in an absolute event horizon?* Indeed, while there are solid mathematical proofs that spacetime singularities actually form at the end state of complete gravitational collapse under many circumstances ('singularity theorems', see Senovilla 1998), no available proof exists that such singularities are not visible to any observers (strongly censored singularities, like the Schwarzschild one) or at least not visible to far away observers (weakly censored singularities, hidden by an event horizon). As a matter of fact the universe could be populated by naked singularities. This prospect is, in a way, enticing. Indeed, if naked singularities could (classically) form in actual, highly asymmetric collapses, the end points of such collapses could allow for a direct observation of quantum gravitational effects. A detailed discussion of the issue of naked singularities is far from the objectives of this introduction, therefore we refer the reader to recent reviews in this field (Joshi 1996, Wald 1997). We would like only to stress that the cosmic censorship conjecture is still a completely open issue. For instance, the singularity occurring at the Kerr ring is always *locally* naked (i.e. visible to an observer who has crossed the horizon). There are, however, some hints of an instability of the Cauchy horizon, which should eventually turn into a censored singularity eliminating the 'unphysical' (for those who believe in censorship) sector of the Kerr spacetime.

In the mid-1960s, as soon as people became convinced that black holes could form in gravitational collapse, there arose the issue of how to search for actual black holes in the galaxy and possibly elsewhere in the universe.

Following some earlier partial suggestions by Zel'dovich and, independently, by Salpeter, an idea that proved to be viable was put forward by Zel'dovich and Novikov in 1966.

Suppose one has a binary system whose components are sufficiently close (of the order of 10^6 km say), one companion being a normal star and the other a compact object. Due to the relative orbital motion of the two bodies, and except in the unfortunate case in which the orbital plane is perpendicular to our line of sight, the spectral lines of the normal companion are periodically Doppler shifted towards the red and the blue (one speaks in this case of a spectroscopic binary). Thus one can measure the period P of the orbit and the maximum value v_* of the component of the velocity of the optical companion along the line of sight. These values can be combined to produce a Keplerian mass function $f(M)$ which depends in a known way on the masses of the two companions and on the sine of the orbital inclination angle (see Treves and Haardt, this volume). Hence, if one is able to estimate the mass M_1 of the normal companion and the inclination angle i, an approximate value for the mass M_X of the compact body can be found. If this value turns out to be larger than the supposed neutron star critical mass (of the order of three solar masses), the compact object is a black hole candidate. Furthermore, the value of the mass function is never larger than M_X, independently of the values of M_1 and of $\sin i$. Therefore, if $f(M_X)$ exceeds the critical mass for a stable neutron star, one can infer the presence of a black hole in the system even without a reliable knowledge of the mass of the optical companion and of the inclination angle.

In addition, the compact companion may accrete matter from the outer atmosphere of the normal star. This matter, as it falls onto the compact object (neutron star or black hole) becomes compressed and heated and emits x-rays.

In conclusion: reliable stellar mass black hole candidates should be sought, in particular, among those spectroscopic binaries in which one companion is optically bright, the other one being x-ray bright (x-ray binaries), and for which the value of the mass function and the estimates of the mass of the optical star and of the orbital inclination angle suggest a value for the mass of the compact body safely higher than the neutron star critical mass.

A thorough account of the present status of the understanding of the astrophysics of stellar mass black holes, both from the theoretical and the observational point of view, is reported in Treves and Haardt, this volume. In particular, this includes elements of the theory of accretion, a detailed discussion of stellar mass black hole candidates, probes of black hole physics such as the hardness of the x-ray spectrum which, if the mass of the black hole is known from the knowledge of the mass function, may provide a measure of the angular momentum of the hole etc. The last section of the contribution is dedicated to the subject of QPOs (quasi-periodic oscillations) in x-ray binaries, which, among other things, can provide an effective probe of general relativistic effects around neutron stars and black holes. Among these, is the putative observation of the Lense–Thirring precession of the inner edge of accretion discs around neutron stars, which one may hope to see in black hole candidates as well.

Stellar mass black holes are believed to be produced by gravitational collapse, following the end of nuclear burning, of stars whose masses are higher than about 45 solar masses. For lower values of the mass, one estimates that a sufficient amount of matter-radiation is ejected during the process of collapse (or contraction) to allow them to settle in the white dwarf or neutron star state.

Besides stellar mass ones, there are other astrophysical black holes whose existence one may envisage. Among these are mini black holes and supermassive black holes. Whereas, from the point of view of classical general relativity, once a black hole is formed its mass can only increase since no matter or radiation can escape from the hole from within the horizon, we shall see below that, due to quantum processes, a black hole is actually expected to lose energy by emission of thermal radiation (Hawking's evaporation process, see Kiefer, this volume). The temperature of the evaporation radiation is inversely proportional to the mass of the hole (more precisely, it is proportional to its surface gravity, see below). The evaporation temperature of a stellar mass black hole is of the order of 10^{-7} K and therefore there is no hope to ever see the evaporation radiation of objects of this kind. Furthermore, any such hole would absorb from the surrounding medium (in particular, from the cosmic microwave background) much more energy than it gives off by Hawking radiation. On the other hand, from an estimate of the power emitted by the evaporation process one can expect from a small black hole with a mass of the order of 10^{15} g (the mass of a typical asteroid) and corresponding Schwarzschild radius of the order of 10^{-13} cm (the size of a nucleon) an evaporation lifetime of the order of the age of our universe.

Such small holes cannot be formed by gravitational collapse but, as suggested by Zel'dovich (1967) and Hawking (1971), they might have been produced in local fluctuations in the early universe under a sufficiently strong external pressure. Should such mini black holes exist, they might go through the last stages of the evaporation process at the present epoch, thus signalling their presence by the gamma rays they emit just before extinction. However, no radiation of this kind has been observed so far (gamma ray bursts are not liable to such an explanation), which sets stringent limits on the average density of such putative primordial black holes.

Regarding supermassive black holes, there are several converging indications that point to the conclusion that, most probably, all types of active galactic nucleus (AGN) such as quasars, radiogalaxies, Seyfert galaxies, BL Lac objects etc, as well as most (and perhaps all) normal galaxies, including our own Milky Way, harbour giant black holes at their centres. Estimates of the masses and of the sizes of AGNs are based on various methods (see e.g. Begelman and Rees 1996). Mass estimates rely on luminosities and on the computation of the total lifetime energy output of an AGN, together with reasonable assumptions regarding the efficiency of the conversion of mass into energy, as well as on the computation of the cumulative energy output of the active nucleus based on the measurement of the energy content of the radio lobes, when these are observed. Sizes can be estimated from the time scales of the variability of luminosities. These estimates invariably point to central mass concentrations which are so dense that they cannot be explained as stable star clusters, so that the black hole hypothesis seems to be the only viable one. The estimated masses of the central black hole range between 10^6 and 10^{10} solar masses.

A widely accepted model for the structure of quasars and radio galaxies is as follows. The central black hole is surrounded by an accretion disc, which is extremely luminous in quasars but rather quiescent in radiogalaxies. Depending on the relative orientations of the corresponding angular momenta, the plane of the disc is generally tilted compared to the equatorial plane of the spinning hole. However, as a consequence of the strong dragging of inertial frames in the vicinity of the hole, due to the hole spin, the central region of the disc, being dragged along by the hole rotation, lies in the latter's equatorial plane. The black hole and possibly the accretion disc power two highly relativistic jets of gas in opposite directions along the spin axis of the hole. These jets race through the galaxy into intergalactic space, feeding energy into the galaxy's radio emitting lobes.

In the case of the Milky Way and of normal nearby galaxies, dynamical methods based on spectroscopic measurements of the orbital velocities of stars and gas close to the central nucleus also point to the central black hole hypothesis as the only convincing one, and the estimated masses of these putative black holes range again essentially between 10^6 and 10^{10} solar masses. For example, in the case of the Milky Way the distribution of the stellar velocities is consistent with the hypothesis of a 2.5×10^6 solar mass black hole.

In particular, recent measurements in a sample of 15 nearby galaxies seem to indicate that there is a definite correlation between the mass of the central 'dormant' black hole and the mass of the bulge of the galaxy, in all these cases the ratio of the two masses being invariably given by $M_{bh}/M_{bulge} = 0.006$.

For a detailed discussion and information on giant black holes in the cores of galaxies, see for example Begelman and Rees (1995), Hehl *et al* (1998).

Due to the intrinsic non-linearity of the Einstein field equations it is really remarkable that, as we have seen, all stationary black hole solutions are known.

There are other important analytical results in the field of exact solutions. For instance, all stationary axisymmetric vacuum solutions are in principle known (see Kramer *et al* 1980 and references therein). However, as soon as the spacetime is no longer required to possess a high degree of symmetry, one has to face the non-linearity of the field equations in full. As a result, for instance, no solution describing the exterior field of a collapsing star is known analytically, apart from spherically symmetric solutions.

On the other hand, of course, there is a wide range of astrophysical processes in which general relativity is expected to play a relevant role, but for which we cannot expect a high degree of symmetry. For instance, Einstein's theory should play a fundamental role in merging events of binary or galactic systems. In the absence of analytical insights we are forced to approach all such processes through a numerical route. Among the most 'dramatic' merging phenomena which are expected to occur in nature, perhaps the most obscure and interesting one is the coalescence of a binary black hole system.

Coalescence of two black holes can be regarded as a succession of three different stages. First, the two objects 'circle' each other with small changes in the orbits on long time scales (inspiral). This can be regarded as a relativistic Kepler problem with the two black holes considered as 'elementary particles' of the theory. Then, the separation becomes smaller and the two objects coalesce rapidly (merger). What is formed is a single dynamical black hole about which very little is known. Finally, the remnant is expected to radiate away all 'perturbations' and settle down to a stationary (i.e. Kerr) state (ring down). The intermediate ('dramatic') period of the above process is expected to produce a strong amount of gravitational radiation. It is worth mentioning that detection of this radiation would serve as the strongest evidence for the existence of black holes, as well as being useful for cosmological measurements. In particular, the observed waveforms would carry clean information about black hole geometry.

It is hoped that the future detectors of gravitational waves, based on the principle of laser interferometry, such as GEO, VIRGO and LIGO, now under construction, as well as the proposed LISA, on a longer term, will reveal events of this kind.

The computational approach to the problem of black hole coalescence has had an enormous development since the foundation of the Binary Black Hole Grand Challenge Alliance, formed by several outstanding scientists in this field. The techniques and the recent results (not only numerical!) of the Alliance as well as the various possible computational approaches to the Einstein field equations are described in detail in the contribution by R Matzner in this book. Here we just mention a few main points to give an idea of the framework in which 'computational black hole physics' operates.

To solve a dynamical problem governed by the vacuum Einstein field equations one has, first of all, to define what 'dynamical' means in mathematical terms. Indeed, one has to take into account that general relativity is a gauge theory, the gauge group being composed by the group of coordinate transformations (diffeomorphisms) of the spacetime. As a result, there is no unique way to introduce a 'space-plus-time' foliation in order to put forward initial data on a spacelike hypersurface and let them evolve numerically. One can, in fact, reparametrize arbitrarily the 'space' coordinates on subsequent hypersurfaces ('shift' freedom) and can also vary the elapsed time between such hypersurfaces in an arbitrary way ('lapse' freedom). As a reflection of this fact, the Einstein field equations split into two parts: constraint equations and evolution equations. The constraint equations do not contain second order derivatives w.r.t. 'time', so that once satisfied on initial data, such equations are always satisfied. This structure is better understood in terms of the Hamiltonian formulation of general relativity (Arnowitt, Deser and Misner, 1962). Although mathematically more complicated, it is conceptually very similar to that of Maxwell electrodynamics in terms of potentials. Therefore, electrodynamics can be used as a useful 'laboratory' to understand canonical general relativity. We refer the reader to Wald's (1984) book for details.

Numerically, one has to choose a specific gauge, and let the dynamical quantities evolve while controlling pointwise the fulfillment of the constraints. The approach has been implemented and developed in many ways from the computational point of view, using for instance splittings based on null advanced-time hypersurfaces instead of spacelike surfaces or spacelike but hyperboloidal hypersurfaces which become null at infinity. It is also possible to cast the evolution equations in a (first order) 'conservative' system, so that the techniques of computational hydrodynamics can be applied to them.

Besides the delicacies involved with the discretization of a constrained theory like general relativity, numerical investigations on black hole dynamics have also to face the 'intrinsic nature' of such objects. Consider, indeed, the merging of two black holes. Initially the two holes are well separated so that two distinct horizons exist. In the merging phenomenon, an enveloping apparent horizon is formed, swallowing the individual horizons until the 'freezing' of the dynamics to a final, stationary black hole state occurs. In practice, one needs to solve stepwise a (generally highly non-linear) partial differential equation which gives the location of the horizon(s) on each spacelike hypersurface.

Even such a brief introduction to the branch of black hole physics related to numerical relativity would not be complete without recalling recent numerical investigations on gravitational collapse and black hole formation. This field was revitalized by the work by Ori and Piran (1990) who found numerical evidence of naked singularity formation in spherically symmetric perfect fluid gravitational collapse, and by Shapiro and Teukolsky (1992), who studied the final stages of a class of non-spherical collapsing objects, for which, again, there is an indication that cosmic censorship does not follow without further specifications. One of the most interesting results that has ever come out from numerical relativity is Choptuik's (1993) discovery of critical behaviour in scalar field spherical collapse. Choptuik investigated numerically the gravitational dynamics of a massless scalar field wave-packet labelled by a scalar parameter, and found that there are black hole forming solutions as well as dispersive solutions. The two branches are separated at a critical value of the parameter at which a sort of phase transition takes place characterized by a universal critical exponent and scaling behaviour. Choptuik's solution has many other interesting features (see Matzner, this volume; a complete review is given by Gundlach, 1998). Here we just want to stress that Choptuik's numerical work has brought up for the first time a new *qualitative* phenomenon contained in the Einstein field equations, giving rise to a new research field.

The beginning of the 1970s marked a leap in black hole research, which ultimately ushered in the era of quantum black hole physics that we are living now. It started in 1970, when Christodoulou first noticed the resemblance of the laws of thermodynamics to those which governed slow changes in the properties of black holes. Shortly thereafter, Stephen Hawking discovered the area-increase theorem, which led Jacob Bekenstein, for the purpose of saving the second law of thermodynamics, to suggest that a black hole should have an entropy proportional to its area. Then, in 1973, Bardeen, Carter and Hawking pointed out clearly the striking analogy between macroscopic thermodynamics and what they called the four laws of black hole mechanics. In this analogy, the role of the temperature is played by the surface gravity κ, which may be looked at as the force at infinity necessary to keep a unit mass immediately outside the horizon of a black hole. The surface gravity has the remarkable property of being constant over the event horizon also for non-spherically symmetric black holes. In analogy with the zeroth law of thermodynamics, this property is referred to as the zeroth law of black hole mechanics, while the area-increase theorem constitutes the second law of black hole mechanics. At that time, however, this analogy was believed to be just a curiosity; even the name 'mechanics' marks a clear-cut distinction between κ and A and the physical temperature and entropy of the black hole; in particular the temperature was thought to be absolute zero, otherwise the hole should emit radiation, a process which is classically impossible since even light cannot escape from it.

Great surprise came about when one year later (1974) Hawking showed that black holes can indeed emit radiation with a thermal spectrum. This changed completely the way of looking at the laws of black hole mechanics, which have since then been regarded as expressing the true thermodynamical properties of black hole physics, through a deep and still undiscovered bridge between quantum geometry and thermodynamics.

It is possible to give a simple picture of Hawking radiation, a phenomenon that arises as a consequence of the existence of the event horizon of the classical (unquantized) black hole, considered as the background where quantized matter evolves according to the paradigm of quantum field theory on curved spacetimes. The radiation can be thought to be produced by quantum vacuum fluctuations, the most typical phenomenon in quantum field theory. Suppose that by one of these fluctuations of the vacuum a pair of photons (or more generally any particle–antiparticle pair) is created, and that one of these photons has negative energy $-E$. If this event happens far from the horizon, then it is necessary that the pair is annihilated within an interval of time of the order $\hbar/E$. In contrast, when the pair is created close to the horizon, then the negative energy particle can be swallowed by the classical black hole by stepping in the horizon before such a time interval has passed; once in the horizon the photon can propagate, since now its negative energy corresponds to spatial momentum, while the other photon can escape away from the hole, reaching an observer at infinity at late times. This observer sees thermal radiation whose temperature is $\hbar\kappa/2\pi k_B$.

It is clear from the previous heuristics that Hawking radiation is not contradicting the classical impossibility of escape from black holes; indeed the photons received at infinity come from outside the horizon. It is worthwhile to stress again that the black hole temperature is a genuine quantum phenomenon; as such it is proportional to Planck's constant $\hbar$ and vanishes when $\hbar$ is set to zero.

In the case of a Schwarschild black hole, the temperature is inversely proportional to the mass M. According to the Stefan–Boltzmann law one obtains a black hole lifetime proportional to M^3. For a stellar mass black hole this gives an unobservably small temperature and a lifetime much longer than the age of the universe. By contrast, putative primordial black holes with masses of the order of 10^{15} g would evaporate now. The evaporation of such small holes could be detected by the hot flash of gamma rays they would emit in their death throes. This phenomenon has not been observed.

Having fixed the temperature, the laws of black hole mechanics then give the following precise value to the Bekenstein–Hawking entropy: $S_{BH} = S_{bh} = Ak_B/4\hbar G$. This entropy is enormous compared to that of ordinary matter. Consider indeed a sphere of radius R filled with thermal radiation at temperature T. The entropy of radiation is the highest possible for ordinary 'matter'; this is proportional to $R^3 T^3$ while the corresponding energy (mass) is proportional to $R^3 T^4$. This amount of radiation will form a black hole when M and R are of the same order (in Planck's units), i.e. T is proportional to $M^{-\frac{1}{2}}$. It follows that the entropy is proportional to $M^{\frac{3}{2}}$, while according to Bekenstein and Hawking the entropy of a black hole should be proportional to M^2.

Where does this entropy come from and where is it located? The question of a microscopic quantum description of the degrees of freedom of a black hole immediately arises. The so-called 'information loss paradox' poses the necessity of such microscopic description in a dramatic way. This paradox stems out from the very existence of Hawking's thermal radiation: indeed, since a radiating black hole may eventually completely evaporate, the information contained in it can be completely lost during the evaporation process, if what comes out is really structureless thermal radiation. Then the time evolution is not unitary, and there is something going wrong with our usual understanding of quantum physics. This is Hawking's point of view which is not shared by other physicists like 't Hooft and Susskind (see Susskind 1997). It is, however, clear that a full answer to these questions cannot be found within the semiclassical approach that led to the discovery of black hole radiation and that it is only a consistent quantum theory of gravity that could solve these paradoxes.

We enter here a highly speculative domain and we confine ourselves to a brief discussion of a few points.

It is fair to say that the search for quantum gravity is still far from achieving its goal. The standard perturbative approach to quantum field theory gives little results in this case since it yields a theory which is not renormalizable. Even worse, it is the idea itself of treating the metric as a relativistic quantum field which fails at the Planck scale: the standard notions on which quantum field theory is based become ill defined or meaningless. This appears clearly by considering for instance the role played in standard quantum field theory by the background spacetime which is endowed with a metric which establishes the causality ordering. These concepts are simply absent from quantum gravity and some radically new ideas are needed. Still, it is difficult to resist the temptation of using an approach that has been so successful in other seemingly similar cases.

One of the most popular attempts is based on the path integral for which, as in standard QFT, it is useful to consider imaginary values of time, a procedure briefly referred to as Wick's rotation. Of course the Euclidean approach to quantum field theory is really justified only for Wightman quantum field theories, being based on the locality, covariance and spectral properties of such theories; in contrast it has little factual support in this context. However, these ideas have been pursued giving rise to the so-called Euclidean quantum gravity. Hawking and Gibbons have obtained again the black hole entropy within this approach, by evaluating the partition function in the zero-loop approximation. One interesting point of the calculation is that only the boundary term in the action is relevant to find the Bekenstein–Hawking formula; this might be suggesting that the degrees of freedom this entropy is counting live on the horizon. However, this calculation is not really counting the quantum states of the black hole and is mathematically equivalent to the purely classical calculation.

On the other side, there exists an attempt to quantize general relativity using a Hamiltonian approach, which is called canonical quantum gravity and is non-perturbative in spirit. Actually, there are several declinations of canonical quantum gravity which are distinguished by different choices of the canonical variables. The major difficulty of this method is that it is based on a split of spacetime as a Cartesian topological product of a three-dimensional spatial manifold times a real line. This seems to be contrary to the whole spirit of general relativity and gives rise to the so-called 'problem of time' in canonical quantum gravity.

The question of a fundamental computation of black hole entropy has been addressed also within canonical quantum gravity. Various expressions for the entropy have been found that agree qualitatively with the Bekenstein–Hawking formula. Some examples are discussed by Kiefer in this book.

Another example has been recently proposed in the context of loop quantum gravity, by considering the number of quantum microstates of the hole which have microscopically distinct effects on its exterior. These states are given by the quantum state of the horizon having the same area, and can be counted using loop quantum gravity. The entropy is again proportional to the hole area. The constant of proportionality is not, however, fixed uniquely (see Rovelli, 1996).

Today, the most widely considered candidate for a theory of quantum gravity is string theory (see e.g. Polchinski 1998). Actually, string theory (or M-theory) has the claim to be much more than a consistent quantum theory of gravity, since it sets out to provide a unified quantum description of all interactions. It contains also a massless spin-2 state whose low energy interactions can be described by general relativity.

One can see by qualitative arguments that string theory owns the right number of degrees of freedom to account for black hole entropy (see e.g. Horowitz 1997). A more quantitative justification needs a non-perturbative treatment of these degrees of freedom. However, until very recently, only a perturbative approach to string theory was available, and therefore the only way to get black holes from strings was through the classical solutions of the various supergravities that emerge in the low energy limit.

The situation changed radically when, in the last few years, new non-perturbative methods (dualities) and objects were discovered in string theory. Most important is in particular the discovery by Polchinski of a new class of solitons that are called 'D-branes'. A D-brane is an object where an open string can end. The fact that D-branes have dynamics means that string theory is actually not only a theory of strings (D-brane democracy).

These new ideas have permitted a microscopic derivation of the black hole entropy which is by now the most successful which has been put forward (a full discussion is given by Dijkgraaf and by Fré and D'Auria, in this book. See also Horowitz 1997 and Peet 1998). Indeed, there exist extremal charged black holes which solve the supergravity obtained in the low energy limit of string theory. In these solutions, the charge is linked to the mass by the relation $M = cQ$ (where c is some constant). All perturbative string states satisfy the bound $M \geq cQ$ for weak coupling. The calculation proceeds, roughly speaking, by counting the number of states saturating the bound. These states are known as Bogomolnyi–Prasad–Sommerfield (BPS) states, and are invariant under a non-trivial subalgebra of the full supersymmetry algebra; their masses do not receive any quantum correction. D-branes enter crucially and it turns out that the number of BPS states corresponds to the Bekenstein–Hawking entropy. It should be remarked that this result makes crucial use of the supersymmetric structure of the theory and it is not clear how to extend it to standard non-supersymmetric black holes.

We would like to add a final remark: whatever the ultimate theory of quantum gravity turns out to be, the puzzles and deep questions 'emerging' from black holes (a typical quantum phenomenon!) may be the best doorway to its understanding.

References

Since most of the results cited here are 'classical', we have limited the references to recent books and reviews. In particular, we benefited very much from reading Thorne's (1994) book.

Begelman M and Rees M 1996 *Gravity's Fatal Attraction: Black Holes in the Universe* (New York: Freeman)

Burko L and Ori A (eds) 1997 *Internal Structure of Black Holes and Spacetime Singularities* (Bristol: Institute of Physics Publishing)

Chandrasekhar S 1983 *The Mathematical Theory of Black Holes* (Oxford: Clarendon)

Gundlach C 1998 Critical phenomena in gravitational collapse *Adv. Theor. Math. Phys.* **2** 1

Hehl F W, Kiefer C and Metzler R J K (eds) 1998 *Black Holes: Theory and Observation (Lecture Notes in Physics 514)* (Berlin: Springer)

Horowitz G T 1997 Quantum states of black holes, gr-qc/9704072

Joshi P S 1996 *Global Aspects in Gravitation and Cosmology* (Oxford: Clarendon)

Kramer D, Stephani H, Herlt E and MacCallum M 1980 *Exact Solutions of Einstein's Field Equations* (Cambridge: Cambridge University Press)

Peet A 1998 The Bekenstein formula and string theory *Class. Quantum Grav.* **15** 3291

Polchinski J 1998 *String Theory* (Cambridge: Cambridge University Press)

Rovelli C 1996 Loop quantum gravity and black hole physics *Helv. Phys. Acta* **69** 582

Senovilla J 1998 *Gen. Rel. Grav.* **30** 701

Susskind L 1997 Black holes and the information paradox *Sci. Am.* **4** 44

Thorne K S 1994 *Black Holes and Time Warps* (New York: Norton)

Wald R 1984 *General Relativity* (Chicago: Chicago University Press)

Wald R 1997 Gravitational collapse and cosmic censorship, gr-qc/9710068

PART 1

Claus Kiefer

Fakultät für Physik, Universität Freiburg, Hermann-Herder-Strasse 3, 79104 Freiburg, Germany

Chapter 2

Thermodynamics of black holes and Hawking radiation

La filosofia è scritta in questo grandissimo libro, che continuamente ci sta aperto innanzi a gli occhi (io dico l'universo), ma non si può intendere se prima non s'impara a intender la lingua, e conoscer i caratteri, ne' quali è scritto. Egli è scritto in lingua matematica, e i caratteri son triangoli, cerchi, ed altre figure geometriche, senza i quali mezi è impossibile a intenderne umanamente parola.

GALILEO GALILEI, *Il Saggiatore*

Modern science is built upon a mathematical description of Nature. Although its characters are no longer only triangles, circles and other geometrical figures, Galileo's characterization of science is as valid as it was in his time.

Black holes—the subject of this school—provide an example *par excellence* for this mathematical description of Nature. Their very existence and simple geometrical properties were predicted by pure theory—Einstein's geometric theory of gravity, the general theory of relativity. By now, their existence has been proven by observations with almost certainty. It is expected, not least with the advent of powerful gravitational wave detectors, that observations of black holes will play a key role in 21st century astronomy.

The subject of my contribution is the thermodynamics of black holes and Hawking radiation. The main focus is therefore on quantum aspects, although I shall elaborate on all the classical aspects of black holes that are a necessary prerequisite for the understanding of their quantum behaviour. Apart from the physics of the early Universe, quantum aspects of black holes will provide the only key towards an understanding of a quantum theory of gravity.

I shall start in the next section with a general introduction into black hole physics. This serves both the purpose to recall the basic notions known from elementary courses as well as to give the material and set the notation needed for the main part. It may also be considered as an introduction to some of the material covered by other lectures.

Section 2.2 then enters the main theme and gives a review of the laws of black hole mechanics. These laws have full analogies to the laws of thermodynamics, but at the classical level of section 2.2 this analogy is purely formal. A physical interpretation is only possible within quantum theory and is given in section 2.3. I there discuss both the derivation of Hawking radiation and its physical interpretation. Section 2.4, then, is devoted to the interpretation of black hole entropy and the problem of information loss. The last two sections enter the most speculative parts of black holes—the possible role they play in and for a quantum theory of gravity. Section 2.5 covers canonical gravity, while section 2.6 covers superstring theory.

To keep the references to a minimum, I put more emphasis on citing books and reviews than on original work. Two recent proceedings that cover all aspects are Hehl *et al* (1998) and Wald (1998). A very recommendable general introduction is Thorne (1994); an excellent short introduction is Luminet (1998).

The metric convention is diag$(1, -1, -1, -1)$; I follow the abstract index notation of Wald (1984).

2.1   Black holes—a general introduction

Black holes have a long and fascinating story, see Israel (1987). Probably the first person who envisaged the existence of 'dark stars' was the Reverend John Michell in 1784. He wanted to develop a method to measure the distance of stars by taking into account the 'diminution of their light' escaping the star's gravitational field. (Recall that the first parallax was measured by Bessel in 1837, so in Michell's time this was an interesting proposal.) At that time, Newton's corpuscular theory of light was prevailing, so it was a sensible problem to study a star with an escape velocity greater than the speed of light. (Recall that the finiteness of the speed of light had already been noted in 1676.) Setting $v = c$ in the expression for the escape velocity v of a spherically symmetric body, $v^2 = 2GM/r$, one finds for the corresponding radius

$$r \equiv R_0 = \frac{2GM}{c^2} \approx 3\frac{M}{M_\odot} \text{ km.} \qquad (2.1)$$

This radius, R_0, is known as the *Schwarzschild radius* (see below). For a body with the mass of the earth, $R_0 \approx 0.9$ cm only.

Michell asked: how big must a star with the density equal to the density of the Sun be, to have $v = c$? Since then the radius of the star is equal to its

Schwarzschild radius R_0, one has ($R_\odot$ is the radius of the Sun)

$$\frac{R_0}{R_\odot} = \frac{2GM}{c^2 R_\odot} = \frac{2G}{c^2} \frac{M_\odot R_0^3}{R_\odot^4}$$

and therefore

$$\frac{R_0}{R_\odot} = \left(\frac{c^2}{2G} \frac{R_\odot}{M_\odot}\right)^{1/2} = \left(\frac{R_\odot}{R_{0\odot}}\right)^{1/2} \approx 483$$

where $R_{0\odot}$ denotes the Schwarzschild radius of the Sun. Note that this corresponds to a mass of about $10^8 M_\odot$—which coincides with the mass of some of the supermassive black holes that are now assumed to exist in the centre of the galaxies (see Treves' lectures)!

A similar discussion can be found in Laplace's famous work *Exposition du Système du Monde* (1796). It also played a role in Soldner's discussion (1801) about light deflection in the neighbourhood of stars and his speculation about a very massive dark object in the centre of our Galaxy.

1801 was also the year where Young discovered the interference properties of light. This gave rise to the advent of the wave theory of light, in which the presence of dark stars did not seem to make much sense. For this reason all reference to them was omitted in the 1808 edition of Laplace's *Exposition*.

The modern theory of black holes started with the advent of general relativity in 1915. Soon after the field equations had been published, Karl Schwarzschild discovered an exact solution to these equations, which describes the gravitational field outside a spherically symmetric mass distribution. This solution is also of the utmost importance for black holes, and it is described below.

The first exact solution that describes the collapse of a body—a pressureless dust cloud—was obtained by Oppenheimer and Snyder in 1939. In this case the dust cloud crunches to a spacetime singularity at $r = 0$ and describes what is now known as a black hole, an object so dense that not even light can escape. However, such objects were not taken seriously by most scientists, including Einstein, for a long time. This was in part due to the ill understood nature of a singularity that occurred at $r = R_0$, see Thorne (1994) for a fascinating account of this story. For this reason it was also not really accepted that black holes could result as the final stage in the life of a star.

This changed in the 1960s, and one could call the period from 1960 to 1975 the *classical period* of black hole research. The nature of the singularity at $r = R_0$ was clarified through the introduction of Kruskal coordinates (see below). The singularity theorems showed that the singularity at $r = 0$ found by Oppenheimer and Snyder is not an artifact of the spherical symmetry assumed in their model, but occurs generally under well defined conditions, see Hawking and Penrose (1996). The uniqueness theorems ('black holes have no hair') exhibit, very surprisingly, that stationary black holes are fully characterized by a small set of parameters (see below).

On the observational side, neutron stars were discovered in 1967. Since neutron stars are possible final states of a star collapse—the others being white dwarfs and black holes, also the most exotic option, the black hole, was then taken more seriously. (1967 was also the year where the name *black hole* was coined by John Wheeler.) In retrospect, the year 1963 is also very important through the discovery of quasars, although this had not been recognized at that time.

Today, there is a general consensus that black holes do exist in Nature and that they have been observed. The best stellar black hole candidate at present is probably the x-ray nova V404 Cygni with a mass $M > 6\,M_\odot$, and the best supermassive black hole candidate is the black hole that lurks in the centre of the Milky Way and has a mass of about $2.6 \times 10^6\,M_\odot$, see the lectures by Treves.

The year 1974 saw the fascinating and surprising—theoretical—discovery that black holes are not really black when quantum theory is taken into account, but radiate with a thermal spectrum like an ordinary 'black body'. This so-called Hawking radiation is the cornerstone of all modern theoretical developments and will play a central role in my review. The year 1974 thus opened the *quantum period* of black hole research, a period where no end is yet in sight. The main open questions are the full quantum picture of black hole evaporation and the interpretation of black hole entropy. Candidates for a quantum theory of gravity, such as canonical quantum gravity or superstring theory, attempt to find a solution for these problems.

In the following I shall give a brief introduction into the Schwarzschild metric describing the gravitational field outside a spherically symmetric mass distribution.[1] This topic is covered in many excellent textbooks including Misner *et al* (1973), Sexl and Urbantke (1983), Straumann (1984), Wald (1984) and—on a more mathematical level—Hawking and Ellis (1973). The Schwarzschild metric is the unique spherically symmetric solution to the vacuum Einstein equations

$$R_{ab} = 0. \tag{2.2}$$

In the standard coordinates, its line element reads

$$ds^2 = \left(1 - \frac{2GM}{r}\right)dt^2 - \left(1 - \frac{2GM}{r}\right)^{-1}dr^2 - r^2\,d\Omega^2 \tag{2.3}$$

where $d\Omega^2 = d\theta^2 + \sin^2\theta\,d\phi^2$ is the line element on the unit two-sphere. It is a direct consequence of spherical symmetry that this metric is also *static* (Birkhoff's theorem). The constant $2GM$ is determined by comparison with the Newtonian limit (this is why, in spite of the vacuum equations (2.2), G comes into play). One easily recognizes the singularities in the metric (2.3) at $r = 0$ and $r = 2GM = R_0$. While the singularity at $r = 0$ is a real one (divergence of curvature invariants), the singularity at $r = R_0$ is a coordinate singularity, see below.

It is of interest to consider also the *interior* solution of a spherically symmetric body whose exterior solution is given by (2.3). Making the *ansatz* of a static spherically symmetric metric,

$$ds^2 = e^{\nu(r)}\,dt^2 - e^{\lambda(r)}\,dr^2 - r^2\,d\Omega^2 \tag{2.4}$$

[1] From now on $c = 1$.

and taking for matter an ideal fluid whose energy–momentum tensor reads

$$T_{ab} = (\rho + p)u_a u_b - g_{ab} p \tag{2.5}$$

where ρ and p are respectively density and pressure, one arrives at the *TOV solution*

$$e^{\lambda(r)} = \left(1 - \frac{2GM(r)}{r}\right)^{-1}$$

$$v(r) = -\lambda(r) + 8\pi G \int_\infty^r dr' \, r'(p + \rho) e^{\lambda(r')}$$

$$M(r) \equiv 4\pi \int_0^r dr' \, r'^2 \rho(r').$$

It is evident that outside the body this is equal to the Schwarzschild solution.

From the covariant conservation equation $T^b_{ab;} = 0$ one finds

$$\frac{dp}{dr} = -\frac{G\left(M(r) + 4\pi r^3 p\right)}{r^2}\left(1 - \frac{2GM(r)}{r}\right)^{-1} (\rho + p). \tag{2.6}$$

This is the general relativistic extension of the well known hydrodynamical equilibrium equation

$$\frac{dp}{dr} = -\frac{GM(r)}{r^2}\rho \tag{2.7}$$

found in Newtonian theory. The modifications due to general relativity are: in addition to $M(r)$ one has a term proportional to p, since also pressure generates gravity; one has $\rho + p$ instead of ρ, since gravity also acts on pressure, and one has $r^{-2} \longrightarrow r^{-2}(1 - 2GM(r)/r)^{-1}$, meaning that gravity increases faster than r^{-2}. As a consequence of this, there is an upper limit to the mass of a neutron star.

For the very important special case $\rho = $ constant, one finds from (2.6) that the pressure in the centre of the star diverges if its radius $R \le \frac{9}{8} R_0$. This is a direct consequence of the general-relativistic feature of nonlinearity that pressure generates more pressure. This lower bound on the radius leads to an upper bound of the mass, $M \le M_{max} = 4[9(3G^3 \pi \rho)^{1/2}]^{-1}$. One can show that the existence of such limits remains for $\rho \ne$ constant (Wald 1984). The above special case of constant density was already discussed by Schwarzschild in 1916; he was pleased by this result, since it suggested to him that the singularity at $R = R_0$ is of no relevance. However, this limit only shows that a static situation can no longer occur for $R \le \frac{9}{8} R_0$; it does not say anything about a dynamical situation, see below.

Can one give an illustration for the geometry of the Schwarzschild metric (2.3)? Taking $t = $ constant (staticity) and $\theta = $ constant $= \pi/2$ (spherical symmetry), one obtains the spatial line element

$$d\sigma^2 = \left(1 - \frac{2GM}{r}\right)^{-1} dr^2 + r^2 \, d\phi^2. \tag{2.8}$$

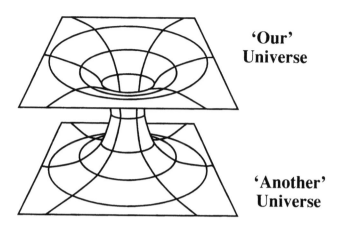

Figure 2.1. Einstein–Rosen bridge.

This can be embedded in an auxiliary three-dimensional Euclidean space with line element $ds^2 = dr^2 + dz^2 + r^2 d\phi^2$ by looking for a rotation surface $z(r)$ that reproduces (2.8). One immediately finds the rotational paraboloid

$$z(r) = \pm\sqrt{8GM}\sqrt{r - 2GM}. \tag{2.9}$$

The corresponding embedding diagram is shown in figure 2.1 (schematically). One recognizes that the region $r < R_0$ is not covered and that there exist two asymptotically flat regions, one corresponding to 'our' Universe. Note also that each point on the surface represents in fact a two-sphere. The geometry in figure 2.1 is often called an *Einstein–Rosen bridge*.

The geometry of the TOV solution with $\rho = $ constant is, for $t = $ constant, given by

$$d\sigma^2 = \left(1 - \frac{8\pi Gr^2\rho}{3}\right)^{-1} dr^2 + r^2 d\Omega^2 \tag{2.10}$$

and describes a space with positive curvature, i.e. a three-sphere with radius $\mathcal{R} = [3/(8\pi G\rho)]^{1/2}$. Taking for the interior part of a star this solution and for the exterior part the Schwarzschild solution, the corresponding embedding diagram looks like figure 2.2. Since the Einstein field equations only determine the local spacetime geometry, the global *topology* is not fixed by them. Instead of figure 2.1, one can identify the two asymptotic regions and arrive at a so-called *wormhole*, as depicted in figure 2.3.

First, the line element (2.3) is written in the form

$$ds^2 = \left(1 - \frac{2GM}{r}\right)\left(dt^2 - dr_*^2\right) - r^2 d\Omega^2 \tag{2.11}$$

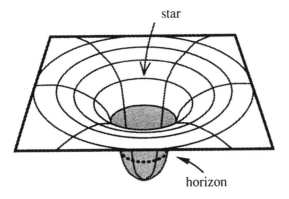

Figure 2.2. Geometry in the vicinity of a star.

through the introduction of the *tortoise coordinate*

$$r_* = r + 2GM \ln \left| \frac{r}{2GM} - 1 \right| . \tag{2.12}$$

This has the advantage that radial light rays propagate as in flat space since in this case $dt = \pm dr_*$.

Second, one looks for a coordinate transformation $(t, r_*) \longmapsto (T, X)$ that preserves these properties, but avoids coordinate singularities. This is achieved by the *Kruskal coordinates*

$$X = \exp\left(\frac{r_*}{4GM}\right) \cosh \frac{t}{4GM}$$

$$= \sqrt{\frac{r}{2GM} - 1} \, \exp\left(\frac{r}{4GM}\right) \cosh \frac{t}{4GM} \tag{2.13a}$$

$$T = \exp\left(\frac{r_*}{4GM}\right) \sinh \frac{t}{4GM}$$

$$= \sqrt{\frac{r}{2GM} - 1} \, \exp\left(\frac{r}{4GM}\right) \sinh \frac{t}{4GM} . \tag{2.13b}$$

The inverse transformation can be given only implicitly,

$$X^2 - T^2 = \left(\frac{r}{2GM} - 1\right) \exp\left(\frac{r}{2GM}\right) \tag{2.14a}$$

$$\frac{T}{X} = \tanh \frac{t}{4GM} . \tag{2.14b}$$

The line element (2.11) then reads

$$ds^2 = \frac{32(GM)^3}{r} \exp\left(-\frac{r}{2GM}\right) \left(dT^2 - dX^2\right) - r^2(T, X) \, d\Omega^2 . \tag{2.15}$$

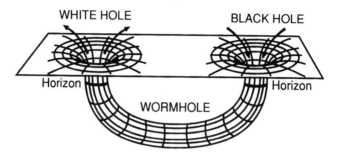

Figure 2.3. Wormhole.

Note that there is no longer any coordinate singularity at $r = R_0 = 2GM$; only the curvature singularity at $r = 0$ remains. The coordinate transformation (2.1) can thus be extended in a straightforward manner to $r < R_0$, and one arrives at the Kruskal diagram, figure 2.4. This is the maximal analytic extension of the Schwarzschild manifold, meaning that every geodesic can be extended either to the value ∞ of its affine parameter *or* encounters a singularity.

To classify the nature of the singularity at $r = R_0$, one looks for a coordinate system that is as nonsingular as possible. This proceeds in two parts:

One recognizes from figure 2.4 that the singularity at $r = 0$ is, in fact, spacelike and therefore distinguishes a certain time, not a space point. An observer present in II cannot 'see' the singularity. Since radial light rays propagate on straight lines inclined by $\pm 45°$ to the axes, the *causal properties* of the Kruskal manifold are evident.

Note in particular the presence of *event horizons* that separate the various regions from each other: by no means whatsoever can any signal emitted in II reach the outside regions I or III. The opposite is true for region IV: no signal emitted from I or III can enter it. The region II never becomes part of the past of an outside observer, as old as he might become.

The Einstein–Rosen bridge shown in figure 2.1 is obtained from figure 2.4 as the cross section $t = \text{constant}$ through the origin. The two asymptotically flat regions of figure 2.1 are thus the regions I and III in figure 2.4. If one investigated instead cross sections $T = \text{constant}$ in figure 2.4, one would obtain changing embedding diagrams suggesting the picture of a dynamical collapse; this demonstrates that the Schwarzschild metric (2.3) is static only in its exterior part, not in its interior part $r < R_0$. The origin in figure 2.4 is often called the *bifurcation two-sphere*.

Instead of the Kruskal diagram (figure 2.4) it is often more convenient to consider a diagram of a conformally related spacetime in which the regions of infinity are mapped to a finite boundary. This so-called *Penrose diagram* is depicted in figure 2.5. It will also play a crucial role in the discussion of Hawking radiation in section 2.3. I want to note that with some topological identifications one can

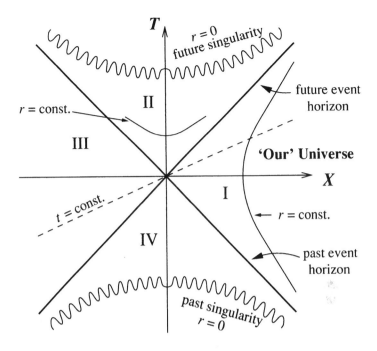

Figure 2.4. Kruskal diagram.

arrive from the Kruskal manifold shown in figures 2.4 and 2.5 at a manifold with only *one* exterior solution. This so-called *geon* is discussed at length by Louko and Marolf (1998); see also the brief discussion in section 2.5 below (figure 2.19).

It is straightforward to study the equations of motion for light rays and observers in the Kruskal spacetime (2.15). The solutions of the corresponding geodesic equation is found in the textbooks cited above. I only want to recall that a particle moving from region I to II reaches the singularity at a finite proper time—in spite of the fact that the coordinate time $t \to \infty$ for $r \to R_0$. For example, the free-fall time s from the horizon to the singularity is given by

$$s = \pi GM \approx 1.54 \times 10^{-5} \frac{M}{M_\odot} \, \text{s} \qquad (2.16)$$

and the proper time for nongeodesic motion is even shorter. The time (2.16) roughly corresponds to the time it would take for light to propagate a distance of R_0 in flat space. We shall encounter this characteristic time at several occasions again.

I want to emphasize that the situation depicted in figures 2.4 and 2.5 is still time symmetric. The region IV (the 'white hole region') is the time reverse of region II (the 'black hole region'). This time-symmetric situation is sometimes

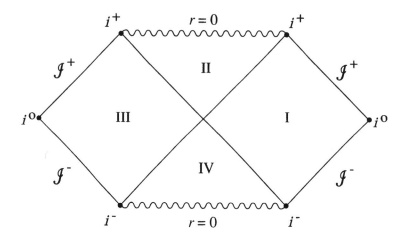

Figure 2.5. Penrose diagram of the Kruskal spacetime.

called an *eternal hole*. A genuine *black hole* is obtained from a time-*asymmetric* situation such as a star collapsing to a singularity (this is the case, for example, in the Oppenheimer–Snyder model). The time reverse of such a situation—a star expanding out of a singularity—is called a *white hole*. Both situations are shown in figure 2.6. It is evident that only two of the four regions of figure 2.4 are present. In particular, the second asymptotically flat region III is absent.

Coordinates that are often used to study a spherically symmetric collapse situation are the ('in-going' version of the) *Eddington–Finkelstein coordinates*. Instead of the original Schwarzschild coordinates (t, r) one uses $(\tilde{v}, r)$ with $\tilde{v} = t + r_*$ and r_* according to (2.12). The line element (2.3) then reads

$$\mathrm{d}s^2 = \left(1 - \frac{2GM}{r}\right)\mathrm{d}\tilde{v}^2 - 2\,\mathrm{d}\tilde{v}\,\mathrm{d}r - r^2\,\mathrm{d}\Omega^2\,. \tag{2.17}$$

The spacetime diagram showing the collapse of a star to form a black hole is shown, using these coordinates, in figure 2.7. This diagram demonstrates in particular that light rays emitted from the surface of the collapsing star are received by the distant observer at later and later times. For the same reason, the emitted light becomes increasingly redshifted according to (for radial rays)

$$\Delta z \equiv \frac{\Delta\lambda}{\lambda} = \exp\left(\frac{t}{4GM}\right)\,. \tag{2.18}$$

For the total decrease of luminosity L one must also take into account nonradial rays, with the result

$$L(t) \approx L_0 \exp\left(-\frac{t}{3\sqrt{3}GM}\right)\,. \tag{2.19}$$

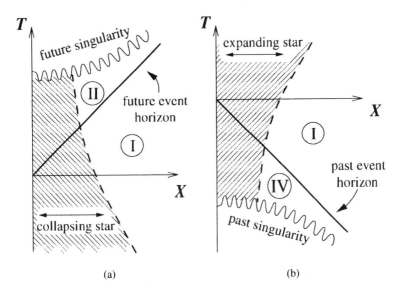

Figure 2.6. (a) Black hole; (b) white hole.

The characteristic time scale τ in this formula is

$$\tau \approx 3\sqrt{3}GM \approx 2.5 \times 10^{-5} \frac{M}{M_\odot} \text{ s} \tag{2.20}$$

and is of the same order of magnitude as (2.16).

That a singularity occurs under a much wider range of conditions than under a spherically symmetric situation is demonstrated by the *singularity theorems* (Hawking and Penrose 1996). One needs a certain energy condition, a condition for the global structure (for example, that there are no closed timelike curves), and the existence of so-called *trapped surfaces*. A trapped surface is a surface where not only in-going light rays converge, but also *out*-going light rays. Such trapped surfaces are for example present in the shaded interior region of figure 2.7. Under these conditions, a singularity necessarily occurs. A spacetime is called *singular* if it is timelike or null geodesically incomplete and cannot be embedded in a bigger spacetime.

A characteristic feature of the collapse situation of figure 2.7 is the occurrence of an *event horizon* that prevents the singularity from being seen from outside. It is unknown whether in a realistic collapse an event horizon always forms; this is the content of the *cosmic censorship hypothesis*: nature abhors naked singularities.

I emphasize that the cosmic censorship hypothesis does *not* exclude the occurrence of spacelike past singularities such as white holes. An important example of a spacelike past singularity is the Big Bang. One can always put a Cauchy surface at a position later than such a past singularity and predict all future evolution from initial data on this Cauchy hypersurface. This is not possible for a timelike

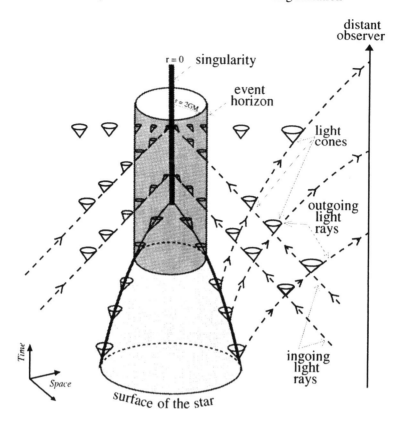

Figure 2.7. Star collapsing to form a black hole.

naked singularity. The apparent nonoccurrence of white holes is related to the second law of thermodynamics (Zeh 1992) and will be briefly discussed in section 2.5. Their presence is often excluded by the *Weyl tensor hypothesis* (Hawking and Penrose 1996).

Can one say anything about the nature of the singularity that occurs in a generic collapse? The general expectation is that upon approaching the singularity so-called BKL oscillations occur—chaotic oscillations in tidal curvature that occur in random directions (Belinsky *et al* 1982). The nature of the singularity itself can only be clarified in a quantum theory of gravity.

A most remarkable development in the mathematical study of black holes has been the proof of various *uniqueness theorems* for black holes (Heusler 1996, 1998). In the Einstein–Maxwell theory, stationary black holes (the asymptotic final stage after the collapse) are uniquely characterized by *three* parameters: mass M, angular momentum J and electric charge q. All other degrees of freedom ('multipoles') are radiated away during the collapse ('black holes have no hair.')

In the presence of other fields (e.g., non-Abelian gauge fields) this theorem no longer necessarily holds, but the corresponding black hole solutions are usually unstable.

I want to conclude this section with a brief description of the stationary black hole solutions that have charge and/or angular momentum.

The spherically symmetric black hole solution with electric charge q is the *Reissner–Nordström* solution :

$$ds^2 = \left(1 - \frac{2GM}{r} + \frac{Gq^2}{r^2}\right) dt^2$$
$$- \left(1 - \frac{2GM}{r} + \frac{Gq^2}{r^2}\right)^{-1} dr^2 - r^2 \, d\Omega^2 . \qquad (2.21)$$

It can be generated from the Schwarzschild metric (2.3) through the substitution

$$M \longrightarrow M - \frac{q^2}{2r} .$$

The metric (2.21) is a solution to the nonvacuum Einstein equations

$$R_{ab} - \frac{1}{2} g_{ab} R = 8\pi G T_{ab} \qquad (2.22)$$

with the energy–momentum tensor corresponding to a point charge, $4\pi T_{ab} = q^2/(2r^4) \, \text{diag}(1, -1, -1, -1)$. In geometrical units, the unit of charge is given by $e\sqrt{G} \approx 1.38 \times 10^{-34}$ cm.

For charge $|q| < \sqrt{G}M$, the metric (2.21) has coordinate singularities at

$$r_\pm = GM \pm \sqrt{(GM)^2 - Gq^2} . \qquad (2.23)$$

The variable r_+ denotes the coordinate radius of the event horizon ($r_+ \to R_0$ for $q \to 0$); r_- characterizes the *Cauchy horizon*—the evolution of fields beyond the Cauchy horizon cannot be predicted from initial data on a spacelike hypersurface. All these features can be immediately recognized from the Penrose diagram of this solution, see figure 2.8.

The interesting feature is that the singularity at $r = 0$ is now timelike—and therefore is a naked singularity for some regions, e.g. regions III—and that there is an infinite repetition of this structure. It should be mentioned, however, that this structure is unstable with respect to small perturbations and that BKL oscillations are expected to occur; they would produce a spacelike singularity. Recent years have seen a tremendous progress in understanding the internal structure of black holes (Israel 1998).

The special case $|q| = \sqrt{G}M$ is referred to as describing an *extremal* black hole; this case plays a crucial role in the quantum gravity sections 2.5 and 2.6. The two horizons now coincide, and the corresponding Penrose diagram is shown in figure 2.9.

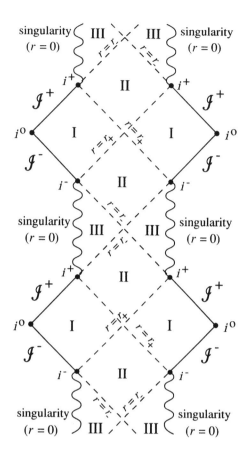

Figure 2.8. Penrose diagram for the Reissner–Nordström solution ($|q| < \sqrt{G}M$).

For $|q| > \sqrt{G}M$, there is no event horizon present: instead of a black hole, there is now a naked singularity. The Penrose diagram is shown in figure 2.10. From an astrophysical point of view, charged black holes are not expected to play any role since they would rapidly attract opposite charges and discharge. They play, however, a useful role as a toy model for the realistic case of rotating black holes to which I now turn.

The solution for a rotating stationary black hole, the so-called *Kerr* solution, is no longer spherically symmetric and static: it is axisymmetric and only stationary. This is characterized by the presence of the two Killing vectors $\xi^a = (\frac{\partial}{\partial t})^a$ and $\psi^a = (\frac{\partial}{\partial \phi})^a$. Apart from its mass M, the solution is characterized by its angular momentum J. It is sometimes more convenient to use the parameter $a = J/(GM)$ that has unit of length. In particular coordinates, so-called Boyer–Lindquist

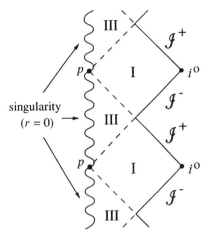

Figure 2.9. Penrose diagram for the extremal Reissner–Nordström solution ($|q| = \sqrt{G}M$). The points p do not belong to the singularity.

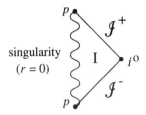

Figure 2.10. Penrose diagram for the Reissner–Nordström solution corresponding to a naked singularity ($|q| > \sqrt{G}M$).

coordinates, the Kerr line element reads

$$ds^2 = \frac{\Delta}{\rho^2} \left(dt - a \sin^2 \theta \, d\phi \right)^2 - \frac{\sin^2 \theta}{\rho^2} \left[(r^2 + a^2) \, d\phi - a \, dt \right]^2$$
$$- \frac{\rho^2}{\Delta} dr^2 - \rho^2 \, d\theta^2 \tag{2.24}$$

with

$$\rho^2 \equiv r^2 + a^2 \cos^2 \theta$$
$$\Delta \equiv r^2 - 2GMr + a^2 . \tag{2.25}$$

The most general solution for a stationary black hole is the Kerr–Newman solution: it possesses in addition an electric charge q and can be obtained from (2.24) through the substitution

$$M \longrightarrow M - \frac{q^2}{2r}.$$

The Penrose diagrams for the Kerr solution show similar features as in figures 2.8, 2.9 and 2.10. (The singularity at $\rho^2 = 0$ is now a ring singularity through which one can 'escape' to a strange anti-gravity universe.) For $|a| < GM$ one has coordinate singularities at $r_\pm = GM \pm \sqrt{(GM)^2 - a^2}$, r_+ referring to the event horizon, and r_- referring to the Cauchy horizon. Again, $|a| = GM$ is the extremal case and $|a| > GM$ describes a naked singularity.

In the next section I shall discuss in which sense these stationary black hole solutions obey formally the laws of thermodynamics.

2.2   The laws of black hole mechanics

In this section I shall discuss the laws of black hole mechanics, laws obeyed by the stationary black hole solutions, that formally mimic the laws of phenomenological thermodynamics. A general reference is Wald (1994), a very readable short introduction is Bekenstein (1980).

For this purpose it is illustrative to follow the historical route and start with some special properties of rotating black holes. The Kerr solution (2.24) exhibits the feature that the Killing field $\xi^a = (\frac{\partial}{\partial t})^a$ becomes spacelike ($\xi^a \xi_a < 0$) in a certain region *outside* the black hole. (For the spherically symmetric black holes, this happens only inside the event horizon.) This region is the so-called *ergosphere*, characterized by

$$r_+ < r < GM + \sqrt{(GM)^2 - a^2 \cos^2 \theta} . \tag{2.26}$$

The ergosphere is shown (for $a \leq GM$) in figure 2.11.

Since the Killing field ξ^a generates time translations at asymptotic infinity, its spacelike nature in the ergosphere means that an observer there would have to travel with more than the speed of light to follow an orbit of ξ^a—he thus cannot remain static and he is forced to rotate with the hole. This is the extreme version of the Lense–Thirring effect that is very weak in the vicinity of the Earth and that only recently has been observed. Quite generally, black holes are characterized by general relativistic effects that under ordinary circumstances are only minor corrections.

For $r \to r_+$, the coordinate angular velocity of this rotation becomes

$$\Omega_H = \frac{a}{r_+^2 + a^2} = \frac{a}{2GMr_+} . \tag{2.27}$$

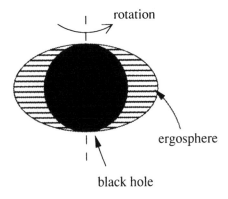

Figure 2.11. Ergosphere of a rotating black hole.

This can be interpreted as the angular velocity of the hole itself. The Killing field

$$\chi^a \equiv \xi^a + \Omega_H \psi^a \tag{2.28}$$

is tangential to the horizon.

Since $\chi^a \chi_a(r_+) = 0$, χ^a is null (and future directed) on the horizon, since the horizon is itself a null surface, χ^a is at the same time tangential and normal to the horizon—the horizon thus also constitutes a so-called *Killing horizon*.

We note that the area of a closed spacelike section through the event horizon is

$$A = \int\limits_{r=r_+} \sqrt{g_{\theta\theta} g_{\phi\phi}}\, d\theta\, d\phi = 4\pi (r_+^2 + a^2) \equiv 16\pi (G M_{\mathrm{irr}})^2 \tag{2.29}$$

where the *irreducible mass* M_{irr} has been introduced. Its significance will be explained below.

As Penrose noted in 1969, the presence of the ergosphere leads to the possibility that energy can be extracted from the black hole. Because ξ^a is spacelike in the ergosphere, the energy of a test particle, defined by $E = p^a \xi_a$, need not be positive. Suppose a projectile with energy E_0 is sent into the ergosphere. Let the projectile disintegrate while in the ergosphere, with one part (with energy E_1) falling down the hole and the other part (with energy E_2) escaping to infinity. If $E_1 < 0$, the recovered fragment has an energy *bigger* than E_0 ($E_0 = E_1 + E_2$). This energy must, of course, have been extracted from the mass of the black hole; $M \rightarrow M - |E_1|$. One also finds for the angular momentum L of the lost fragment that $L < E_1/\Omega_H$. So a negative-energy particle has negative angular momentum.

How much energy can be extracted in this way from the black hole? It is clear that the black hole parameters should obey $\delta J < \delta M / \Omega_H$ after the above fragment

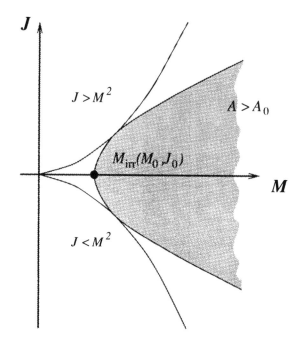

Figure 2.12. The region $A > A_0$ reachable via the Penrose process.

has been swallowed. Christodoulou was able in 1970 to rewrite this condition as a condition on the irreversible mass introduced in (2.29),

$$\delta M_{\text{irr}}^2 = \frac{M r_+}{\sqrt{(GM)^2 - a^2}} (\delta M - \Omega_{\text{H}} \delta J) > 0. \qquad (2.30)$$

Since the irreducible mass of the black hole can thus not be lowered by the Penrose process, the mass itself cannot be reduced below $M_{\text{irr}}(M_0, J_0)$, where M_0, J_0 denote the initial mass and angular momentum. Since $M_{\text{irr}}(M_0, J_0) \geq M_0/\sqrt{2}$, at most 29% of the mass can be extracted by the Penrose process. The parameter space allowed by the Penrose process is shown in figure 2.12.

I should note that there exists also a wave analogue of the Penrose process, the so-called *superradiance*. Consider an incident wave of the form $\phi_0(r, \theta)$ $\exp(-i\omega t) \exp(im\phi)$: for $0 < \omega < m\Omega_{\text{H}}$, the reflected wave has an energy greater than the incident wave, as was noted by Starobinsky and Zel'dovich in the 1970s.

The Penrose process could become of utmost importance in astrophysics: rotational energy of a black hole immersed in a magnetic field can be transferred to an escaping jet ('Blandford–Znajek process') as observed in active galaxies.

From M_{irr} one also finds, see (2.29),

$$\delta A \geq 0 \tag{2.31}$$

i.e. the area of the horizon cannot decrease.

Surprisingly, this *area law* holds much more generally than just for the Penrose process. This generalization is the content of Hawking's famous theorem that he proved in 1971:

For a predictable black hole satisfying $R_{ab}k^a k^b \geq 0$ for all null k^a, the surface area of the *future* event horizon *never* decreases with time.

A 'predictable' black hole is one for which the cosmic censorship hypothesis holds (see Wald 1984)—this is thus a major assumption for the area law. I emphasize that the time asymmetry comes into play because a statement is made about the future horizon, not the past horizon; the analogous statement for white holes would then be that the past event horizon never increases. I also emphasize that the area law only holds in the classical theory, not in the quantum theory (see section 2.3).

The area law seems to exhibit a close formal analogy to the second law of thermodynamics—there the *entropy* can *never* decrease with time (for a closed system). However, the conceptual difference could not be more drastic: while the second law is related to statistical behaviour, the area law is just a theorem in differential geometry. That the area law is in fact directly related to the second law will become clear in the course of this section.

Further support for this analogy is given by the existence of analogies to the other laws of thermodynamics. The zeroth law states that there is a quantity, the temperature, that is constant on a body in thermal equilibrium. Does there exist an analogous quantity for a black hole? We introduced in (2.28) the Killing field χ^a that is null and future directed on the horizon. Since then in particular $\chi^a \chi_a$ is constant (zero) on the horizon, its gradient is normal to the horizon and therefore parallel to χ^a. One has

$$\nabla^a(\chi^b \chi_b) = -2\kappa \chi^a \tag{2.32}$$

where κ is the so-called *surface gravity* that plays an important role in both classical and quantum black hole physics. For a Kerr black hole, κ is given by

$$\kappa = \frac{\sqrt{(GM)^2 - a^2}}{2GMr_+} \xrightarrow{a \to 0} \frac{1}{4GM} = \frac{GM}{R_0^2} \tag{2.33}$$

and one recognizes in the Schwarzschild limit the well known expression for the Newtonian gravitational acceleration. Note that κ also sets the time scale for the redshift during collapse, see (2.18). One can show that κ is the limiting force that must be exerted at infinity to hold a unit test mass in place when approaching the horizon. This justifies the name surface gravity.

It is now possible to prove (see Wald 1984) that κ is in fact *constant* on the event horizon. This is the desired formal analogy to the constancy of temperature on a body in thermal equilibrium.

The surface gravity also enters the relation between two important parameters: the horizon is generated by null geodesics that are conveniently parametrized by their affine parameter λ. On the other hand, there exists the so-called Killing parameter, v, that parametrizes the orbits of the Killing field (2.28). The two parameters are related by

$$\lambda \propto e^{\kappa v} \tag{2.34}$$

a relation that will be of crucial relevance in the discussion of the Hawking effect in section 2.3.

With a tentative formal relation between surface gravity and temperature, and between area and entropy, the question arises whether a first law of thermodynamics can be proved. This can in fact be done and the result for a Kerr–Newman black hole is

$$dM = \frac{\kappa}{8\pi G} dA + \Omega_H dJ + \Phi dq \tag{2.35}$$

where Φ denotes the electrostatic potential. As Wald (1994) emphasizes, this relation can be obtained by conceptually different methods: a *physical process version* whereby a stationary black hole is altered by infinitesimal physical processes, and an *equilibrium state version* whereby the areas of two infinitesimally nearby stationary black hole solutions to Einstein's equations are compared. Both methods lead to the same result (2.35).

Since M is the energy of the black hole, (2.35) is the analogue to the first law of thermodynamics given by

$$dE = T dS - p dV + \mu dN. \tag{2.36}$$

'Modern' derivations of (2.35) make use of both Hamiltonian and Lagrangian methods of general relativity (Wald 1998). For example, a first law follows from an arbitrary diffeomorphism invariant theory of gravity whose field equations can be derived from a Lagrangian.

What about the third law of thermodynamics? A 'physical process version' was proved by Israel—it is impossible to reach $\kappa = 0$ in a finite number of steps. This corresponds to the 'Nernst version' of the third law. Whether the stronger 'Planck version' holds is a matter of dispute and will be discussed at some length in section 2.5. The 'Planck version' states that the entropy goes to zero (or a material-dependent constant) if the temperature approaches zero. The above analogies are summarized in table 2.1.

The identification of the horizon area with an entropy can be obtained from a conceptually different point of view. If a box with, say, thermal radiation of entropy S is thrown into the black hole, it seems as if the second law could be violated, since the black hole is characterized only by mass, angular momentum and charge, and nothing else. The rescue of the second law immediately leads to the concept of a black hole entropy, as will be discussed now (Bekenstein 1980; Sexl and Urbantke 1983).

Consider a box with thermal radiation of mass m and temperature T lowered from a spaceship far away from a spherically symmetric black hole towards the hole (figure 2.13). As an idealization, both the rope and the walls are assumed to

Table 2.1. Analogies between laws of thermodynamics and black hole mechanics.

Law	Thermodynamics	Stationary black holes
Zeroth	T constant on a body in thermal equilibrium	κ constant on the horizon of a black hole
First	$dE = T\,dS - p\,dV + \mu\,dN$	$dM = \dfrac{\kappa}{8\pi G}\,dA + \Omega_H\,dJ + \Phi\,dq$
Second	$dS \geq 0$	$dA \geq 0$
Third	$T = 0$ cannot be reached	$\kappa = 0$ cannot be reached

have negligible mass. At a coordinate distance r from the black hole, the energy of the box is given by

$$E_r = m\sqrt{1 - \frac{2GM}{r}} \quad \overset{r \to R_0}{\longrightarrow} \quad 0. \tag{2.37}$$

If the box is lowered down to the horizon, the energy gain is thus given by m. The box is then opened and thermal radiation of mass δm escapes into the hole. If the box is then closed and raised again to the spaceship, the energy loss is $m - \delta m$. In total the energy δm of the thermal radiation can be transformed into work with a degree of efficiency $\eta = 1$. This looks as if one possessed a *perpetuum mobile* of the second kind.

The key to the resolution of this apparent paradox lies in the observation that the box must be big enough to contain the wavelength of the enclosed radiation. This, in turn, leads to a lower limit on the distance that the box can approach the horizon. Therefore, only part of δm can be transformed into work, as I shall show now.

According to Wien's law, one must have a linear extension of the box of at least

$$\lambda_{\max} \approx \frac{\hbar}{k_B T}. \tag{2.38}$$

I emphasize that at this stage Planck's constant $\hbar$ comes into play. The box can then be lowered down to the coordinate distance δr (assumed to be $\ll 2GM$) from the black hole, where according to the Schwarzschild metric (2.3) the relation between δr and $\lambda_{\max}$ is

$$\lambda_{\max} \approx \int_{2GM}^{2GM+\delta r} \left(1 - \frac{2GM}{r}\right)^{-\frac{1}{2}} dr \approx 2\sqrt{2GM\delta r} \implies \delta r \approx \frac{\lambda_{\max}^2}{8GM}.$$

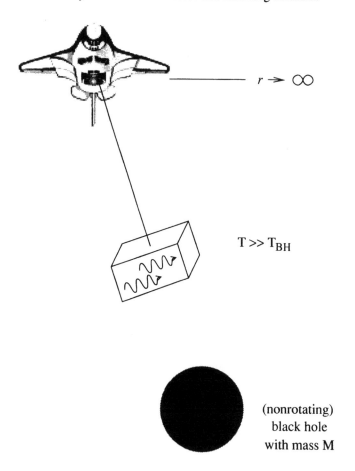

$r \to \infty$

$T \gg T_{BH}$

(nonrotating)
black hole
with mass M

Figure 2.13. Gedankenexperiment to demonstrate the second law of thermodynamics for black holes.

According to (2.37), the energy of the box at $r = 2GM + \delta r$ is

$$E_{2GM+\delta r} = m\sqrt{1 - \frac{2GM}{2GM + \delta r}} \approx \frac{m\lambda_{\max}}{4GM} \approx \frac{m\hbar}{4Gk_B T M}.$$

Recalling that according to (2.35) the formal temperature of the black hole, T_{BH}, is proportional to the surface gravity $\kappa = 1/(4GM)$, the energy of the box before opening is

$$E_{2GM+\delta r}^{(before)} \approx m\frac{T_{BH}}{T}$$

while after opening it is

$$E_{2GM+\delta r}^{(\text{after})} \approx (m - \delta m)\frac{T_{BH}}{T}.$$

The degree of efficiency of transforming thermal radiation into work is thus given by

$$\eta \approx \left(\delta m - \delta m \frac{T_{BH}}{T}\right)\bigg/ \delta m = 1 - \frac{T_{BH}}{T} < 1$$

which is just the well known Carnot limit for the efficiency of heat engines. From the first law (2.35) one then finds for the entropy of the black hole $S_{BH} \propto A = 16\pi (GM)^2$. It is this agreement of conceptually different approaches to black hole thermodynamics that gives rise to the confidence in the results. In the next section I shall show how all these formal results can be physically interpreted in the context of quantum theory.

2.3   Hawking radiation

We have already seen in the gedankenexperiment discussed in the last section that $\hbar$ enters the scene, see (2.38). That Planck's constant has to play a role, can be seen also from the First Law (2.35). Since $T_{BH}\,dS_{BH} = \kappa/(8\pi G)\,dA$, one must have

$$T_{BH} = \frac{\kappa}{G\zeta} \qquad S_{BH} = \frac{\zeta A}{8\pi}$$

with an undetermined factor ζ. What is the dimension of ζ? Since S_{BH} has the dimension of Boltzmann's constant k_B, k_B/ζ must have the dimension of a length squared. There is, however, only one fundamental length available, the Planck length

$$l_p = \sqrt{G\hbar} \quad \approx \quad 10^{-33}\,\text{cm}. \tag{2.39}$$

(We do not yet consider string theory, see section 6.)[2] Therefore,

$$T_{BH} \propto \frac{\hbar\kappa}{k_B} \qquad S_{BH} \propto \frac{k_B A}{G\hbar}. \tag{2.40}$$

The determination of the precise factors in (2.40) is the content of this section. To this purpose, it is necessary to briefly introduce the framework of quantum theory in curved spacetime. A standard reference for this is Birrell and Davies (1982), see also Wald (1994). An excellent brief introduction is Wipf (1998). Before entering this topic, I want to briefly recapitulate some basic notions of quantum field theory in flat Minkowski space. There, the Poincaré symmetry—together with Wightman axioms—select an invariant *vacuum state* and therefore a well defined notion of *particles*. All inertial observers agree on these notions.

---

[2] It is amusing to note that Planck found this length before the 'invention' of $\hbar$, since $\hbar$ is implicitly contained in Wien's law, cf. (2.38).

Consider for simplicity a free massive scalar field $\phi(x)$ satisfying the Klein–Gordon equation $(\Box + (m/\hbar)^2)\phi(x) = 0$. It can be decomposed into positive and negative frequencies with respect to the distinguished Killing time t according to

$$\phi(x) = \frac{1}{(2\pi)^{3/2}} \int \frac{d^3k}{\sqrt{2\omega_k}} \left[a_k \exp(i k \cdot x - i\omega_k t) + a_k^\dagger \exp(-i k \cdot x + i\omega_k t) \right]$$

$$\equiv \phi^{(+)}(x) + \phi^{(-)}(x) \tag{2.41}$$

where $\omega_k = \sqrt{k^2 + (m/\hbar)^2}$. In quantum theory, $\phi(x)$ is a field operator that satisfies

$$\left[\phi(x, t), \pi_\phi(y, t) \right] = i\hbar \, \delta^{(3)}(x - y) \tag{2.42}$$

where π_ϕ is the canonical momentum operator. From (2.41) and (2.42) one has

$$[a_k, a_{k'}^\dagger] = \delta^{(3)}(k - k') \tag{2.43}$$

the well known relations for the annihilation and creation operators of harmonic oscillators. From these relations the standard *Fock space* can be constructed; the vacuum state $|0\rangle$ is defined by

$$a_k|0\rangle = 0 \quad \forall_k \tag{2.44}$$

and excited states are found through application of the creation operators $a_k^\dagger$. In the presence of interactions, these concepts only hold in asymptotic 'free' regions. Since I shall be concerned with linear quantum fields only, it will not be necessary to elaborate on this.

I shall now consider quantum field theory on a globally hyperbolic spacetime with metric $g_{ab}(\det g_{ab} \equiv g)$. Restricting again attention to (minimally coupled) scalar fields, the curved-space version of the Klein–Gordon equation reads

$$\Box_g \phi + (m/\hbar)^2 \phi = 0 \tag{2.45a}$$

with

$$\Box_g = \frac{1}{\sqrt{-g}} \partial_a [\sqrt{-g}\, g^{ab} \partial_b]. \tag{2.45b}$$

Consider now two solutions, u_1 and u_2, of (2.45a, 2.45b). Their conjugate momenta are $\pi_1 = n^a \nabla_a u_1$ and $\pi_2 = n^a \nabla_a u_2$, where n^a denotes the normal vector with respect to some spacelike hypersurface Σ. One can define the following inner product for such solutions:

$$(u_1, u_2) \equiv i \int_\Sigma (u_1^* \pi_2 - \pi_1^* u_2)\, d^3x = (u_2, u_1)^*. \tag{2.46}$$

The Klein–Gordon equation guarantees that this inner product is independent of the choice of Σ ('independent of time'), but it is not positive definite.

Choose now a complete set of solutions $\{u_k, u_k^*\}$ (k can stand here for an arbitrary index, not necessarily k) normalized according to

$$(u_k, u_{k'}) = \delta(k, k') \quad \Rightarrow \quad (u_k^*, u_{k'}^*) = -\delta(k, k') \tag{2.47}$$

and

$$(u_k, u_{k'}^*) = 0. \tag{2.48}$$

The $\{u_k\}$ are the generalization of the plane-wave solutions

$$u_k = \frac{1}{(2\pi)^{3/2}} \frac{\exp(i\boldsymbol{k} \cdot \boldsymbol{x} - i\omega_k t)}{\sqrt{2\omega_k}}$$

used in (2.41) to curved spacetime. Since $\{u_k, u_k^*\}$ are a complete set of solutions, any field operator ϕ can—analogous to the flat-space case (2.41)—be expanded with respect to them,

$$\phi(x) = \int d\mu(k) \left(a_k u_k + a_k^\dagger u_k^* \right). \tag{2.49}$$

Here, $d\mu(k)$ is an abbreviation for the used measure. From (2.49) follows $a_k = (u_k, \phi)$ and $a_k^\dagger = -(u_k^*, \phi)$. As in the flat case, a Fock space can be constructed from the vacuum $|0\rangle_u$, where

$$a_k |0\rangle_u = 0 \quad \forall_k. \tag{2.50}$$

The vacuum state is supposed to be normalized according to $_u\langle 0|0\rangle_u$, where $\langle\,|\,\rangle$ denotes the positive definite inner product in the constructed Hilbert space (not to be confused with the inner product (2.46)).

The crucial point is now that in a general spacetime—in contrast to inertial coordinates for flat space—there is no distinguished set of coordinates, in particular no distinguished time, with respect to which (2.49) can be uniquely defined. This is of course a consequence of the 'general covariance' of general relativity. The definition of the vacuum therefore depends on the chosen set of solutions—this fact has already been taken into account by adjoining the index 'u' to $|0\rangle$ in (2.50).

One can therefore expand the field into a *different* set of complete solutions $\{v_p, v_p^*\}$,

$$\phi(x) = \int d\mu(p) \left(b_p v_p + b_p^\dagger v_p^* \right). \tag{2.51}$$

One can also expand one basis with respect to the other,

$$v_p = \int d\mu(k) \left(\alpha(p, k) u_k + \beta(p, k) u_k^* \right) \tag{2.52}$$

where α and β are the so-called *Bogolubov coefficients*:

$$\alpha(p, k) = (u_k, v_p) \qquad \beta(p, k) = -(u_k^*, v_p). \tag{2.53}$$

In an obvious matrix notation (suppressing the indices (p, k)), the Bogolubov coefficients obey the following conditions

$$\alpha\alpha^\dagger - \beta\beta^\dagger = 1 \tag{2.54}$$

$$\beta\alpha^T - \alpha\beta^T = 0. \tag{2.55}$$

Comparing the alternative expressions (2.49) and (2.51), one can also express the 'old' creation and annihilation operators with respect to the 'new' ones,

$$(a \quad a^\dagger) = (b \quad b^\dagger) \begin{pmatrix} \alpha & \beta \\ \beta^* & \alpha^* \end{pmatrix}. \tag{2.56}$$

For a given Fock space, the operator $a_k^\dagger a_k$ 'measures' the particle content of type k in a given state and is therefore called the *particle number operator*. Its expectation value with respect to the vacuum is of course zero. If, however, the expectation value of the 'new' particle number operator $b_p^\dagger b_p$ with respect to the 'old' vacuum $|0\rangle_u$ is calculated, the result does *not* vanish in general:

$$_u\langle 0|b_p^\dagger b_p|0\rangle_u = \int d\mu(k)|\beta(p, k)|^2. \tag{2.57}$$

The 'old' vacuum thus contains 'new' particles! Note that the integral in (2.57) may even be divergent, in which case both Fock spaces cannot be related by a unitary transformation (this is a possibility that exists in the case of infinitely many degrees of freedom). For quantum field theory on a curved spacetime (this is also true for general external fields) the definition of a vacuum—and therefore the whole particle concept—is ambiguous if $\beta(p, k)$ is nonvanishing; as Paul Davies once noted: 'particles don't exist'.

How can one define a sensible vacuum state? In general, no set of solutions to (2.45a, 2.45b) is distinguished. An exception holds if the external spacetime exhibits certain symmetries. For a *stationary* spacetime—a spacetime that has a timelike Killing vector $\xi^a \equiv (\partial/\partial t)^a$—there are distinguished modes of *positive frequency* that obey

$$\frac{\partial u_k}{\partial t} = -i\omega_k u_k. \tag{2.58}$$

(Strictly speaking, the left-hand side is the Lie derivative with respect to ξ^a.) Such solutions are a natural generalization of the plane waves in Minkowski space, see (2.41). If a different set of solutions $\{v_p\}$ is a linear combination of the $\{u_k\}$ only, i.e. independent of the $\{u_k^*\}$, the Bogolubov coefficient $\beta(p, k)$ is zero and both set of modes share a common vacuum state. For $\beta(p, k) \neq 0$, $\{v_p\}$ contains a mixture of positive frequencies $\{u_k\}$ and negative frequencies $\{u_k^*\}$ and the 'v-vacuum' contains 'u-particles' (and vice versa).[3]

---

[3] Even if there are no Killing fields present, there exists for a globally hyperbolic spacetime a distinguished vacuum state if the two-point functions obey the so-called *Hadamard condition* (Wipf 1998).

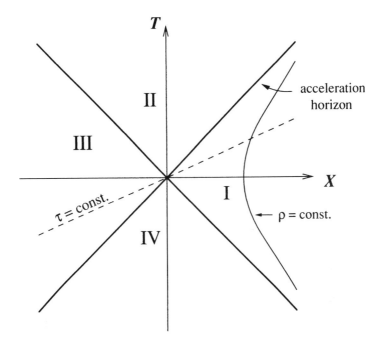

Figure 2.14. Uniformly accelerated observer in Minkowski space.

Before discussing the Hawking effect, I want to address briefly an analogous effect discovered by Unruh (1976) that already exists for *noninertial* observers in flat space. As remarked above, only inertial observers share the standard Minkowski vacuum. What happens for noninertial observers?

Consider an observer that is uniformly accelerating along the X-direction in (1+1)-dimensional Minkowski spacetime (figure 2.14). To emphasize the analogy with the Kruskal situation (figure 2.4), the Minkowski cartesian coordinates are labelled *here* by upper-case letters. The orbit of this observer is the hyperbola shown in figure 2.14. One recognizes that, as in the Kruskal situation, the observer encounters a horizon; there is, however, no singularity behind this horizon. The region I is a globally hyperbolic spacetime on its own—the so-called *Rindler spacetime*. This spacetime can be described by coordinates (τ, ρ) that are connected to the Cartesian coordinates via the coordinate transformation

$$\begin{pmatrix} T \\ X \end{pmatrix} = \rho \begin{pmatrix} \sinh a\tau \\ \cosh a\tau \end{pmatrix} \tag{2.59}$$

where a is a constant (the orbit in figure 2.14 describes an observer with acceleration a, who has $\rho = 1/a$).

Since

$$ds^2 = dT^2 - dX^2 = a^2 \rho^2 \, d\tau^2 - d\rho^2 \tag{2.60}$$

the orbits $\rho = $ constant are also orbits of a timelike Killing field $\partial/\partial\tau$. It is clear that τ corresponds to the external Schwarzschild coordinate t and that ρ corresponds to r in figure 2.4. Like in the Kruskal case, $\partial/\partial\tau$ becomes spacelike in regions II and IV.

The analogy with Kruskal becomes even more transparent if the Schwarzschild metric (2.3) is expanded around the horizon at $r = 2GM$. Introducing $\rho^2/(8GM) = r - 2GM$ and recalling (2.33), one has

$$ds^2 \approx \kappa^2 \rho^2 \, dt^2 - d\rho^2 - \frac{1}{4\kappa^2} \, d\Omega^2 . \tag{2.61}$$

Comparison with (2.60) shows that the first two terms on the right-hand side of (2.61) correspond exactly to the Rindler spacetime (2.60) with the acceleration a replaced by the surface gravity κ. The last term in (2.61) describes a two-sphere with radius $(2\kappa)^{-1}$.[4]

An inertial observer in Minkowski spacetime would of course employ the quantization of a massless scalar field according to (2.41). In (1+1) dimensions one has, with $\omega_k = |k|$,

$$\phi(T, X) = \int dk (a_k u_k + a_k^\dagger u_k^*) \tag{2.62a}$$

$$u_k(T, X) = \frac{1}{\sqrt{4\pi|k|}} e^{-i|k|T + ikX} . \tag{2.62b}$$

The accelerated observer is restricted to region I and employs a quantization scheme that is adapted to the 'Rindler coordinates' τ and ρ. Instead of the plane waves (2.62b) one has to use the corresponding set of solutions to (2.45a, 2.45b) rewritten in terms of τ and ρ.

This leads to (Birrell and Davies 1982)

$$\phi(\tau, \rho) = \int dp (b_p v_p + b_p^\dagger v_p^*) \tag{2.63a}$$

$$v_p(\tau, \rho) = \frac{1}{\sqrt{4\pi|p|}} e^{-i|p|\tau} \rho^{ip/a} . \tag{2.63b}$$

Calculating the Bogolubov coefficient $\beta(p, k) = -(u_k^*, v_p)$, see (2.53), one finds for the expectation value of the particle number operator $b_p^\dagger b_p$ with respect to the standard Minkowski vacuum $|0\rangle_M$ the expression

$$_M\langle 0|b_p^\dagger b_p|0\rangle_M = \int dk |\beta(p, k)|^2$$

$$= (\text{volume}) \times \frac{1}{e^{2\pi|p|/a} - 1} . \tag{2.64}$$

[4] It is this term that is responsible for the nonvanishing curvature of (2.61) compared to the flat-space metric (2.60) whose extension into the (neglected) other dimensions would be just $-dY^2 - dZ^2$.

(The volume term becomes infinite if the orbit is infinitely long.) Equation (2.64) describes a *Planckian distribution* at a temperature

$$T_U = \frac{\hbar a}{2\pi k_B} \approx 4 \times 10^{-23} a \left[\frac{cm}{s^2} \right] K. \tag{2.65}$$

An observer that is accelerating uniformly through Minkowski space thus sees a *thermal* distribution of particles. This is an important manifestation of the nonuniqueness of the vacuum state in quantum field theory, even for flat spacetime.

A more detailed investigation of an accelerated detector makes use of the so-called *response function* $\mathcal{F}(E)$ as evaluated along the spacetime path $x(\tau) \equiv (T(\tau), X(\tau))$,

$$\mathcal{F}(E) = \int_{-\infty}^{\infty} d\tau \int_{-\infty}^{\infty} d\tau' \, e^{-iE(\tau-\tau')/\hbar} \,_M \langle 0 | \phi(x(\tau)) \phi(x(\tau')) | 0 \rangle_M \tag{2.66}$$

where E denotes the detector's energy.

For the situation of the uniformly accelerating observer, the two-point function appearing in (2.66) again contains a Planck factor with the temperature (2.65), see Birrell and Davies (1982). The vacuum two-point function for a uniformly accelerated detector corresponds to the thermal two-point function for an inertial detector.

Can the Unruh temperature (2.65) be observed? Although T_U is tiny for most accelerations, it might be noticeable for electrons in accelerators where spin precession is used as 'detector'. Unruh (1998) has argued that the well known depolarization effect of an electron in a storage ring can be interpreted as a vacuum effect of this kind; however, due to the *circular* nature of the acceleration, the effect is not a thermal one.[5]

I shall now turn to the case of black holes. From the form of the line element near the horizon, (2.61), one can already anticipate that—according to the equivalence principle—there is a black hole radiation with temperature (2.65) in which a is replaced by κ. This is in fact what we shall find.

The following discussion follows the original calculation performed by Hawking (1975). We consider a spherically symmetric star that collapses to form a black hole, see figure 2.15. I shall again treat the case of a scalar field, see (2.3). Because the background is spherically symmetric, a solution of the Klein–Gordon equation may be separated according to

$$\phi(x) = \frac{f(t, r)}{r} Y_{lm}(\theta, \phi). \tag{2.67}$$

Inserting this *ansatz* into (2.3) and using the coordinate r_*, see (2.12), one finds

$$\frac{\partial^2 f}{\partial t^2} - \frac{\partial^2 f}{\partial r_*^2} + V(r_*)f = 0. \tag{2.68}$$

---

[5] A different, but related, effect is the radiation produced by an accelerating mirror through Minkowski space. It has been argued that the observed sonoluminescence is a manifestation of this effect, see e.g. Liberati *et al* (1998).

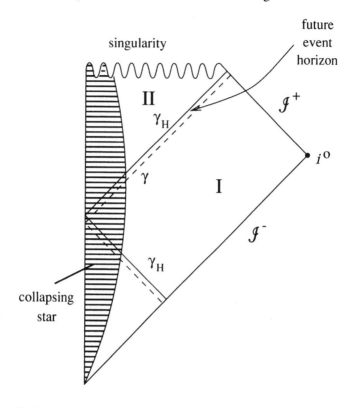

Figure 2.15. Penrose diagram showing the collapse of a star to form a black hole; γ denotes a light ray that is traced back from $\mathfrak{J}^+$ through the collapsing star to $\mathfrak{J}^-$.

This is just a two-dimensional wave equation with a potential

$$V(r_*) = \left(1 - \frac{2GM}{r}\right)\left[\frac{l(l+1)}{r^2} + \frac{2GM}{r^3} + \left(\frac{m}{\hbar}\right)^2\right]. \qquad (2.69)$$

The task now is to solve this equation with appropriate boundary conditions. For r approaching the horizon, $r \to 2GM$, the potential vanishes and (2.68) becomes a free wave equation. For $r \to \infty$, the potential approaches $(m/\hbar)^2$ for $m \neq 0$ and zero for $m = 0$. For simplicity, I shall restrict myself to $m = 0$, but this qualitatively also reflects the case $m \neq 0$.

It is convenient to use the null coordinates

$$u = t - r_* \qquad v = t + r_* \qquad (2.70)$$

that play the role of a retarded and an advanced time, respectively. In figure 2.15, v runs along $\mathfrak{J}^-$ from $-\infty$ to $+\infty$, and u runs along $\mathfrak{J}^+$ from $-\infty$ to $+\infty$.

Considering for the moment the full Kruskal spacetime, figure 2.5, the solution to (2.68) is for $m = 0$ uniquely fixed by either specifying $f(t, r)$ on the union of future horizon and $\mathfrak{J}^+$ or on the union of past horizon and $\mathfrak{J}^-$.

On $\mathfrak{J}^+$ and $\mathfrak{J}^-$ there is a well defined notion of positive frequency in the sense of (2.58) ($f \propto \exp(-i\omega t)$), while there is some ambiguity in this definition on the horizon. One can there, for example, use either the affine parameter or the Killing parameter, which are related by (2.34). The exact definition is, however, irrelevant as long as one restricts attention only to 'measurements' in region I, since the definition of creation and annihilation operators of particles escaping to $\mathfrak{J}^+$ is independent of the notion used (Wald 1984).

We now consider solutions in region I of the full Kruskal manifold that have positive frequency, $f_\omega \propto \exp(-i\omega t)$. Since for $r \to 2GM(r_* \to -\infty)$ the potential (2.69) vanishes, such solutions approach close to the horizon 'plane waves' in (t, r_*)-coordinates,

$$f_\omega(t, r_*) = a\, e^{-i\omega t}\, e^{i\omega r_*} + b\, e^{-i\omega t}\, e^{-i\omega r_*}$$

$$= a\, e^{-i\omega u} + b\, e^{-i\omega v}$$

$$\equiv f_\omega^{(\text{out})} + f_\omega^{(\text{in})}\,. \tag{2.71}$$

The solution $f_\omega^{(\text{out})}$ is referred to as 'outgoing', since $u = $ constant corresponds to 'light rays' escaping to $\mathfrak{J}^+$; analogously, the solution $f_\omega^{(\text{in})}$ is 'ingoing', since $v = $ constant corresponds to 'light rays' penetrating the horizon.

It is now important to notice that close to the horizon $u \to \infty$ and that therefore $f_\omega^{(\text{out})}$ rapidly oscillates. This becomes especially transparent by using Kruskal coordinates: defining from the standard coordinates (T, X) in (2.1) the corresponding null coordinates

$$U = T - X \qquad V = T + X \tag{2.72}$$

one has

$$U = -\exp\left(-\frac{u}{4GM}\right) = -e^{-\kappa u} \tag{2.73a}$$

$$V = \exp\left(\frac{v}{4GM}\right) = e^{\kappa v} \tag{2.73b}$$

and therefore

$$f_\omega^{(\text{out})} = a\, e^{i\omega\kappa^{-1}\ln(-U)}\,. \tag{2.74}$$

Note that $U \to 0$ as the horizon is approached. Because of the rapid oscillation of $f_\omega^{(\text{out})}$, the approximation of geometric optics should be excellent for $r \to 2GM$; this is why it is justified to talk of 'light rays' (more precisely, rays corresponding to the scalar field).

Consider a null geodesic that is entering the black hole region II from region I; let λ be its affine parameter ($\lambda = 0$ corresponding to the crossing point with the horizon). Then $U = -\lambda$ (since $ds^2 \sim dU \, dV + \cdots$, U is the affine parameter, not u) and one has

$$f_\omega^{(\text{out})} = a \, e^{i\omega\kappa^{-1}\ln\lambda}.$$ (2.75)

The frequency of each mode thus *diverges* at the horizon—an extreme manifestation of the gravitational redshift (compare (2.18)) that is also responsible for the behaviour of the modes of a quantum field. In fact, this redshift lies at the heart of the Hawking effect.

Hawking (1975) now noticed that it is most convenient to consider in the collapse diagram (figure 2.15) an outgoing ray at $\mathfrak{I}^+$ and trace it *back* to $\mathfrak{I}^-$; one part is directly scattered back to $\mathfrak{I}^-$, the other part passes *through* the collapsing matter and reaches $\mathfrak{I}^-$—it is this part that is of interest for our analysis. In figure 2.15, γ is an example of such a ray; the passage through the collapsing star corresponds in the diagram to a reflection at the origin. The ray γ_H denotes a limiting ray that stays on the future horizon and is traced back to $\mathfrak{I}^-$. Since the considered rays (such as γ) are close to γ_H, the potential V in (2.68) is negligible.

The propagator of γ back to $\mathfrak{I}^-$ reaches $\mathfrak{I}^-$ at a distance λ in the affine parameter along null geodesics on $\mathfrak{I}^-$ (this is because $ds^2 = -du \, dv + \cdots$ along $\mathfrak{I}^-$, so that v is there the affine parameter). If the crossing of γ_H with $\mathfrak{I}^-$ is at $v = 0$, γ reaches $\mathfrak{I}^-$ at $v = -\lambda$.

The propagation of (2.75) therefore leads to the following solution in the vicinity of $v = 0$ on $\mathfrak{I}^-$:

$$f_\omega(v) = \begin{cases} a \, e^{i\omega\kappa^{-1}\ln(-v)} & v < 0 \\ 0 & v > 0. \end{cases}$$ (2.76)

Since γ_H is the limiting ray, $f_\omega(v)$ vanishes for $v > 0$. To obtain the frequency content of (2.76), its Fourier transform is calculated:

$$\tilde{f}_\omega(\omega') = \frac{1}{\sqrt{2\pi}} \int\limits_{-\infty}^{\infty} dv \, e^{i\omega'v} f_\omega(v)$$

$$\sim \int\limits_{-\infty}^{0} dv \, e^{i\omega'v} \, e^{i\omega\kappa^{-1}\ln(-v)} = \int\limits_{0}^{\infty} dv \, e^{-i\omega'v} \, v^{i\omega\kappa^{-1}}.$$ (2.77)

(Note the similarity of the integrand to (2.63b) for the Unruh effect.) The integral (2.77) can be evaluated if use is made of the integral formula

$$\int\limits_{0}^{\infty} dx \, x^{\nu-1} \exp[-(A + iB)x] = \Gamma(\nu)(A^2 + B^2)^{-\nu/2} \exp\left(-i\nu \arctan \frac{B}{A}\right).$$

(2.78)

One then easily recognizes that, taking $\omega' > 0$,

$$\tilde{f}_\omega(-\omega') = -e^{\omega\pi\kappa^{-1}} \, \tilde{f}_\omega(\omega') \neq 0$$ (2.79)

so that a mode of positive frequency on $\mathfrak{I}^+$ is a *mixture of positive and negative frequency* on $\mathfrak{I}^-$! To find the exact amount of particle creation, the Bogolubov coefficient β has to be calculated. Decomposing in the manner of (2.52) the solution $f_\omega(v)$, (2.76), which is of positive frequency on $\mathfrak{I}^+$, into positive and negative frequencies on $\mathfrak{I}^-$, one has up to numerical factors

$$f_\omega(v) \sim \int\limits_0^\infty d\omega \left(\alpha_{\omega\omega'} \frac{e^{-i\omega' v}}{\sqrt{\omega'}} + \beta_{\omega\omega'} \frac{e^{i\omega' v}}{\sqrt{\omega'}} \right). \tag{2.80}$$

On the other hand, using (2.79),

$$f_\omega(v) = \frac{1}{\sqrt{2\pi}} \int\limits_{-\infty}^\infty d\omega' \, e^{-i\omega' v} \, \tilde{f}_\omega(\omega')$$

$$\sim \int\limits_0^\infty d\omega' \left(\tilde{f}_\omega(\omega') e^{-i\omega' v} - e^{-\omega\pi\kappa^{-1}} \tilde{f}_\omega(\omega') e^{i\omega' v} \right). \tag{2.81}$$

Comparing (2.80) with (2.81), one finds

$$\beta_{\omega\omega'} = -e^{-i\omega\pi\kappa^{-1}} \alpha_{\omega\omega'}. \tag{2.82}$$

Using (2.76) and (2.80) one can then evaluate α and β in terms of Γ-functions. If one now attempted to calculate the particle number expectation value according to (2.57), one would find a diverging result. This is due to the fact that the collapsing body produces particles infinitely long, as can be recognized from figure 2.15 (again an effect of the infinite redshift at the horizon). One can instead calculate the number of emitted particles per unit time (Birrell and Davies 1982)—either by considering wave packets or confining the system into a box—to obtain for the number of particles per unit time in the frequency range ω to $\omega + d\omega$ the expression

$$\frac{d\omega}{2\pi} \frac{1}{e^{2\omega\pi\kappa^{-1}} - 1}. \tag{2.83}$$

This is a Planck distribution with the temperature

$$T_{BH} = \frac{\hbar\kappa}{2\pi k_B}. \tag{2.84}$$

One immediately notes that this 'Hawking temperature' follows from the Unruh temperature in (2.65) through the substitution $a \to \kappa$, as anticipated.

An alternative derivation of this result can be made through the use of the energy–momentum tensor (DeWitt 1975). This treatment also allows a straightforward implementation of the back-scattering effect: if the potential (2.69) is fully taken into account, some of the modes are scattered back into the black hole instead of escaping to $\mathcal{J}^+$. This leads to the following expression for the total luminosity of the black hole:

$$L = -\frac{dM}{dt} = \frac{1}{2\pi} \sum_{l=0}^{\infty} (2l+1) \int_0^{\infty} d\omega\, \omega \frac{\Gamma_{\omega l}}{e^{2\omega\pi\kappa^{-1}} - 1}. \qquad (2.85)$$

The term $\Gamma_{\omega l}$—called 'greybody factor' because it encodes a deviation from the blackbody spectrum—is the fraction of the incoming mode that enters the black hole; it depends explicitly on ω and l. Its calculation requires a detailed discussion of (2.68) and (2.69) (and similar equations for higher spins).

For the special case of the Schwarzschild metric, $\kappa = (4GM)^{-1}$, and (2.84) becomes

$$T_{\text{BH}} = \frac{\hbar}{8\pi G k_B M} \approx 10^{-6} \frac{M_\odot}{M}\, \text{K}. \qquad (2.86)$$

For solar-mass black holes, this is of course utterly negligible—the black hole absorbs much more from the ubiquitous 3 K microwave background radiation than it radiates.

One can, however, estimate the lifetime of a black hole by making the plausible assumption that the decrease in mass is equal to the energy radiated to infinity and using the Stefan–Boltzmann law:

$$\frac{dM}{dt} \propto -A T_{\text{BH}}^4 \propto -M^2 \times \left(\frac{1}{M}\right)^4 = -\frac{1}{M^2}$$

which integrated yields

$$t(M) \propto (M_0^3 - M^3) \approx M_0^3 \qquad (2.87)$$

where M_0 is the initial mass, and it has been assumed that after the evaporation $M \ll M_0$. Very roughly, the lifetime of a black hole is thus given by

$$\tau_{\text{BH}} \approx \left(\frac{M_0}{m_p}\right)^3 t_p \approx 10^{65} \left(\frac{M_0}{M_\odot}\right)^3 \text{years} \qquad (2.88)$$

(m_p and t_p denote Planck mass and Planck time: $m_p = \hbar/l_p$, $t_p = l_p$.) If in the early universe primordial black holes with $M_0 \approx 5 \times 10^{14}$ g were created, they would evaporate at the present age of the universe.

A very detailed investigation into black hole evaporation was made by Page (1976). He found that for $M \gg 10^{17}$ g the power emitted from an (uncharged, nonrotating) black hole is

$$P \approx 2.28 \times 10^{-54} L_\odot \left(\frac{M}{M_\odot}\right)^{-2}$$

of which 81.4% is in neutrinos (he considered only electron and muon neutrinos), 16.7% in photons and 1.9% in gravitons, assuming of course that there are no other massless particles. Since a black hole evaporates *all* existing particles in Nature, this result would of course be changed by the existence of massless supersymmetric or other particles. In the range 5×10^{14} g $\ll M \ll 10^{17}$ g, Page found

$$P \approx 6.3 \times 10^{16} \left(\frac{M}{10^{15} \text{ g}} \right)^{-2} \frac{\text{erg}}{\text{s}}$$

of which 45% is in electrons and positrons, 45% in neutrinos, 9% in photons and 1% in gravitons. Massive particles with mass m are only suppressed if $k_B T_{BH} < m$. For $M < 5 \times 10^{14}$ g, also higher-mass particles are emitted.

All of the above derivations use the approximation where the spacetime background remains classical.[6] In a theory of quantum gravity, however, such a picture cannot be maintained, see sections 2.5 and 2.6. Since the black hole becomes hotter while radiating, see (2.86), its mass will eventually enter the quantum-gravity domain $M \approx m_p$, where the semiclassical approximation breaks down. The evaporation then enters the realm of speculation, see the following sections. As an intermediate step one might consider the heuristic 'semiclassical' Einstein equations,

$$R_{ab} - \tfrac{1}{2} g_{ab} R = 8 \pi G \langle T_{ab} \rangle \tag{2.89}$$

where on the right-hand side the quantum expectation value of the energy–momentum tensor appears. The evaluation of $\langle T_{ab} \rangle$—which requires regularization and renormalization—is a difficult subject on its own (Birrell and Davies 1982). The renormalized $\langle T_{ab} \rangle$ is essentially unique (its ambiguities can be absorbed in coupling constants) if certain sensible requirements are imposed (Wald 1984). Technically, it is most convenient to handle $\langle T_{ab} \rangle$ through the use of the *effective action* (Wipf 1998). Evaluating the components of the renormalized $\langle T_{ab} \rangle$ near the horizon, one finds that there is a flux of *negative energy* into the hole. Clearly this leads to a decrease in the black hole's mass. These negative energies are a typical quantum effect and are well known from the—accurately measured—Casimir effect. This occurrence of negative energies is also responsible for the breakdown of the classical area law discussed in the last sections.

The negative flux near the horizon lies also at the heart of the 'pictorial' representation of Hawking radiation that is often used (see figure 2.16). In vacuum, virtual pairs of particles are created and destroyed. However, close to the horizon, one partner of this virtual pair might fall into the black hole, thereby liberating the other partner to become a real particle and escaping to infinity as Hawking radiation.

It is interesting to note that the quantum fields exhibit entanglement ('EPR correlations') between the interior and exterior of the event horizon (Wald 1986). This was shown for both the 'eternal hole' (figure 2.5) and the Rindler spacetime (figure 2.14), see Birrell and Davies (1982). The global vacuum state comprising

---

[6] This limit is referred to as the semiclassical approximation to quantum gravity (see e.g. Kiefer 1994).

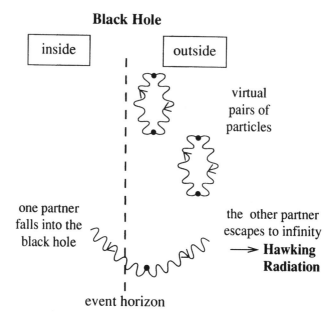

Figure 2.16. Heuristic 'visualization' of the Hawking effect.

the regions I and II in these diagrams can be written in the form

$$|0\rangle = \prod_{\omega} \sqrt{1 - e^{-2\pi\omega\kappa^{-1}}} \sum_{n} e^{-n\pi\omega\kappa^{-1}} |n_{\omega}^{\mathrm{I}}\rangle \otimes |n_{\omega}^{\mathrm{II}}\rangle \qquad (2.90)$$

where $|n_{\omega}^{\mathrm{I}}\rangle$ and $|n_{\omega}^{\mathrm{II}}\rangle$ are n-particle states with frequency ω in regions I and II, respectively; in the situation of figure 2.14, κ has to be replaced by a. The expression (2.90) is just the Schmidt expansion for two entangled quantum states, see e.g. Giulini *et al* (1996); note the analogy of (2.90) to a BCS state in the theory of superconductivity.

Since in the presence of an event horizon, observations are restricted to the outside region, the state (2.90) cannot be distinguished by operators with support in I only from a density matrix that is found from (2.90) by tracing out all degrees of freedom in region II,

$$\rho_{\mathrm{I}} \equiv \mathrm{Tr}_{\mathrm{II}} \, |0\rangle\langle 0| = \prod_{\omega} \left(1 - e^{-2\pi\omega\kappa^{-1}}\right) \sum_{n} e^{-2\pi n\omega\kappa^{-1}} |n_{\omega}^{\mathrm{I}}\rangle\langle n_{\omega}^{\mathrm{I}}|. \qquad (2.91)$$

Note that ρ_{I} describes a canonical ensemble with the temperature (2.84). The thermal nature of Hawking radiation is thus a consequence of the fact that observations are restricted to region I—and this is a consequence of the presence of an event horizon!

I want to end this section by giving the explicit expressions for the Hawking temperature (2.84) in the case of rotating and charged black holes. For the Kerr solution (2.24), one has

$$k_B T_{BH} = \frac{\hbar \kappa}{2\pi} = 2 \left(1 + \frac{M}{\sqrt{M^2 - a^2}} \right)^{-1} \frac{\hbar}{8\pi M} < \frac{\hbar}{8\pi M}. \qquad (2.92)$$

Rotation thus reduces the Hawking temperature. The integrand in (2.85) then becomes

$$\frac{\omega \Gamma_{\omega l}}{e^{2\pi \kappa^{-1}(\omega - m\Omega_H)} - 1} \qquad (2.93)$$

where Ω_H is given by (2.27) and m is here the azimuthal number of the incident wave. For $\omega - m\Omega_H < 0$ and $\kappa \to 0$ (i.e., $T_{BH} \to 0$), (2.93) goes to $-\omega \Gamma_{\omega l}$: this is just the classical phenomenon of superradiance mentioned in section 2.2 (see the paragraph above (2.31)), see also DeWitt (1975).

For the Reissner–Nordström solution (2.21) one has

$$k_B T_{BH} = \frac{\hbar}{8\pi M} \left(1 - \frac{(Gq)^4}{r_+^4} \right) < \frac{\hbar}{8\pi M}. \qquad (2.94)$$

Thus, also electric charge reduces the Hawking temperature. For an extremal black hole, $r_+ = GM = \sqrt{G}|q|$, and thus $T_{BH} = 0$. The question whether its entropy is also zero or proportional to $A \neq 0$ plays a crucial role in the quantization of black holes, see sections 2.5 and 2.6.

2.4 Interpretation of entropy and the problem of information loss

We have seen in the last section that—if quantum theory is taken into account—black holes emit thermal radiation with the temperature (2.84). Consequently, the laws of black hole mechanics discussed in section 2.2 have indeed a physical interpretation as thermodynamical laws—black holes *are* thermodynamical systems.

One can therefore from the first law (2.35) also infer the expression for the black hole entropy. From $dM = T_{BH} dS_{BH}$ one finds the 'Bekenstein–Hawking entropy'

$$S_{BH} = \frac{k_B A}{4G\hbar} \qquad (2.95)$$

in which the unknown factor in (2.40) has now been fixed. For the special case of a Schwarzschild black hole, this yields

$$S_{BH} = \frac{k_B \pi R_0^2}{G\hbar}. \qquad (2.96)$$

It can easily be estimated that S_{BH} is much bigger than the entropy of the star that collapsed to form the black hole. A physical interpretation of S_{BH} must therefore be based on other principles—but on which? Certainly, up to now the laws of black hole mechanics are only phenomenological thermodynamical laws. The central open question is: can S_{BH} be derived from quantum-statistical considerations? This would mean that S_{BH} could be calculated from a Gibbs-type formula according to

$$S_{BH} \overset{?}{=} -k_B \, \text{Tr}(\rho \ln \rho) \equiv S_{SM} \qquad (2.97)$$

where ρ denotes an appropriate density matrix; S_{BH} would then somehow correspond to the number of quantum microstates that are consistent with the macrostate of the black hole that is—according to the no hair theorem—uniquely characterized by mass, angular momentum, and charge. Some important questions are:

- Does S_{BH} correspond to states hidden behind the horizon?
- Does S_{BH} corresponds to the number of possible initial states?
- Where is S_{BH} located (if it is located at all)?
- What happens to S_{BH} after the black hole has evaporated?

There have been some attempts to calculate S_{BH} by counting internal states of freedom (see the review in Kiefer 1998). However, although one could derive $S_{BH} \propto A$ in this way, the factor of proportionality was divergent and needed some regularization. At least these derivations seem to indicate that the entropy is somehow located at or near to the event horizon. Preliminary results in the theory of induced gravity show that a regularization can indeed be invoked such that the desired result (2.95) can be obtained, although at the price of introducing nonminimally coupled fields. The only clear-cut microscopic derivation of S_{BH} was done in string theory (see section 2.6), although the applicability is as yet restricted to extremal (or near-extremal) black holes.

The attempts to calculate S_{BH} by state counting are usually done in the 'one-loop limit' of quantum field theory in curved spacetime—this is the limit where gravity is classical but nongravitational fields are fully quantum, and it is the limit where the Hawking radiation (2.84) has been derived. Surprisingly, however, the expression (2.95) can already be calculated from the so-called 'tree level' of the theory, where only the gravitational degrees of freedom are taken into account (Hawking 1979, Hawking and Penrose 1996). This is already reviewed in Kiefer (1998), and I shall be brief on this in the following.

The derivations employ the analogy of Euclidean path integrals and partition sums. Calculating the Euclidean path integral to quantum gravity at highest order (tree-level or saddle-point approximation), one finds for the partition sum in the Schwarzschild case

$$Z \approx \exp\left(-\frac{\hbar \beta^2}{16\pi G}\right) \qquad (2.98)$$

with $\beta = (k_B T)^{-1}$. One can then calculate the standard thermodynamical quantities in the usual manner. In this way one arrives at the mean energy

$$E = -\frac{\partial \ln Z}{\partial \beta} = \frac{\hbar \beta^2}{8\pi G} \tag{2.99}$$

from which (since E must be equal to the mass M of the black hole) the Hawking temperature (2.86) can be found, $T = T_{BH}$. For the entropy one then finds

$$S = k_B(\ln Z + \beta M) = \frac{\hbar \beta^2}{16\pi G} = \frac{k_B A}{4G\hbar} = S_{BH} \tag{2.100}$$

and for the specific heat

$$C = -\beta \frac{\partial S}{\partial \beta} = -\frac{\hbar \beta^2}{8\pi G} < 0 \tag{2.101}$$

whose negativity signals instability; this just expresses the fact that the black hole in asymptotically flat space becomes hotter by radiating—a typical thermodynamical feature of gravitational systems (Zeh 1992). One can try to stabilize the black hole by putting it into a box or embedding it into an asymptotically anti-de Sitter space, but I shall not elaborate on this here.

It is also instructive to see how the Hawking temperature in the Schwarzschild case can be found from the *Euclidean* line element of the Schwarzschild metric (2.3). Writing $\tau = it$, one obtains

$$ds^2 = \left(1 - \frac{2GM}{r}\right) d\tau^2 + \left(1 - \frac{2GM}{r}\right)^{-1} dr^2 + r^2 d\Omega^2. \tag{2.102}$$

Introducing the new radial coordinate

$$R = 4GM\sqrt{1 - \frac{2GM}{r}} \tag{2.103}$$

this assumes the form

$$ds^2 = R^2 d\left(\frac{\tau}{4GM}\right)^2 + \left(\frac{r(R)}{2GM}\right)^4 dR^2 + r^2(R) d\Omega^2. \tag{2.104}$$

This metric has a coordinate singularity at $R = 0$ (corresponding to $r = 2GM$). *Regularity* is obtained if $\tau/(4GM)$ is interpreted as an angular coordinate with periodicity 2π; τ itself has then periodicity $8\pi GM$ (figure 2.17) which, when set equal to $\beta\hbar$, yields the Hawking temperature (2.86). This result suggests the existence of a thermal equilibrium state on the eternal black hole spacetime (figure 2.5) at a temperature $T = T_{BH}$; this state is the so-called *Hartle–Hawking state* (a general formulation is achieved in terms of KMS states).

Since R in (2.103) is only defined outside the horizon ($r > 2GM$), the Euclidean Schwarzschild line element (2.102) does not penetrate into the horizon—this is an expression of the fact that the interior is never classically forbidden, see section 2.5.

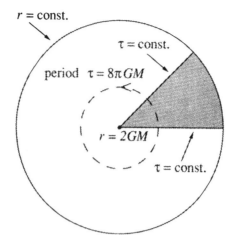

Figure 2.17. Coordinates for the Euclidean Schwarzschild solution; the Euclidean time is identified with period $\tau = 8\pi GM$. All slices $\tau =$ constant meet at the origin.

Hawking put forward the conjecture that gravitational entropy is connected with a nontrivial *topology* of the Euclideanized spacetime (see e.g. his article in Wald 1998). The Euclidean line element (2.104) has the boundary $S^2 \times S^1$ at infinity, where S^2 is a large two-sphere at $r \to \infty$ and S^1 corresponds to the periodically identified imaginary time coordinate. In the Euclidean Schwarzschild case, the topology is $\mathbb{R}^2 \times S^2$ (figure 2.17), while in the case of filling this boundary with flat Euclidean space, one would obtain the topology $\mathbb{R}^3 \times S^1$. A nonvanishing entropy is obtained only in the first case. Other examples suggest the same connection between entropy and topology. Obviously, in this interpretation gravitational entropy would be a truly *global* concept and cannot be assumed to be localized near the horizon. The important question, however, is: how fundamental are *Euclidean* concepts? An answer can probably only be found in a quantum theory of gravity, see sections 2.5 and 2.6.

If the entropy (2.95) is to make sense, there should be a generalized second law of thermodynamics in the sense that

$$\frac{d}{dt}(S_{BH} + S_M) \geq 0 \qquad (2.105)$$

where S_M denotes all contributions to nongravitational entropy. The validity of (2.105), although far from being proven in general, has been shown in a variety of gedankenexperiments. One of the most instructive of such experiments has been devised by Unruh and Wald (see Wald 1994). It makes use of the box shown in figure 2.13 that is adiabatically lowered towards a (spherically symmetric) black hole.

At asymptotic infinity $r \to \infty$, the black hole radiation is given by (2.84). However, for finite r the temperature is modified by the occurrence of a redshift factor $\chi(r) \equiv (1 - 2GM/r)^{1/2}$ in the denominator. Since the box is not in free fall, it is accelerated with an acceleration a. From the relation (Wald 1984)

$$\kappa = \lim_{r \to R_0} (a\chi) \tag{2.106}$$

one has

$$T_{\mathrm{BH}}(r) = \frac{\hbar\kappa}{2\pi k_{\mathrm{B}}\chi(r)} \quad \overset{r \to R_0}{\longrightarrow} \quad \frac{\hbar a}{2\pi k_{\mathrm{B}}} \tag{2.107}$$

which is just the Unruh temperature (2.65)! This means that a freely falling observer near the horizon observes no radiation at all, and the whole effect (2.107) comes from the observer (or box) being noninertial with acceleration a.

The analysis of Unruh and Wald, which is a generalization of the gedanken-experiment discussed at the end of section 2.2, shows that the entropy of the black hole increases at least by the entropy of the Unruh radiation displaced at the floating point—this is the point where the gravitational force (pointing downwards) and the buoyancy force from the Unruh radiation (2.107) are in equilibrium. Amusingly, it is just the application of 'Archimedes' principle' to this situation that rescues the generalized Second Law (2.105).

An inertial, i.e. free-falling, observer does not see any Unruh radiation. How does he interpret the above result? For him the box is accelerated and therefore the interior of the box fills up with negative energy and pressure—a typical quantum effect that occurs if a 'mirror' is accelerated through the vacuum, cf. footnote 5. The 'floating point' is then reached after this negative energy is so large that the total energy of the box is zero.

I want to conclude this section with some speculations about the final stages of black hole evolution and the information loss problem. The point is that— in the semiclassical approximation used by Hawking—the radiation of a black hole is purely thermal. If the black hole evaporates completely and leaves only thermal radiation behind, one would have a conflict with established principles in quantum theory: any initial state (in particular a pure state) would evolve into a mixed state. In ordinary quantum theory, because of the unitary evolution of the total system,[7] this cannot happen. Formally, $\mathrm{Tr}\,\rho^2$ remains *constant* under the von Neumann equation; the same is true for the entropy $S_{\mathrm{SM}} = -k_{\mathrm{B}}\,\mathrm{Tr}(\rho \ln \rho)$: for a unitarily evolving system, there is no increase in entropy. If these laws are violated during black hole evaporation, information would be destroyed. This is indeed the speculation that Hawking made after his discovery of black hole radiation. The attitudes towards this *information loss problem* can be roughly divided into the following classes (see e.g. Page 1994 and Preskill 1993 for reviews):

- The information is indeed lost during black hole evaporation, and the quantum-mechanical Liouville equation is replaced by an equation of the form

$$\rho \longrightarrow \$\,\rho \neq S\rho S^{\dagger}. \tag{2.108}$$

---

[7] For an *open* quantum system, a state will in general become more mixed and its entropy will increase (Giulini *et al* 1996).

- The full evolution is in fact unitary; the black hole radiation contains subtle quantum correlations that cannot be seen in the semiclassical approximation.
- The black hole does not evaporate completely, but leaves a 'remnant' with mass in the order of the Planck mass that carries the whole information.

All of these options have advantages and disadvantages that are discussed at length in the above-cited literature.[8] From a conservative point of view, it looks reasonable to stick to the second option as long as possible. In fact, as long as the black hole has not evaporated and there exists a horizon, it is always possible to assume that the information is 'hidden' behind the horizon. As (2.90) and (2.91) demonstrate, although the global vacuum state is pure, it appears outside the black hole as a thermal state, since there are nonlocal correlations with inaccessible states behind the horizon. The entropy (2.95) would then be just an expression of this lack of information about the global quantum state. Its appearance would signal the presence of nonlocal entanglement, and S_{BH} would thus be itself a *global* quantity.

The relations (2.90) and (2.91) have only been shown for the 'eternal hole' (figure 2.5) and the Rindler spacetime (figure 2.14), but not for a collapse situation. One might, however, expect that—independently of the unknown details of the final stage—all quantum correlations would become accessible again, demonstrating that the full evolution is unitary.[9]

One argument in favour of the first option above is the observation that in the Euclideanized spacetime constant time surfaces meet at a common point, leading to a zero point of the time translation vector field (see Hawking's article in Wald 1998). An example of this can be seen in the geometry of the Euclidean Schwarzschild geometry, see figure 2.17. In such a case one cannot use a Hamiltonian to obtain a unitary evolution from the initial to the final state, and a transition from a pure to a mixed state can be expected to occur. It is, however, by no means evident that such a Euclidean viewpoint is more fundamental than the Lorentzian viewpoint. In particular, the constant-time surfaces in figure 2.17 all meet at the origin—the place where the horizon sits. Their crossing at this point might be only an indication for the presence of quantum entanglement behind the horizon in the Lorentzian framework.

2.5   Black holes in canonical quantum gravity

The framework for the discussion in the last two sections is the limit where the gravitational background is classical, but all nongravitational fields are quantum. The central results are that black holes radiate with a thermal spectrum and that they possess an intrinsic gravitational entropy.

---

[8] A recent vote among experts at a meeting in Utrecht showed an overwhelming majority for the second option and a practical exclusion of the third option.

[9] The accelerated observer in figure 2.14 would only have to go over to inertial motion in order to gain all quantum correlations again.

According to (2.86), black holes become hotter through emitting radiation, while losing mass. The framework of the fixed gravitational background is expected to break down if the hole's mass approaches the *Planck mass*

$$m_{\mathrm{p}} = \sqrt{\frac{\hbar}{G}} \approx 10^{-5} \text{ g} \tag{2.109}$$

that is related to the Planck length (2.39) by $m_{\mathrm{p}} = \hbar/l_{\mathrm{p}}$. The reason for this expectation is that m_{p} should set the scale for the occurrence of quantum gravitational effects—effects where the classical picture of spacetime breaks down. (That the approximation of section 2.3 must break down somewhere is by itself evident, since otherwise $T_{\mathrm{BH}} \to \infty$.)

The final stages of black hole evaporation can thus only be understood within a *quantum theory of gravity*. Unfortunately, such a theory is not yet available. Nevertheless, there exist quite advanced approaches towards such a theory, within which sensible questions can be asked and partially answered. The two most popular approaches at present are *canonical quantum gravity* and *superstring theory*. I shall devote the last two sections to them. Canonical quantum gravity, the topic of this section, is the framework that is found if standard quantization rules are applied to the general theory of relativity. This is a rather conservative approach, since no unification of fundamental forces is attempted. However, this approach exhibits in a most transparent way the basic conceptual features that a quantum theory of gravity should contain. The starting point for canonical quantization is a Hamiltonian formulation on a classical spacetime $\mathcal{M}$ that is globally hyperbolic, i.e. that can be written in the form

$$\mathcal{M} = \Sigma \times \mathbb{R} \tag{2.110}$$

where Σ is a three-dimensional manifold (Wald 1984). Depending on the topological structure of Σ one obtains different configuration spaces. In addition, the canonical framework can be further subdivided into the following classes that are characterized by the configuration variable that is used:

- quantum geometrodynamics: three-metric h_{ab} on Σ;
- quantum connection dynamics: SU(2) connection A_a^i on Σ;
- quantum loop dynamics: the trace of the holonomy of the connection along a loop γ, $\mathrm{tr}\, \mathcal{P} \exp \oint_\gamma \mathcal{A}$.

Although much progress has been made in the last two approaches (Ashtekar 1997), I restrict myself to the first approach that uses the three-dimensional metric as the fundamental variable. The main feature is the existence of *constraints*, one Hamiltonian constraint (per space point),

$$\mathcal{H} \approx 0 \tag{2.111}$$

and three diffeomorphism constraints (per space point),

$$\mathcal{D}_a \approx 0. \tag{2.112}$$

The sign $\approx$ denotes *weak equality* in the sense of Dirac: the constraints define a subspace in the full phase space.

If Σ is compact without a boundary (a situation often used in cosmological models), the total Hamiltonian is a combination of these constraints only and thus (weakly) vanishes. If Σ is asymptotically flat (this is the case relevant for black holes), the total Hamiltonian has in addition boundary terms. A comprehensive introduction into all aspects of canonical gravity can be found in Ehlers and Friedrich (1994).

The crucial feature is the treatment of the constraints (2.111), (2.112) in the quantum theory. Here I shall follow Dirac's approach and implement the constraints—at least formally—as constraints on physically allowed wave functionals:

$$\hat{\mathcal{H}}\Psi = 0 \qquad\qquad (2.113)$$

$$\hat{\mathcal{D}}_a\Psi = 0. \qquad\qquad (2.114)$$

The wave functional Ψ depends (apart from nongravitational fields) on the three-metric $h_{ab}(x)$; (2.114) guarantees that Ψ remains invariant under coordinate transformations, so it *de facto* depends only on the three-dimensional geometry—this is often emphasized by writing $\Psi[^{(3)}\mathcal{G}]$, but one must remember that Ψ is always given as $\Psi[h_{ab}(x)]$.

On the fundamental level, there is only a collection of spaces (of three-dimensional geometries), but no spacetime. The latter has only meaning in a semiclassical approximation (Kiefer 1994). This lack of spacetime on a fundamental level is often referred to as the 'problem of time in quantum gravity' (see e.g. Kiefer 1997). Since the event horizon of a black hole plays a crucial role in the derivation of Hawking radiation, and since the horizon is a genuine *classical* spacetime concept, this drastically demonstrates that the semiclassical picture of section 2.3 *must* be modified in quantum gravity.[10]

In the following I shall discuss the quantization of spherically symmetric 'eternal' black holes. This can either be interpreted as an exact quantization of the matter-free case or as the first step in the semiclassical approximation to the case where also matter is present (see Kiefer 1998 for details and references).

The starting point is the general spherically symmetric metric,

$$ds^2 = N^2(r, t)\,dt^2 - \Lambda^2(r, t)(dr + N^r\,dt)^2 - R^2(r, t)\,d\Omega^2 \qquad (2.115)$$

where $d\Omega^2$ is the metric on the unit two-sphere, Λ and R are the dynamical variables (the only components that are left in $h_{ab}(x)$ after spherical symmetry is imposed) and N and N^r are Lagrange multipliers (the so-called lapse and shift functions). To encompass also the case of charged black holes, I shall include a spherically symmetric electromagnetic one-form

$$A = \Phi(r, t)\,dt + \Gamma(r, t)\,dr. \qquad (2.116)$$

---

[10] Figure 21.4 in Misner *et al* (1973) shows a foliation of the Kruskal diagram (figure 2.4) into three-geometries. In quantum gravity, Ψ depends on both this set of three-geometries (classically allowed ones for this case) and other three-geometries (that are classically not allowed).

The classical constraints (2.111) and (2.112) then read

$$\mathcal{H} = \frac{G}{2}\frac{\Lambda P_\Lambda^2}{R^2} - G\frac{P_\Lambda P_R}{R} + \frac{\Lambda P_\Gamma^2}{2R^2} + G^{-1}\left(\frac{RR''}{\Lambda} - \frac{RR'\Lambda'}{\Lambda^2} + \frac{R'^2}{2\Lambda} - \frac{\Lambda}{2}\right)$$
$$\approx 0 \tag{2.117}$$

$$\mathcal{D}_r = P_R R' - \Lambda P_\Lambda' - \Gamma P_\Gamma' \approx 0. \tag{2.118}$$

In addition, we have Gauß's law,

$$\mathcal{G} = -P_\Gamma' \approx 0. \tag{2.119}$$

(The explicit form of $\mathcal{D}_r$ is different from Kiefer (1998), since we have redefined in the action $\Phi \to \Phi - N^r\Gamma$.) The variables P_Λ, P_R, P_Γ are the momenta canonically conjugate to Λ, R, Γ, respectively.

I consider for Σ a three-space that in the classical picture of figure 2.4 would correspond to a hypersurface that starts at the bifurcation point of the horizons (taken to be $r \to 0$) and spatial infinity ($r \to \infty$). A crucial role in the whole procedure is played by the careful discussion of *boundary conditions* at $r \to 0$ and $r \to \infty$.

I shall first consider boundary conditions at $r \to \infty$. To avoid the unwanted conclusion that the Lagrange multipliers N and Φ vanish there one has to add the boundary term (Kiefer 1998)

$$-G\int dt\, N_+ M - \int dt\, \Phi_+ q \tag{2.120}$$

in the action; N_+ (Φ_+) is the limiting value of N (Φ); M is the ADM mass and q is the electric charge. A canonical transformation then exhibits that (except for variables at $r \to 0$, see below) the only dynamical variables are M and q.

What about boundary conditions at the bifurcation point $r \to 0$? I wish to adopt boundary conditions that enforce every classical solution to be (part of) an exterior region of a Reissner–Nordström black hole, see (2.21); the constant t hypersurfaces are asymptotic to the constant Killing time hypersurfaces as $r \to 0$. It turns out that the extremal case ($\sqrt{G}M = |q|$) has boundary conditions *different* from the nonextremal case ($\sqrt{G}M > |q|$), see Kiefer and Louko (1998).

Consider first the nonextremal case. The variables N, Λ, R then exhibit the following asymptotic behaviour at $r \to 0$:

$$N(r, t) = N_1(t)r + \mathcal{O}(r^3) \tag{2.121a}$$

$$\Lambda(r, t) = \Lambda_0(t) + \mathcal{O}(r^2) \tag{2.121b}$$

$$R(r, t) = R_0(t) + R_2(t)r^2 + \mathcal{O}(r^4). \tag{2.121c}$$

To avoid the unwanted conclusion that $N_1 = 0$, a boundary term similar to (2.120) must be added at $r \to 0$,

$$(2G)^{-1}\int dt\, N_0 R_0^2 \tag{2.122}$$

where $N_0 \equiv N_1/\Lambda_0$. The quantity

$$\alpha \equiv \int_{t_1}^{t} dt\, N_0(t) \tag{2.123}$$

plays the role of a 'rapidity' because it boosts the normal vector to the constant t hypersurfaces at $r \to 0$.

For the extremal case, one has instead of (2.5) the boundary conditions

$$N(r,t) = \Lambda^{-1} R'(\tilde{N}_0(t) + \mathcal{O}(r)) \tag{2.124a}$$

$$\Lambda(r,t) = \Lambda_{-1}(t) r^{-1} + \mathcal{O}(1) \tag{2.124b}$$

$$R(r,t) = R_0(t) + R_1(t) r + \mathcal{O}(r^2). \tag{2.124c}$$

The falloff (2.124b), in particular, encodes the fact that in the extremal case the point at $r \to 0$ is infinitely far away, $\int_0^{r_1} g_{rr}\, dr \to \infty$. It then turns out from the action that *no* term of the form (2.122) has to be added. The geometrical reason for this lies in the fact that the boundary term is proportional to the surface gravity κ, and one has $\kappa = 0$ for the extremal case, cf. (2.94).

For the canonical formalism, one needs canonical *pairs* of variables at the boundaries. This is achieved by the introduction of the variables τ, λ and α, and *parametrization* (Kiefer 1998),

$$N_+(t) = \dot{\tau}(t) \tag{2.125a}$$

$$\Phi_+(t) = \dot{\lambda}(t) \tag{2.125b}$$

$$N_0(t) = \dot{\alpha}(t). \tag{2.125c}$$

(In the extremal case, the last condition is absent.) The physical interpretation of these variables is as follows: τ is the proper time at $r \to \infty$, λ the gauge parameter and α is the rapidity (2.123). While τ and λ are conjugate to mass and charge, respectively, α is conjugate to the area A of the event horizon. The remaining quantum constraints can than be solved, and a plane-wave-like solution reads

$$\Psi(\alpha, \tau, \lambda) = \chi(M, q) \exp\left[\frac{i}{\hbar} \left(\frac{A(M,q)\alpha}{8\pi G} - M\tau - q\lambda \right) \right]. \tag{2.126}$$

$\chi(M, q)$ is an arbitrary function of M and q; one can construct superpositions of the solutions (2.126) in the standard way by integrating over M and q.

Varying the phase in (2.126) with respect to M and q yields the classical equations

$$\alpha = 8\pi G \left(\frac{\partial A}{\partial M} \right)^{-1} \tau = \kappa\tau \tag{2.127}$$

$$\lambda = \frac{\kappa}{8\pi G} \frac{\partial A}{\partial q} \tau = \Phi\tau. \tag{2.128}$$

The solution (2.126) holds for nonextremal holes. If one made a similar quantization for extremal holes on their own, the first term in the exponent of (2.126) would be absent.

An interesting analogy with (2.126) is the plane-wave solution for a free nonrelativistic particle,

$$\exp(ikx - \omega(k)t) . \tag{2.129}$$

As in (2.126), the number of parameters is one less than the number of arguments, since $\omega(k) = k^2/2m$. A quantization for extremal holes on their own would correspond to choosing a particular value for the momentum, say p_0, at the classical level, and demanding that no dynamical variables (x, p) exist for $p = p_0$. This is, however, not the usual way to find classical correspondence—this is gained not from the plane-wave solution (2.129) but from *wave packets* that are obtained by superposing different wave numbers k. This then yields quantum states that are sufficiently concentrated around classical trajectories such as $x = p_0t/m$.

It seems therefore appropriate to proceed similarly for black holes—construct wave packets for nonextremal holes that are concentrated around the classical values (2.127), (2.128) and then *extend* them by hand to the extremal limit. This would correspond to 'extremization after quantization', in contrast to the 'quantization after extremization' made above. Expressing in (2.126) M as a function of A and q and using Gaussian weight functions, one has

$$\Psi(\alpha, \tau, \lambda) = \int\limits_{A>4\pi q^2} dA\, dq\, \exp\left[-\frac{(A - A_0)^2}{2(\Delta A)^2} - \frac{(q - q_0)^2}{2(\Delta q)^2}\right]$$

$$\times \exp\left[\frac{i}{\hbar}\left(\frac{A\alpha}{8\pi G} - M(A, q)\tau - q\lambda\right)\right]. \tag{2.130}$$

The result of this calculation is given and discussed in Kiefer and Louko (1998). As expected, one finds Gaussian packets that are concentrated around the classical values (2.127), (2.128). As for the free particle, the wave packets exhibit dispersion with respect to Killing time τ. Using for ΔA the Planck length squared, $\Delta A \propto G\hbar \approx 2.6 \times 10^{-66}$ cm^2, one finds for the typical dispersion time in the Schwarzschild case

$$\tau_* = \frac{128\pi^2 R_0^3}{G\hbar} \approx 10^{65}\left(\frac{M}{M_\odot}\right)^3 \text{ years} . \tag{2.131}$$

Note that this is just of the order of the black hole evaporation time (2.88)! The dispersion of the wave packet just gives the time scale after which the semiclassical approximation breaks down.

Coming back to the charged case, and approaching the extremal limit $\sqrt{G}M = |q|$, one finds that the widths of the wave packet (2.130) are *independent* of τ for large τ. This is due to the fact that for the extremal black hole $\kappa = 0$ and therefore no evaporation takes place. If one takes, for example, $\Delta A \propto G\hbar$ and

$\Delta q \propto \sqrt{G\hbar}$, one finds for the α-dependence of (2.130) for $\tau \to \infty$ the factor

$$\exp\left(-\frac{\alpha^2}{128\pi^2}\right) \tag{2.132}$$

which is independent of both τ and $\hbar$. It is clear that this packet, although concentrated at the value $\alpha = 0$ for extremal holes, has support also for $\alpha \neq 0$ and is qualitatively not different from a wave packet that is concentrated at a value $\alpha \neq 0$ close to extremality.

An interesting question is the possible occurrence of a naked singularity (cf. figure 2.10) for which $\sqrt{G}M < |q|$. Certainly, both the boundary conditions (2.5) and (2.5) do not comprise the case of a singular three-geometry. However, the wave packets discussed above also contain parameter values that would correspond to the 'naked' case. Such geometries could be avoided if one imposed the boundary condition that the wave function vanishes for such values. But then continuity would enforce the wave function also to vanish on the boundary, i.e. at $\sqrt{G}M = |q|$. This would mean that extremal black holes could not exist at all in quantum gravity—an interesting speculation.

A possible thermodynamical interpretation of (2.126) can only be obtained if—analogous to section 2.4—an appropriate transition into the Euclidean regime is performed. This transition is achieved by the 'Wick rotations' $\tau \to -i\beta\hbar$, $\alpha \to -i\alpha_E$ (from (2.123) it is clear that α is connected to the lapse function and must be treated similarly to τ) and $\lambda \to -i\beta\hbar\Phi$. With an argument analogous to the one used below (2.103)—regularity of the Euclidean line element—one arrives at the conclusion that $\alpha_E = 2\pi$. But this means that the Euclidean version of (2.127) just reads $2\pi = \kappa\beta\hbar$, which with $\beta = (k_B T_{BH})^{-1}$, is just the expression for the Hawking temperature (2.84)! Alternatively, one could use (2.84) to derive $\alpha_E = 2\pi$.

The Euclidean version of the state (2.126) then reads

$$\Psi_E(\alpha, \tau, \lambda) = \chi(M, q) \exp\left(\frac{A}{4G\hbar} - \beta M - \beta\Phi q\right). \tag{2.133}$$

One recognizes in the exponent of (2.133) the occurrence of the Bekenstein–Hawking entropy (2.95). Of course, (2.133) is still a pure state and should not be confused with a partition sum. However, the transition to a partition sum is straightforward, and one then finds indeed from the standard thermodynamical relations in section 2.4 the expression for S_{BH}. Moreover, the factor $\exp[A/(4G\hbar)]$ in (2.133) directly gives the enhancement factor for the rate of black hole pair creation relative to ordinary pair creation. It must be emphasized that S_{BH} fully arises from a boundary term at the horizon ($r \to 0$). This is similar to the path-integral approaches discussed in section 2.4 (the entropy arises there from a boundary term at the centre of the disc in figure 2.17).

It is now clear that a quantization scheme that treats extremal black holes as a limiting case gives $S_{BH} = A/(4G\hbar)$ also for the extremal case. On the other hand, quantizing extremal holes on their own would yield $S_{BH} = 0$. From this point of view it is also clear why the extremal (Kerr) black hole that occurs in the transition from the disc-of-dust solution to the Kerr solution has entropy $A/(4G\hbar)$, see Neugebauer's article in Hehl *et al* (1998). If $S_{\text{BH}} \neq 0$ for the extremal hole (that has temperature zero), the stronger version of the third law of thermodynamics (that would require $S \to 0$ for $T \to 0$) mentioned in section 2.2 apparently does not hold. This is not particularly disturbing, since many systems in ordinary thermodynamics violate the strong form of the third law; it just means that the system does not approach a unique state for $T \to 0$.

The topological difference of the classical charged black hole solutions between extremal and nonextremal cases can immediately be inferred from figures 2.8 and 2.9. If the extremal case were quantized on its own, the reason for its vanishing entropy could be understood as follows. Consider first the nonextremal case (figure 2.8) and a spacelike hypersurface that starts at one of the bifurcation two-spheres and extends through the right part of region I up to spatial infinity i^0. Initial data on such hypersurfaces can be evolved only in the right part of region I; one could thus interpret the occurrence of the entropy $A/(4G\hbar)$ as signalling a 'lack of information' about the left part of region I and region II. (In this interpretation, 'full' knowledge would refer to the evolution up to the Cauchy horizon at $r = r_-$.) From a hypersurface that extends from the left i^0 up to the right i^0 one could infer the whole evolution up to r_-; in fact, no boundary term appears in the canonical analysis that would give rise to a term $A/(4G\hbar)$.

On the other hand, for a hypersurface that passed from point p in the extremal case (figure 2.9) through region I up to i^0 one could infer the whole evolution in region I from initial data on this hypersurface (see figure 2.18). Consequently, its entropy should be zero since there is no 'lack of information'. However, the situation shown in figure 2.18 cannot be reached by any continuous limit from figure 2.8. Moreover, it must be emphasized again that spacetime is a *classical* concept and that in particular a singular case such as the extremal hole may play no role in quantum gravity.

An interesting example that I want to mention only in passing is the $\mathbb{RP}^3$ geon (Louko and Marolf 1998). This is an eternal black hole that is locally isometric to the Kruskal spacetime (figure 2.4), but contains only one exterior region. For this spacetime it was found that it possesses an entropy that is half of the Bekenstein–Hawking value, i.e. $A/(8G\hbar)$. The Penrose diagram for the $\mathbb{RP}^3$ geon is shown in figure 2.19; in which sense the result $S_{\text{BH}}/2$ is related to some 'lack of information' is not yet clear.

In the case where additional fields are present, the above-discussed quantization of black holes is only valid at the highest order of a semiclassical approximation (Kiefer 1998). Even at that order, the solution (2.126) is augmented by a factor $\exp(iS_0/(G\hbar))$, where S_0 is a solution to the functional Hamilton–Jacobi equation that follows from the constraints (2.113), (2.114) in this limit. From a discussion of S_0 one can also infer that the interior of the black hole horizon is always a classically allowed region—this is why the horizon shrinks to a point in the Euclidean version that exhibits classically forbidden regions, see figure 2.17.

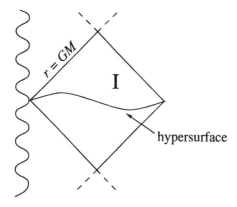

Figure 2.18. Spacelike hypersurface in the extremal Reissner–Nordström solution.

The next order of the semiclassical approximation makes it clear why the system can no longer be reduced to a system with finitely many degrees of freedom—field theoretic aspects play an important role. It is the level where Hawking radiation becomes manifest in quantum gravity. Unfortunately, the equations are much too complicated (in particular a viable regularization is needed) to be solved. Therefore, the full evolution of black holes in canonical gravity remains unknown.

Even at the semiclassical level, nonclassical black hole quantum states can easily be constructed by using the superposition principle. However, most such states *decohere*, i.e. become indistinguishable from a classical stochastic ensemble, through their own Hawking radiation (Giulini *et al* 1996). Such a decoherence only follows for macroscopic (semiclassical) black holes; it does not occur for microscopic (virtual) black holes (more properly called black-and-white holes). The time symmetry of such microscopic states remains thus unbroken, and the loss of quantum coherence that is claimed to happen by scattering off vacuum fluctuations in which virtual black-and-white holes appear and disappear (see Hawking's article in Wald 1998) is spurious: as for the corresponding situation in QED (Giulini *et al* 1996), no loss of quantum coherence should occur.

As far as the quantum state (2.126) is concerned, all variables and parameters are of a continuous nature, as for the free particle. It is, however, often speculated that mass and area are quantized (Bekenstein 1998). This can be found heuristically from (2.126) if it is assumed that the range of τ (or α) is compact—in a similar way one can find momentum quantization on finite spaces, e.g. on a circle. Since $\Psi(\alpha, \tau, \lambda) = \Psi(\alpha, \tau + \Delta\tau, \lambda)$, one arrives at (Kastrup 1996)

$$\frac{M\Delta\tau}{\hbar} = 2\pi n. \tag{2.134}$$

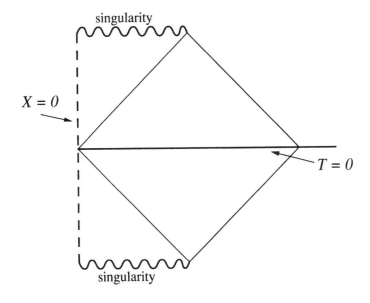

Figure 2.19. Spacelike hypersurface in the $\mathbb{RP}^3$ geon.

Restricting to vanishing charge, one can assume that $\Delta\tau$ is proportional to the Schwarzschild radius,

$$\Delta\tau = \gamma R_0 = 2\gamma GM \qquad (2.135)$$

with an unknown constant γ that is probably of order unity. This then yields for mass and area, respectively,

$$M_n = \sqrt{\frac{\pi n}{\gamma}} m_{\mathrm{p}} \quad \Rightarrow \quad A = \frac{16\pi^2}{\gamma} n l_{\mathrm{p}}^2 . \qquad (2.136)$$

Since one would expect that then also α has a finite range, one is led to

$$\frac{A\Delta\alpha}{8\pi G\hbar} = 2\pi \quad \Rightarrow \quad \Delta\alpha = \gamma \qquad (2.137)$$

and therefore $\Delta\tau = 2\Delta\alpha GM$ (which apart from a factor 2 would follow from the classical relation (2.127)).

A similar quantization would follow if one imposed an *ad hoc* Bohr–Sommerfeld quantization rule in the Euclidean version (recall $\alpha_{\mathrm{E}} = 2\pi = \gamma$)

$$nh = \oint \pi_{\alpha_{\mathrm{E}}} \, d\alpha_{\mathrm{E}} = \int_0^{2\pi} \frac{A \, d\alpha_{\mathrm{E}}}{8\pi G} = \frac{A}{4G} . \qquad (2.138)$$

Whether these results survive a rigorous derivation remains at present open. If true, however, this area quantization would modify the thermal spectrum of black hole radiation found in the semiclassical limit—even for black holes much bigger than the Planck length (Bekenstein 1998)!

I want to emphasize finally that canonical quantum gravity can also address the issue of black holes in quantum cosmology (Kiefer and Zeh 1995), but this goes beyond the scope of this review.

2.6   Black holes in string theory

A much more ambitious framework for a quantum theory of gravity is *superstring theory*. In fact, this theory is not constructed through the application of standard quantization rules to well known classical theories, but through the use of a very different route that leads directly to a fundamental unified quantum theory. The various interactions—such as gravity—can then only be distinguished within this theory in certain limits. The role of black holes in string theory has been extensively discussed in the lectures by Dijkgraaf and by Fré and D'Auria. For this reason I shall be rather brief in this section and try mainly to discuss how this topic fits into the general scheme outlined in my earlier sections; for details I refer to the other lectures as well as to reviews such as Horowitz (1998) and Peet (1998).

Superstring theory (or 'M-theory') on its most fundamental level does not have any notion of spacetime, although it contains one fundamental length scale, $l_s = \sqrt{\alpha'\hbar}$ (α' being the inverse string tension). Spacetime only emerges in a semiclassical approximation, very similar to canonical quantum gravity (Kiefer 1994). This is most conveniently expressed through an *effective action* that is found by an expansion with respect to l_s (or $\sqrt{\hbar}$); it is also a low energy expansion since higher orders in l_s lead to higher order spacetime derivatives in the effective action. Usually, only low energy effective actions are considered in which terms of l_s and higher are neglected; their classical solutions correspond to a gravitational theory including novel fields such as a dilaton, axion fields etc. The presence of the dilaton field, in particular, gives rise to the effective string coupling constant g_s that connects Planck length and string scale,

$$l_p \propto g_s l_s . \tag{2.139}$$

In the low energy approximation, one can in particular address either the weak coupling ($g_s \ll 1$) or the strong–coupling limits ($g_s \gg 1$). As has been shown in recent years, *dualities* connect these limits in different (or even the same) string theories. For $g_s \ll 1$, perturbation theory can be applied. For $g_s \gg 1$ one can (still on the semi-classical level!) find nonperturbative classical solutions to the effective actions—in particular black holes with various charges that are generalizations of the Reissner–Nordström solution (2.21). For $g_s \gg 1$ one has $l_p \gg l_s$ and one would thus expect that canonical quantum gravity should be a good approximation.

The presence of dualities, together with (2.139), allows us to address the issue of black hole entropy in string theory. A special role is played by BPS states, states where mass is equal to charge (in a precise sense). Certain 'nonrenormalization theorems' guarantee that, while the coupling g_s is varied, this relation and the degeneracy of the states remain unchanged. Now, in the weak coupling limit $l_p \to 0$ and one is effectively in a flat spacetime; the theory predicts the existence of bound states of certain extended objects ('D-branes') in flat space whose degeneracy can easily be calculated from standard considerations. A corresponding entropy is then defined as the logarithm of this degeneracy.

As g_s increases, the (effective) theory yields extremal black hole solutions. They have nonvanishing horizon area, and the interesting result is that $S_{BH} = A/(4G\hbar)$ is exactly equal to the D-brane entropy. It is in this sense that a 'microscopic derivation' of S_{BH} has been provided; one must, however, emphasize that this is merely a consistency check, since the D-brane state has no resemblance to a black hole (there is, in particular, no horizon), and the connection is established only via duality arguments. Perhaps more surprising is the fact that for nonextremal black holes (but close to extremality), the full spectrum including the greybody factor $\Gamma_{\omega l}$, see (2.85), is recovered for the corresponding D-brane states. Unfortunately, for the general case (black holes far from extremality) no rigorous calculation exists as yet; in particular, the case of the ordinary Schwarzschild black hole remains elusive.

In the string calculations, unitarity is always preserved. This seems to indicate that no information would be lost during black hole evaporation. Since Hawking radiation corresponds, in the string picture, to the emission of closed strings from D-branes, the mixed character of Hawking radiation would solely result from decoherence (Giulini *et al* 1996)—the closed strings are quantum entangled with the D-brane states and integrating out the latter would then lead to a mixed state. It would be an interesting exercise to find out how the *thermal* nature of Hawking radiation arises in this picture.

I must emphasize again that the string calculations are only made on the semiclassical level—the same level as the canonical treatment in section 2.5. Going beyond would necessitate taking higher order corrections in l_s into account; this is, however, up to now as untractable as the treatment of the higher order l_p-corrections in section 2.5. In particular, the full black hole evaporation and the final decision about the information-loss problem remain elusive.

Can anything of the quantum aspects of black holes be observed in the foreseeable future? As was mentioned after (2.88), Hawking radiation can only be measured for primordial black holes (PBHs) that were created with initial mass $\approx 5 \times 10^{14}$ g in the early Universe. Black holes that result from stellar collapse are much too heavy to show noticeable radiation. Still, attempts have been and are still being made to look for the existence from a contribution of PBH distribution to the diffuse γ-background and to look directly for the final evaporation of a single PBH (Halzen *et al* 1991). The first method yields an upper limit onto the density of PBHs of $\mathcal{N} \lesssim 10^4$ pc^{-3}, while the second method yields a (conservative) evaporation rate $dn/dt < 4.4 \times 10^5$ pc^{-3} year^{-1} (Funk 1997). PBH explosions could in the future be observed, for example, by the MILAGRO project[11] that measures

---

[11] http://hana-mana.lanl.gov/milagro/MGRO_Homepage.html

secondary showers arising from primary γ-rays. If the final evaporation of black holes could be observed, this would open the first window towards an experimental test of quantum gravity.

Acknowledgments

I would like to thank the organizers of the school for inviting me to give lectures in a very pleasant atmosphere. I am also grateful to Jorma Louko and Jörg Thorwart for discussions and a critical reading of this manuscript.

References

Ashtekar A 1997 *General Relativity and Gravitation* ed M Francaviglia, G Longhi, L Lusanna and E Sorace (Singapore: World Scientific) p 3

Bekenstein J D 1980 *Phys. Today* January 24

——1999 *Proc. VIII Marcel Grossmann Meeting* ed R Ruffini and T Piran (Singapore: World Scientific)

Belinsky V A, Khalatnikov I M and Lifshitz E M 1982 *Adv. Phys.* **19** 525

Birrell N D and Davies P C W 1982 *Quantum Fields in Curved Space* (Cambridge: Cambridge University Press)

DeWitt B S 1975 *Phys. Rep.* **19** 295

Ehlers J and Friedrich H (eds) 1994 *Canonical Gravity: from Classical to Quantum* (Berlin: Springer)

Funk B 1997 Search for gamma-ray burst counterparts with the HEGRA air shower array *Dissertation* Wuppertal *Report* WUB-DIS 97-7

Giulini D, Joos E, Kiefer C, Kupsch J, Stamatescu I-O and Zeh H D 1996 *Decoherence and the Appearance of a Classical World in Quantum Theory* (Berlin: Springer)

Halzen F, Zas E, MacGibbon J H and Weekes T C 1991 *Nature* **353** 807

Hawking S W 1975 *Commun. Math. Phys.* **43** 199

——1979 *General Relativity* ed S W Hawking and W Israel (Cambridge: Cambridge University Press) p 746

Hawking S W and Ellis G F R 1973 *The Large Scale Structure of Space–Time* (Cambridge: Cambridge University Press)

Hawking S W and Penrose R 1996 *The Nature of Space and Time* (Princeton: Princeton University Press)

Hehl F W, Kiefer C and Metzler R (eds) 1998 *Black Holes: Theory and Observation* (Berlin: Springer)

Heusler M 1996 *Black Hole Uniqueness Theorems* (Cambridge: Cambridge University Press)

——1998 *Black Holes: Theory and Observation* ed F W Hehl *et al* (Berlin: Springer) p 157

Horowitz G T 1998 *Black Holes and Relativistic Stars* ed R M Wald (Chicago: University of Chicago Press) p 241

Israel W 1987 *Three Hundred Years of Gravitation* ed S W Hawking and W Israel (Cambridge: Cambridge University Press) p 199

——1998 *Black Holes: Theory and Observation* ed F W Hehl *et al* (Berlin: Springer) p 364

Kastrup H A 1996 *Phys. Lett.* B **385** 75

Kiefer C 1994 *Canonical Gravity: from Classical to Quantum* ed J Ehlers and H Friedrich (Berlin: Springer) p 170

——1997 *Time, Temporality, Now* ed H Atmanspacher and E Ruhnau (Berlin: Springer) p 227

——1998 *Black Holes: Theory and Observation* ed F W Hehl *et al* (Berlin: Springer) p 416

Kiefer C and Louko J 1999 *Ann. Phys., Lpz.* **8** 67, gr-qc/9809005

Kiefer C and Zeh H D 1995 *Phys. Rev.* D **51** 4145

Liberati S, Belgiorno F, Visser M and Sciama D W 1998 Sonoluminescence as a QED vacuum effect *Report* quant-ph/9805031

Louko J and Marolf D 1998 *Phys. Rev.* D **58** 24 007

Luminet J P 1998 *Black Holes: Theory and Observation* ed F W Hehl *et al* (Berlin: Springer) p 3

Misner C W, Thorne K S and Wheeler J A 1973 *Gravitation* (New York: Freeman)

Page D N 1976 *Phys. Rev.* D **13** 198

——1994 *Proc. 5th Canadian Conf. on General Relativity and Relativistic Astrophysics* ed R Mann and R McLenaghan (Singapore: World Scientific) p 1

Peet A W 1998 *Class. Quantum Grav.* **15** 3291

Preskill J 1993 *Int. Symp. on Black Holes, Membranes, Wormholes, and Superstrings* ed J Kalara and D V Nanopoulos (Singapore: World Scientific) p 22

Sexl R U and Urbantke H K 1983 *Gravitation und Kosmologie* (Mannheim: Bibliographisches Institut)

Straumann N 1984 *General Relativity and Relativistic Astrophysics* (Berlin: Springer)

Thorne K S 1994 *Black Holes and Time Warps* (New York: Norton)

Unruh W G 1976 *Phys. Rev.* D **14** 870

——1998 Acceleration Radiation for Orbiting Electrons *Report* hep-th/9804158

Wald R M 1984 *General Relativity* (Chicago: University of Chicago Press)

——1986 *Quantum Concepts in Space and Time* ed C J Isham and R Penrose (Oxford: Clarendon) p 293

——1994 *Quantum Field Theory in Curved Spacetime and Black Hole Thermodynamics* (Chicago: University of Chicago Press)

——1998 *Black Holes and Relativistic Stars* (Chicago: University of Chicago Press)

Wipf A 1998 *Black Holes: Theory and Observation* ed F W Hehl *et al* (Berlin: Springer) p 385

Zeh H D 1992 *The Physical Basis of the Direction of Time* (Berlin: Springer)

PART 2

Robbert Dijkgraaf

Departments of Mathematics and Physics, University of Amsterdam, Plantage Muidergracht 24, 1018 TV Amsterdam, The Netherlands

Chapter 3

Strings, matrices and black holes

3.1 Introduction

Our lack of understanding of black holes has been the main motivation to develop theories of quantum gravity. Indeed, the various paradoxes surrounding quantum black holes form a rich proving ground for every theory of quantum gravity, and string theory is no exception. The remarkable progress made in our understanding of the structure of string theory in the last years has made it possible to address some issues of black hole physics in a comprehensive model of quantum gravity, and the results have been very promising. It seems that we now have for the first time a possible consistent microscopic theory of quantum black holes—although still in a very preliminary phase.

Instead of giving a general survey of the field, which deserves a much more systematic review, I will mostly concentrate on one particular example, that has been studied in most detail and carries all the significant features: the five-dimensional black hole that was first studied by Strominger and Vafa in their seminal work on quantum black hole states in string theory [1]. This system can be studied from various points of view: classical solutions, sigma models, Yang–Mills gauge theories, conformal field theory, matrix theory and finally gravity on anti-de Sitter spacetimes. Thereby it presents a microcosmos of the ideas that permeate modern day string theory. It should be emphasized that there are various extensions of these ideas to black holes in other dimensions (in particular four) and with more or less supersymmetry. These generalizations contain some new ingredients but are, in my opinion, essentially based on the same framework. Therefore we do not lose much by concentrating on this one example. For further reading on the by now extensive topic of strings in black holes we refer to [2] and references therein.

3.2 What is string theory?

Before we delve into the concrete model it is perhaps appropriate to make some remarks about string theory of a more general nature. For a good modern introduction to these topics see [3].

3.2.1 The magic triangle

It is slowly appearing that string theory is not (only) a theory of strings. This is sometimes stressed by calling it M-theory. It is better defined as the unique theory that combines the following ingredients: gravity, gauge theory, quantum mechanics and supersymmetry. Of these four principles only the last one is contested, and much of the fate of string theory as the theory of our universe depends on confirmation that our world is indeed supersymmetric, although the symmetry must be broken at a certain scale (hopefully within reach of the coming accelerators).

In fact, not only does string theory combines these fundamental principles, it more or less shows that they are equivalent. It leads to what I like to call the magic triangle of M-theory. Under the benevolent influence of supersymmetry the following three independent physical ideas seem to be essentially equivalent: gravity, gauge theory and strings (i.e. one-dimensional extended objects, little loops). This is quite an astonishing and unbelievable claim, but all the recent progress seems to point in that direction. And we are right in the middle of making a three-way dictionary translating between gravity, gauge theory and string theory.

Of the three equivalences, two are rather well understood: the ways in which strings relate to gravity and gauge theory. The relation to gravity appears because the massless mode of a closed string represents a spin two particle that can be identified with the graviton, a perturbation of the metric. This is already a remarkable phenomenon. Of course a string, just as a point-particle or a higher-dimensional extended object, couples to the background metric. But it is quite another thing that this background is actually produced by the string itself, and that a consistent propagation of strings in a space–time metric $G_{\mu\nu}$ requires through conformal invariance and vanishing of the beta-functions that this metric solves (to leading order) Einstein's equation in vacuum

$$R_{\mu\nu} = 0. \tag{3.1}$$

So string theory naturally combines general relativity and quantum mechanics and as such is a possible model for a theory of quantum gravity.

There are several relations between strings and gauge theory. For example, the closed heterotic string produces space–time gauge fields through affine Kac–Mody algebras, and recently more exotic appearances of gauge symmetries in closed strings have been discovered (for example when the space–time develops a particular singularity). However, the most straightforward relation with gauge theory is through open strings. At the massless level an open string represents a gauge boson A_μ. Open strings can carry additional labels (Chan–Paton indices) and lead thus (depending on the details) to the classical gauge symmetry groups $U(N)$, $SO(N)$ and $Sp(N)$.

The importance of open strings has been forcefully demonstrated by Polchinski [4] with his discovery of D-branes as the 'missing links' in our picture of string dynamics. D-branes are submanifolds of space–time where strings can end. Completeness of string theory requires that these objects are included, and are treated to some extent at the same level as the 'fundamental' strings. Because D-branes are formulated (in perturbative string theory) in terms of open strings and thus gauge fields, they serve as a dictionary between string theory and gauge theory.

D-branes are remarkable objects. Not only do they show us that string theory is not exclusively a theory of one-dimensional extended objects (D-brane democracy), but there are particular regimes of string theory where the D-branes can dominate the perturbative strings. This is most spectacularly shown in the matrix theory limit [5]. In this description of M-theory the relevant degrees of freedom are entirely carried by the D0-branes or D-particles. These are pointlike solitonic excitations of IIA string theory that probe distances that can be much smaller than the string scale—what was usually considered the fundamental small-distance cut-off.

3.2.2 Holography

The final link, that between gravity and gauge theory, seems to be more mysterious but might prove to be the deepest. As we will discuss in these lecture notes, there are now various examples where a quantum gravity theory is *equivalent* to a (large-N) gauge theory. This is quite remarkable since a theory of quantum gravity should carry as its fundamental symmetry group general covariance, a property not obviously shared by gauge theories.

However, the equivalence relation is not between theories on the same space–time or even in the same dimensions. Typically one finds that a theory of gravity in d dimensions is related to a gauge theory in $d-1$ dimensions! This is an example of what is called the holographic principle as formulated by 't Hooft and Susskind [6, 7]: theories of quantum gravity carry essentially the same number of degrees of freedom of a non-gravitational QFT in one dimension less.

The motivating fact behind this idea is the Bekenstein–Hawking formula for black hole entropy, that states that the number of quantum states of a black hole has an entropy that grows (in four dimensions) with the *area* of the horizon, not with the volume, as would be standard for a conventional quantum field theory that is composed of local degrees of freedom. Indeed, a quantum gravity theory is supposed not to carry exactly localizable observables due to the fluctuating properties of the space–time metric.

A toy model for such a relation can be found in the realm of topological field theories. Topological field theories are a much simpler class of objects. They share with models of quantum gravity the property of being generally covariant. They achieve this not by averaging over all metrics, but simply because these models do not depend on the metric at all. In a TFT the stress tensor, that measures the coupling to a background metric, simply vanishes identically as a quantum operator.

A good example, that we will return to at the end of these notes, is Chern–Simons gauge theory in three dimensions [8]. This has a topological Lagrangian

$$\int \mathrm{Tr}\,(A \wedge \mathrm{d}A + \tfrac{2}{3} A \wedge A \wedge A) \tag{3.2}$$

that can be written just using differential forms, without picking a metric.

This topological invariance persists at the quantum level, and leads to a QFT without local observables. The only measurable operators are the Wilson loops or holonomies around non-trivial cycles or knots. However, if we put this model on a 3-manifold M with a boundary Σ, we do obtain local degrees of freedom at the boundary [9]. The model becomes equivalent to a WZW group manifold sigma-model, with the group valued field g essentially the trivialization of the gauge field $A = g^{-1} \, dg$ on the boundary Σ.

So in this model a three-dimensional general covariant topological field theory without local observables is equivalent to a two-dimensional (chiral) conformal field theory at the boundary. The conformal field theory has an infinite set of local operators, that correspond to fixing locally the boundary value of the gauge field A.

3.2.3 Gravity as an effective field theory

Many recent discoveries seem to point in the direction that gravity may not be a fundamental microscopic phenomenon: that metrics and curvature only appear after we have 'integrated out' the small-scale UV degrees of freedom. As stated in [10] 'gravity is thermodynamics, gauge theory is statistical physics'.

This is particularly clear in D-brane dynamics [11]. If we probe the geometry of one D-brane using another D-brane, by computing the effective world-volume action that encodes the 'curved space' in which the probe brane moves, we find that such a description only emerges after we have integrated out the (massive) open strings that connect the two branes. This only makes sense if the separation of the probe and the source is sufficiently large, so that the massive modes can safely be integrated out.

Schematically a system of two branes is described by a 2×2 matrix of coordinates x_{IJ}^{μ} with a potential of the form

$$V = \sum_{\mu, \nu} \mathrm{Tr}[x^{\mu}, x^{\nu}]^2. \tag{3.3}$$

The diagonal entries x_{11}^{μ} and x_{22}^{μ} can be considered as the positions of the two branes, but the off-diagonal elements $x_{12}^{\mu} = (x_{21}^{\mu})^*$ have no direct geometric interpretation. A typical classical configuration minimizes the potential V and satisfies

$$[x^{\mu}, x^{\nu}] = 0. \tag{3.4}$$

In this case the commuting matrices can be brought into diagonal form and the separation

$$r = |x_{11} - x_{22}| \tag{3.5}$$

is well defined. The small fluctuations of the off-diagonal fields now become massive, since the potential induces a term

$$r^2 |\delta x_{12}^{\mu}|^2 \tag{3.6}$$

in the Lagrangian. For r large enough these extra degrees of freedom can be integrated out, and they will, through quantum loops, modify the kinetic term of the diagonal coordinates to be of a more general form

$$g_{\mu\nu}(x)\partial x^\mu \partial x^\nu \tag{3.7}$$

thereby in effect curving the background space.

But this procedure clearly becomes ill defined if r becomes small, since then the off-diagonal components of the matrix x_{IJ}^μ become just as relevant as the diagonal components. So apparent singularities at $r = 0$ in the space–time metric (as seen by the D-brane probe) are just artifacts of integrating out (almost) massless degrees of freedom. A good analogue is the Seiberg–Witten solution of $\mathcal{N} = 2$ supersymmetric gauge theories, where the singularities in the moduli space of vacua of the effective Abelian theory are explained by massless monopoles and dyons. At these singularities the physics becomes non-singular as soon as one remembers to include these massless fields in the dynamics.

Indeed since the fundamental description of D-branes is in terms of gauge theories, space–time appears as a moduli space with a non-trivial metric induced by quantum corrections—a typical effective quantum field theory phenomenon. The fact that the underlying microscopic description is in terms of non-Abelian gauge theories, where the coordinates x^μ no longer commute, strongly suggest that quantum geometry is in an essential way non-commutative. It is indeed satisfying that in the specific case of toroidal compactifications of M-theory the general framework of non-commutative geometry can be encompassed [12].

These deep relations between gravity and gauge theory suggest that many of the present paradoxes surrounding black hole physics might be solved in a perhaps disappointing way. Since gravity can turn out to be only an effective phenomenon, that manifests itself in perhaps a large class of microscopic theory, issues of space–time singularities, space–time foam etc might just all be ill-posed questions. At these small distances the world can appear to be genuinely non-commutative and a picture of a quantized geometry can be a wrong framework to phrase questions!

3.3 Strings, supersymmetry and branes

3.3.1 Extremal black holes and BPS states

By now there is a considerable literature on black holes in string theory. Although very much is known, I think it is fair to say that we do not yet have a comprehensive picture, and most likely the present ideas are only a first step in the direction of a full understanding of the nature of quantum black holes. For example much more is known about charged black holes than the conventional neutral Schwarzschild-type black holes. The typical reason for this preference of charged objects is that a charged black hole with mass M and charge Q satisfies an equality of the form

$$M \geq |Q|. \tag{3.8}$$

So, for a fixed charge, there is a minimum mass, that is attained by a so-called extremal solution with

$$M = |Q|. \tag{3.9}$$

These extremal black holes are stable objects. They cannot lower their mass by Hawking radiation, indeed the Hawking temperature is zero. But often these extremal solutions do have a non-vanishing Bekenstein–Hawking entropy S_{BH}, so we do expect that after quantization they are described by a non-trivial Hilbert space $\mathcal{H}$ of quantum states with

$$\log \dim \mathcal{H} \sim S_{BH} \tag{3.10}$$

in the limit of large charge or mass.

In supergravity theories these stable solutions preserve a subgroup of the super-Poincaré group. The classical fields ϕ of a supersymmetric soliton or black hole satisfy an equation of the form

$$\delta_\epsilon \phi = \epsilon_\alpha G^\alpha(\phi) = 0 \tag{3.11}$$

for some particular set of supersymmetry variations ϵ_α. Since for a fermionic field the supersymmetry variation $G(\phi)$ will be typically some first-order equation in the bosonic fields, these supersymmetric or BPS solutions satisfy much simpler equations of motion than generic classical solutions. A non-gravitational example is supersymmetric Yang–Mills theory, where the supervariation of the fermion gives the first-order self-duality condition $F_+ = 0$ instead of the second-order Maxwell equation $d^*F = 0$.

Quantum mechanically, the black hole quantum states will be annihilated by these particular supersymmetry transformations. That is, we have an equation of the form

$$\epsilon_\alpha Q^\alpha |\text{BPS}\rangle = 0. \tag{3.12}$$

Since a number of supersymmetry generators act trivially, these BPS states will therefore correspond to smaller representations of the supersymmetry algebra. The fact that the dimension of a BPS supermultiplet is smaller than a generic one is a quantum mechanical reflection of the stability that we just discussed. An extremal BPS state is unable to develop in a non-extremal state by a simple dimensional argument: the resulting multiplet would contain more states than the original one.

More precisely, suppose we are dealing with some supersymmetry algebra with a set of n supercharges Q^α (n will always be even.) The Hilbert space will then decompose in irreducible representations of this algebra. The supersymmetry algebra will be of the general form

$$\{Q^\alpha, Q^\beta\} = \omega_i^{\alpha\beta} K^i \tag{3.13}$$

with

$$[Q^\alpha, K^i] = 0 \qquad [K^i, K^j] = 0. \tag{3.14}$$

Here the K^i are some set of bosonic charges, consisting of the translation operator P_μ and some extra set of central charges that commute with P_μ and that are usually denoted as Z. The symmetric bilinear forms ω_i will always be non-degenerate. Therefore, when we consider a representation where the operators K^i have fixed generic eigenvalues k^i, so that the total bilinear form

$$\omega = \omega_i k^i \tag{3.15}$$

is non-degenerate, we are essentially dealing with a representation of an n-dimensional Clifford algebra. The dimension of the representation will therefore be $2^{n/2}$.

However, for special values of the charges k^i, it might be the case that the bilinear form ω becomes accidentally degenerate. In that case there are certain linear combinations of supercharges that annihilate the representation. This representation is then a BPS representation. If it satisfies the conditions

$$\epsilon_\alpha Q^\alpha |\text{BPS}\rangle = 0 \tag{3.16}$$

for m independent spinors ϵ, the rank of the Clifford algebra will be $n - m$ and therefore the dimension of the representation will be $2^{(n-m)/2}$. Furthermore, by a well known argument, for such a BPS object the mass M of the state can be related to the eigenvalues of the central charge Z.

So in supergravity theories there is a preferred role for extremal black holes. And we will see that their quantum description is indeed much more straightforward. Furthermore, given our understanding of the extremal situation we will be able to also study the so-called near-extremal regime defined by

$$\delta M = M - |Q| \ll M. \tag{3.17}$$

In this regime the relevant degrees of freedom can be described by small perturbations around the BPS states.

3.3.2 Type IIB strings

Although string duality transformations are able to connect all perturbative string theories, it will be useful to start in a definite weak-coupling regime. Our starting point in this course will be the type IIB string theory. This theory reduces at low energy to IIB supergravity, which has $\mathcal{N} = (2, 0)$ spacetime supersymmetry. The bosonic low-energy fields come in two families: Neveu–Schwarz and Ramond.

The Neveu–Schwarz fields contain the metric $G_{\mu\nu}$ together with the two-form field $B_{\mu\nu}$ and the dilaton field Φ. The two-form B-field is an example of a generalized gauge field. Gauge transformations act as

$$B \rightarrow B + d\Lambda \tag{3.18}$$

with Λ a one-form. (Actually, this is not quite right. Technically a two-form B field gives what is called a connection on a gerbe. The gauge transformation only takes locally the above form. In general two B-field can differ by a closed two-form F that has integer periods, just as we can have large gauge transformation in a $U(1)$ gauge theory defined on a complex line bundle.)

The low-energy Lagrangian for these fields reads

$$\int d^{10}x \, e^{-2\Phi} \sqrt{-G}(R + 4|d\Phi|^2 - |dB|^2). \tag{3.19}$$

An important property of the NS fields is that fundamental strings form sources for all the fields. In particular the B-field has a 'minimal coupling' to the string world-sheet Σ

$$\int_\Sigma B_{\mu\nu} \partial x^\mu \wedge \bar{\partial} x^\nu. \tag{3.20}$$

So the fundamental string is the natural electric source for the NS B-field. If we compute the flux of the dual field strength

$$H^D = *e^{-2\Phi} H \qquad H = dB \tag{3.21}$$

through a S^7 surrounding a string in nine spatial dimensions, we find

$$\int_{S^7} H^D = 1. \tag{3.22}$$

We further note that the mass, or more properly the tension, of a fundamental strings is given by

$$T = 1/\alpha' \tag{3.23}$$

and is therefore independent of the string coupling g_s that is set by the expectation value of the dilaton

$$g_s = \langle e^\Phi \rangle. \tag{3.24}$$

3.3.3   The NS five-brane

Both in the IIA and IIB string there is also an object that is magnetically charged with respect to the NS B-field: the so-called NS five-brane. A solution including a single five-brane satisfies

$$\int_{S^3} H = 1 \tag{3.25}$$

for an S^3 that surrounds the five-brane in the nine spatial dimensions. It can be thought of as the analogue of a Dirac monopole in an Abelian gauge theory. Locally we write $H = dB$ but on the equator of the S^3, which is now a two-sphere S^2, the B-fields on the upper and lower hemisphere are related through a gauge transformation

$$B^+ = B^- + F \tag{3.26}$$

with F the field strength of a Dirac monopole of the S^2 satisfying

$$\int_{S^2} F = 1. \tag{3.27}$$

The NS five-brane is a typical soliton: a localized classical solution of the quations of motion. The metric and dilaton read in terms of the four transversal coordinates $x_\perp$ and the $5+1$ longitudinal or tangent coordinates $x_\parallel$ along the world-volume of the five-brane

$$ds^2 = f(r)\,dx_\perp^2 + dx_\parallel^2 \qquad e^{2\Phi} = f(r) \tag{3.28}$$

with

$$f(r) = g_s^2 + \frac{1}{2\pi^2 r^2} \qquad r^2 = x_\perp^2. \tag{3.29}$$

Note that this geometry describes an asymptotically flat space–time at $r = \infty$ connected to an infinite throat at $r \approx 0$. From any finite value of the transversal separation r the total distance to the point $r = 0$, where one would expect to find the five-brane located, is actually infinite in the five-brane metric. Close to the point $r = 0$ we can approximate the transversal part of the metric as

$$ds_\perp^2 \sim \frac{dr^2}{r^2} + d\Omega_3^2 \tag{3.30}$$

and we see that the size of the three-sphere surrounding the five-brane attains a constant value. So the local geometry looks like an infinite long cylinder $\mathbf{R}^+ \times S^3$, often referred to as the 'throat'.

However, the dilaton blows up logarithmically at $r = 0$ so string perturbation theory breaks down deep in the throat and the above metric cannot be trusted for the sigma model describing a single non-interacting string. We will see later what the relevant degrees of freedom are that describe string theory in the background of a NS five-brane at these small values of r.

The characteristic property of a NS five-brane, that gives away its nature of a conventional soliton, is that its tension is of the form

$$T \propto \frac{1}{g_s^2 \alpha'^3}. \tag{3.31}$$

Why is $M \sim 1/g_s^2$ the right behaviour for a soliton in string theory? Recall that Newton's constant G_N is proportional to g_s^2, so the gravitational field generated by a NS five-brane will behave as

$$G_N M \sim g_s^2 \frac{1}{g_s^2} \sim 1 \tag{3.32}$$

and is therefore independent of the string coupling. In particular in the weak-coupling, semi-classical regime $g_s \to 0$ the gravitational field will be non-zero and there is an actual extended curved space–time solution to Einstein's equations. Indeed there is exact conformal field theory that describes perturbative string theory around the five-brane solution [13].

3.3.4 D-branes

The second set of massless bosonic fields in type II string theory are the Ramond–Ramond fields. These are all generalized Abelian gauge fields: differential forms A of degree $p+1$ with a field strength $F = \mathrm{d}A$ that is a $(p+2)$-form. Such an object naturally couples electrically to a p-brane, and magnetically to a $(6-p)$-brane. In the type IIB the RR gauge fields are of even degrees and therefore we have odd-dimensional D-branes.

Because the low-energy action for the RR gauge fields is of the form

$$\int \mathrm{d}^{10}x \sqrt{-G} \, |F|^2 \qquad (3.33)$$

with no explicit powers of the string coupling, D-branes have the defining property that their tensions scale as

$$T \sim \frac{1}{g_s \ell_s^{p+1}}. \qquad (3.34)$$

Therefore the strength of the gravitational field scales as

$$G_N M \sim g_s^2 \frac{1}{g_s} \sim g_s. \qquad (3.35)$$

So in the weak-coupling limit $g_s \to 0$ the gravitational field vanishes. In string perturbation theory a D-brane does not curve the surrounding space–time, apart from a delta-function singularity at the location of the brane. This is very uncharacteristic for solitons, and D-branes should therefore not be considered as such. This is closely related to the fact that a D-brane solution contributes to the classical action as

$$S \sim 1/g_s \qquad (3.36)$$

instead of the familiar behaviour of solitons $S \sim 1/g_s^2$.

A Dp-brane of charge Q gives rise to the following classical supergravity solution

$$\mathrm{d}s^2 = f(r)^{1/2} \, \mathrm{d}x_\perp^2 + f(r)^{-1/2} \, \mathrm{d}x_\parallel^2 \qquad e^{2\Phi} = f(r)^{(3-p)/2} \qquad (3.37)$$

with

$$f(r) = 1 + \frac{cg_s Q\alpha'^{(7-p)/2}}{r^{7-p}} \qquad r^2 = x_\perp^2 \qquad (3.38)$$

with c some numerical constant.

These supergravity solutions are only to be trusted when the curvatures (in string units) are small. Since the characteristic length of a D-brane is given by

$$\ell \sim (g_s Q)^{1/(7-p)} \ell_s \qquad (3.39)$$

this gives the condition

$$g_s Q \gg 1. \qquad (3.40)$$

Furthermore absence of string loop corrections to this metric requires $g_s \ll 1$.

The description of D-branes in terms of perturbation theory of open strings and their low-energy Yang–Mills reduction requires that

$$g_s Q = g_{YM}^2 N \ll 1. \tag{3.41}$$

So, there seems to be no overlap between the supergravity and perturbative gauge theory regime.

There is, however, a possibility to relate the supergravity solution to a *non-perturbative* gauge theory description. We can take the low-energy limit of the open string theory on the D-brane world-volume. In terms of the space–time physics this means considering distances to the D-brane or black hole that are so small that the masses set by the lowest modes of the stretched open strings match these energies. So a UV-limit or near-horizon limit in space–time makes contact with the gauge theory description. We will make ample use of this fact.

There is of course one obvious case in which all regimes can be brought together, namely an extremal solution. For such an extremal black hole, that gives rise to BPS states, the physics is essentially independent of the string coupling g_s, and we can vary it from one regime to another.

3.4 Five-dimensional black holes

In these lecture we will mostly concentrate on the most simple model which is built up from D-branes. We will start our discussion by describing the classical supergravity solutions.

3.4.1 Supergravity solutions

The solutions of five-dimensional black holes take the following form [14, 15]. They are parametrized by four real quantities $r_0, \alpha_1, \alpha_2, \alpha_3 \geq 0$. The metric is then given by

$$ds^2 = -f(r)^{-2/3}\left(1 - \frac{r_0^2}{r^2}\right)dt^2 + f(r)^{1/3}\left[\left(1 - \frac{r_0^2}{r^2}\right)^{-1}dr^2 + r^2\, d\Omega_3^2\right] \tag{3.42}$$

with (in units where the five-dimensional Planck length $\ell_p = 1$)

$$f(r) = \prod_{i=1}^{3}\left(1 + \frac{r_0^2}{r^2}\sinh^2\alpha_i\right). \tag{3.43}$$

This solution carries three independent Abelian charges

$$Q_i = r_0^2 \sinh 2\alpha_i. \tag{3.44}$$

These charges are quantized as

$$Q_i = \lambda_i N_i \qquad N_i \in \mathbf{Z} \tag{3.45}$$

where the parameters λ_i depend on the specific moduli of the compactification and satisfy

$$\prod \lambda_i = 1. \tag{3.46}$$

The total ADM mass is given in terms of the variables r_0, α_i as

$$M = r_0^2 \sum_i \cosh 2\alpha_i. \tag{3.47}$$

The Bekenstein–Hawking entropy, defined by the volume A of the horizon (which is in this five-dimensional case an S^3) can be written as

$$S_{BH} = \frac{A}{4} = 2\pi r_0^3 \prod_i \cosh \alpha_i. \tag{3.48}$$

Many of these formulas obtain a very suggestive form if we introduce a new notation by introducing 'left-moving' and 'right-moving' charges N_i^L, N_i^R, which are no longer quantized in integers, but are defined as

$$N_i = N_i^L - N_i^R \tag{3.49}$$

with

$$N_i^R = N_i^L \, e^{-4\alpha_i}. \tag{3.50}$$

Then the formula for the geometric entropy takes the form

$$S = 2\pi \prod_i \left(\sqrt{N_i^L} + \sqrt{N_i^R} \right) \tag{3.51}$$

(note that it does not depend on the moduli λ_i) and the mass formula becomes

$$M = \sum_i \lambda_i \left(N_i^L + N_i^R \right). \tag{3.52}$$

This last formula is particularly suggestive, since it seems to point to six independent contributions to the mass, measured by the 'occupation numbers' N_i^L, N_i^R of left- and right-moving degrees of freedom. Although a complete quantum description of these black holes is still to be found, we will see that in particular limits this decomposition in left- and right-movers indeed makes sense.

The ten-dimensional origin of this solution becomes clear, if we build it out of D-branes in a compactification of the type IIB theory from ten to five dimensions along a T^5. (We will return shortly to the specifics of this compactification.) We can put $Q_5 = N$ D5-branes wrapped around the T^5 with coordinates $(x^5, x^6, x^7, x^8, x^9)$, and $Q_1 = m$ D1-branes in the x^5-direction, and also turn on a momentum $p_5 = n2\pi/R_5$ in the same x^5-direction. The three independent charges are now

$$N_1 = Q_1 = m \qquad N_2 = Q_5 = N \qquad N_3 = n. \tag{3.53}$$

In this way we obtain the following (string) metric:

$$ds^2 = (f_1 f_2)^{-1/2}\left[-dt^2 + dx_5^2 + \frac{r_0^2}{r^2}(\cosh\alpha_3\, dt + \sinh\alpha_3\, dx_5)^2\right]$$

$$+ f_1^{-1/2} f_2^{1/2}[dx_6^2 + \cdots + dx_9^2]$$

$$+ (f_1 f_2)^{1/2}\left[\frac{dr^2}{1 - r_0^2/r^2} + r^2\, d\Omega_3^2\right] \tag{3.54}$$

with

$$f_1(r) = \left(1 + \frac{r_0^2 \sinh^2\alpha_1}{r^2}\right)$$

$$\tag{3.55}$$

$$f_2(r) = \left(1 + \frac{r_0^2 \sinh^2\alpha_2}{r^2}\right).$$

Note that the moduli of the T^5 that appear in this metric are given by

$$r_0^2 \sinh^2\alpha_1 = (2\pi)^4 g_s m \ell_s^6 / V$$
$$r_0^2 \sinh^2\alpha_2 = g_s N \ell_s^2 \tag{3.56}$$
$$r_0^2 \sinh^2\alpha_3 = (2\pi)^5 g_s^2 n \ell_s^8 / V.$$

Here V is the volume of the T^4 spanned by the coordinates (x^6, x^7, x^8, x^9). From this formula one easily reads off the concrete expressions for the parameters λ_i, that we used in the general formulas for the mass and the physical charges.

3.4.2 Extremal black holes

Extremal solutions are recovered in the limit $r_0 \to 0$ and $\alpha_i \to \infty$ while keeping the physical charges Q_i fixed. In that case the mass formula degenerates into

$$M = \sum_i Q_i = \sum_i \lambda_i N_i. \tag{3.57}$$

The formula for the entropy also simplifies since in terms of the 'left-moving and right-moving' charges N_i^L, N_i^R we have

$$N_i^R = 0 \qquad N_i^L = N_i \tag{3.58}$$

so that

$$S = 2\pi\sqrt{N_1 N_2 N_3} = 2\pi\sqrt{Q_1 Q_5 n}. \tag{3.59}$$

This formula is one of the simplest forms of a non-trivial geometric BH entropy in string theory, and in these lectures we will mostly concentrate on deriving this fundamental result from a microscopic point of view following [1].

3.4.3 String compactification

Let us now turn to a more detailed description of a specific model. We compactify the type IIB string on a five-dimensional Euclidean manifold down to five uncompactified Lorentzian dimensions $\mathbf{R}^{4,1}$. This five-dimensional compactification space we take to be

$$X^4 \times S^1 \tag{3.60}$$

where the four-manifold X^4 should be Ricci flat. This leaves the option of a four-torus T^4 or a $K3$ manifold.

Let us start with the case of a compactification on T^5. In that case there are the following 27 independent charges or, equivalently, five-dimensional Abelian gauge fields. (Compactified coordinates carry an i-index, uncompactified coordinates carry a μ-index.)

- NS fields: five momenta p_i and five winding numbers w^i, corresponding to the Kaluza–Klein gauge fields $G_{\mu i}$ and $B_{\mu i}$. Furthermore there is a NS five-brane charge q_{ijklm} corresponding to the dual field $B^D_{\mu ijklm}$.
- RR fields: five D1-branes r_i, ten D3-branes s_{ijk}, one D5-brane t_{ijklm}, corresponding to RR gauge fields $\tilde{B}_{\mu i}$, $C_{\mu ijk}$ and $\tilde{B}^D_{\mu ijklm}$.

The IIB string compactified on T^5 is supposed to have a U-duality symmetry group [16]

$$U = E_{6(6)}(\mathbf{Z}) \tag{3.61}$$

an integer form of the corresponding continuous Lie group. The 27 charges combine into one irreducible representation of this group: an 8×8 pseudo-real, antisymmetric and traceless matrix Z [17]. The group $E_{6(6)}(\mathbf{Z})$ can be defined as those transformations in $Sp(8, \mathbf{Z})$ that preserve the cubic invariant

$$\text{Tr } Z^3. \tag{3.62}$$

Under the T-duality subgroup $SO(5, 5, \mathbf{Z}) \subset E_{6(6)}(\mathbf{Z})$ the 27 charges decompose as

$$\mathbf{27} = \mathbf{10} \oplus \mathbf{1} \oplus \mathbf{16} \tag{3.63}$$

where the defining vector representation $\mathbf{10}$ of $SO(5, 5, \mathbf{Z})$ is the momenta and winding numbers (p_i, w^i) carried by the fundamental strings and waves; the singlet $\mathbf{1}$ corresponds to the NS five-brane and the $\mathbf{16}$ spinor representation is given by the D-brane charges that can be thought to lie in the homology lattice

$$H_{odd}(T^5, \mathbf{Z}) \cong \mathbf{Z}^{16}. \tag{3.64}$$

The values of the background fields parametrize the moduli space of T^5 compactifications of type II string theory, which is locally given by the Grassmannian

$$E_{6(6)}/Sp(8). \tag{3.65}$$

For weak string coupling this modulus space degenerates to the CFT modulus space, which is locally the familiar Narain modulus space

$$O(5,5)/O(5) \times O(5) \tag{3.66}$$

together with the flat RR gauge fields, that take values in

$$H^{even}(T^5, \mathbf{R}/\mathbf{Z}). \tag{3.67}$$

More precisely we have the local isomorphism

$$E_{6(6)}/Sp(8) \cong (O(5,5)/O(5) \times O(5)) \times \mathbf{R}^+ \times \mathbf{R}^{16}. \tag{3.68}$$

In the case of a compactification on $S^1 \times K3$ we again have a charge lattice of rank 27, which now carries the structure

$$\Gamma = \Gamma^{5,21} \oplus \mathbf{Z}. \tag{3.69}$$

This is built up as follows. First of all, the S^1 factor gives a two-dimensional lattice $\Gamma^{1,1}$ of momenta and winding numbers. The D-branes give rise to a copy of

$$\Gamma^{4,20} \cong H_*(K3, \mathbf{Z}). \tag{3.70}$$

Finally, there is the NS five-brane, wrapped over the total five-manifold. In this case we have a U-duality group

$$U = SO(5, 21, \mathbf{Z}) \tag{3.71}$$

that acts in the obvious way on the sublattice $\Gamma^{5,21}$ that consists of all charges except the momentum along the S^1. In fact, even on a compactification of type IIB on $K3$ we meet this U-duality group, where the lattice $\Gamma^{5,21}$ has an interpretation as the charge lattice of strings in the uncompactified six dimensions.

3.5   D5–D1-brane system

The essential ingredient of the stringy black hole will be a system of a D5-brane and a D1-brane. Of course, by various U-dualities such a system can be mapped into many other, equivalent combinations, in particular to a bound state of a fundamental string and a NS five-brane. We will now study this system from two points of view: from the point of view of the D5-brane and from the point of view of the D1-brane.

3.5.1   The D5-brane action

The D5-brane world-volume theory is at low energies described by a six-dimensional super-Yang–Mills (SYM) theory with $\mathcal{N} = (1, 1)$ supersymmetry [18]. The bosonic fields are a six-dimensional $U(N)$ gauge field A_μ together with four scalar fields U^i. The theory has a

$$SO(4) \cong Sp(1) \times Sp(1) \tag{3.72}$$

R-symmetry. Six-dimensional gauge theory is not a renormalizable field theory. Its coupling constant g_{YM}^2 has dimension [length]^2, and therefore the theory becomes strongly coupled in the UV. At short distance new degrees of freedom have to appear, and indeed they do. We will see later that the five-brane world-volume theory is a non-Abelian six-dimensional *string* theory, with a characteristic string length given by the Yang–Mills scale [19, 20]. However, for many applications, in particular for our considerations of near-extremal black holes, the effective field theory description in terms of the low-energy SYM theory is good enough, since we will only probe the IR physics.

Ignoring the fermions, the leading part of the D5-brane world-volume action on a six-dimensional world-volume Y embedded in ten-dimensional space–time (we will assume with a trivial normal bundle) is of the form

$$S = \int_Y \frac{1}{g_s \alpha'} \operatorname{Tr} \left(\mathcal{F} \wedge *\mathcal{F} + |D_\mu U^i|^2 + [U^i, U^j]^2 \right) + \mathcal{C} \wedge v'(\mathcal{F}). \quad (3.73)$$

Let us explain the various objects in this Lagrangian: g_s is the type IIB string coupling. The covariant field strength $\mathcal{F}$ is defined as [18]

$$\mathcal{F} = F - 2\pi i B \quad (3.74)$$

with F the usual curvature of the $U(N)$ connection on the rank N vector bundle $\mathcal{E}$ over Y, and B the background NS tensor field, a harmonic two-form on Y (actually, the pull-back to Y of the ten-dimensional space–time B-field.) Note that B is a singlet under $U(N)$, so it only couples to the $U(1)$ piece $\operatorname{Tr} F$ of the YM curvature.

The background RR gauge field $\mathcal{C}$ is the pull-back to Y of an arbitrary harmonic form of even degree in space–time and it can be decomposed as

$$\mathcal{C} = \theta + \widetilde{B} + G \quad (3.75)$$

with θ a scalar, $\widetilde{B}$ the RR two-form field and G a four-form with a field strength that satisfies the self-duality constraint in ten dimensions, $dG = *dG$. These RR fields couple to the five-brane through the generalized Mukai vector v' given by [21, 22, 23, 24]

$$v' = \operatorname{Tr} \exp \left(\frac{i\mathcal{F}}{2\pi} \right) \wedge \widehat{A}(Y)^{1/2}. \quad (3.76)$$

Here the first term is a generalization of the usual Chern character ch($\mathcal{E}$) including the NS B-field. The expression $\widehat{A}(Y)$ that figures in the second term is the so-called A-roof genus of the manifold Y. It appears in the index theorem of the Dirac operator and can be expressed as a particular combination of Pontryagin classes of Y. For a Calabi–Yau threefold (or twofold) the $\widehat{A}$ genus equals the Todd genus

$$Td(Y) = 1 + \frac{c_2}{12} \quad (3.77)$$

and we can write the RR charge vector as

$$v' = \operatorname{Tr} \exp \left(\frac{iF}{2\pi} + B + \frac{c_2}{24} \right). \quad (3.78)$$

From this coupling of the RR gauge fields we see that a D5-brane can carry charges with respect to other fields than the six-form gauge field if non-trivial topological sectors are included. In particular it can carry a charge with respect to the RR B-field when the second Chern class of the gauge bundle is non-zero. From the Lagrangian we see that the total D1-brane charge is

$$ch_2(E) + \frac{c_2(Y)}{24} = \frac{1}{8\pi^2}(-\text{Tr } F \wedge F + \tfrac{1}{24}\text{Tr } R \wedge R). \qquad (3.79)$$

Physically speaking, instantons in the gauge theory carry D1-brane charge. Indeed, in six dimensions an instanton is a stringlike object. It should be thought of as a D1-brane 'dissolved' into the D5-brane.

In the same way a D3-brane is represented by a non-trivial first Chern class

$$c_1(E) = \frac{i}{2\pi}\text{Tr } F. \qquad (3.80)$$

A D3-brane should have a Minkowskian world-volume, so the corresponding Poincaré-dual two-form Tr F should be completely spatial. Therefore the *magnetic* fluxes of the gauge theory represent the D3-branes.

Finally, we also see that the fundamental strings can be represented. They couple to the NS B-field, and from the action we read off that they correspond to *electric* fluxes. The coupling to the NS five-brane is a little bit more mysterious, since it corresponds to a finite energy density.

Summarizing and specializing to the relevant case: at low energies the D5–D1-brane system is described by a $(5 + 1)$-dimensional gauge theory of rank $N = Q_5$ with total instanton number $m = Q_1$.

We can reduce this system further if we assume that the six-dimensional world-volume factors are in the form $Y^6 = X^4 \times \Sigma^2$, where the four-manifold X is of much smaller volume than the two-dimensional surface Σ. Then, for fixed position in Σ we can solve the classical equations for the fields on X. In this case, these solutions will satisfy the ASD equations of the gauge instantons

$$\mathcal{F}_+ = 0. \qquad (3.81)$$

Note that if the NS B-field is non-zero, this equation gives the usual ASD relation of the $SU(N)$ part of the curvature, but the $U(1)$ part satisfies the modified relation $F_+ = B$.

So, in this adiabatic limit we are led to a sigma model, describing maps from the surface Σ to the moduli space of $U(N)$ m-instantons on the four-manifold X. We will indicate this moduli space by

$$\mathcal{M}_{N,m}(X). \qquad (3.82)$$

3.5.2 The D1-brane action

An alternative starting point is the world-sheet theory for a set of $m = Q_1$ D1-branes, see [25, 26]. According to the general philosophy this is a two-dimensional

$\mathcal{N} = (8, 8)$ $U(m)$ SYM theory with eight adjoined scalar fields U^i that represent the transverse oscillations. This system has an $SO(8)$ R-symmetry.

Including a collection of N D5-branes, say in the $(6, 7, 8, 9)$ directions, breaks the R-symmetry to

$$SO(4)_\perp \times SO(4)_\parallel.$$

We will denote the vector representations of these groups by the indices i and μ respectively. So the eight scalar fields split into four Y^i transverse to the five-brane and four X^μ parallel to the five-brane. Similarly, we will denote the chiral spinor representations of these groups by indices $\alpha, \dot{\beta}$ and $a, \dot{b}$. The supersymmetry is broken down to $\mathcal{N} = (4, 4)$ and in terms of this supersymmetry algebra the two-dimensional gauge field combines with the scalars X^μ to form a vector multiplet. The parallel scalars Y^i form a hypermultiplet. It will be sometimes convenient to write these fields in terms of spinor indices as $X^{\alpha\dot{\beta}}$, $Y^{a\dot{b}}$.

We also have to add 1–5 strings, that stretch from the D1-brane to the D5-brane. From the perspective of the two-dimensional gauge theory on the D1-branes these 1–5 strings correspond to N hypermultiplets in the fundamental representation of $U(m)$. We will denote the scalar components of these hypermultiplets as H^a where we suppressed the $U(N)$ and $U(m)$ indices.

The full Lagrangian contains three potential terms for these scalar fields, namely (schematically)

$$V_1 = \text{Tr}\, [X^\mu, X^\nu]^2 \tag{3.83}$$

$$V_2 = |X^\mu H^a|^2 + \text{Tr}\, [X^\mu, Y^i]^2 \tag{3.84}$$

$$V_3 = |[Y^i, Y^j]_+ + H^a \sigma^{ij}_{ab} H^b|^2. \tag{3.85}$$

There are two important zero sets of these scalar potentials: the Coulomb branch and the Higgs branch.

3.5.3 The Coulomb branch

In the Coulomb branch the expectation values of the hypermultiplets are zero, whereas the vector multiplet can have a non-zero vev

$$\langle H \rangle = 0 \qquad \langle X^\mu \rangle \neq 0. \tag{3.86}$$

(We ignore for the moment the hypermultiplets Y^i.) In that case the potential term V_1 requires

$$[X^\mu, X^\nu] = 0. \tag{3.87}$$

Therefore the scalar fields X^μ can be simultaneously diagonalized and brought into the form

$$X^\mu = \begin{pmatrix} x^\mu_1 & & 0 \\ & \ddots & \\ 0 & & x^\mu_k \end{pmatrix}. \tag{3.88}$$

The eigenvalues x_I^μ are only uniquely determined up to a permutation in S_m. The classical Coulomb modulus space is therefore given by the symmetric product

$$\mathcal{M}_C^{cl} = (\mathbf{R}^4)^m / S_m. \tag{3.89}$$

There is an obvious interpretation of this modulus space. The eigenvalues x_I denote the positions of the D1-branes as they move away from the five-brane, which is located at $x = 0$. So we are left with the configuration space of m unordered points in the transversal $\mathbf{R}^4$. Separating the one-branes and five-branes stretches the 1–5 strings which gives a mass to the hypermultiplets H. Indeed from the potential term V_2 we see that such a mass is indeed induced

$$V_2 \sim (x_I^\mu H)^2. \tag{3.90}$$

However, this is not the end of the story, because quantum corrections modify the classical modulus space (3.89). Because of the $\mathcal{N} = (4, 4)$ supersymmetry there is only a one-loop correction to the kinetic term of the scalars x_I. This contribution is easily seen to be of the form $\sim m/x^2$ and this gives the following full quantum metric [27]

$$ds^2 = dx^2 \left(\frac{1}{g_s^2} + \frac{m}{x^2} \right). \tag{3.91}$$

Here we recognize the transversal part of the metric of the D5-brane (3.29). (Note that the D5-brane is simply S-dual to the NS five-brane.) It is precisely the infinite throat that we discussed at length in the case of the NS five-brane.

So at large distances away from the D5-brane, where the hypermultiplets H can be considered heavy and can therefore be harmlessly integrated out, the description of the Coulomb branch of the gauge theory reduces to a sigma-model describing a gas of strings moving in the five-brane background geometry. Although we are dealing with D-strings in the background of a D5-brane, a simple S-duality will relate this system to a gas of fundamental strings moving in the background of an NS five-brane. It is quite remarkable how the world-sheet sigma model, that is the fundamental starting point of perturbative string theory, arises here as an effective CFT describing the IR dynamics of a particular branch of a gauge theory.

3.5.4 The Higgs branch

There is a second set of field configurations that minimalize the potential of the D1–D5-brane system: the Higgs branch. Here we give expectation values to the hypermultiplets and not to the vector multiplets,

$$\langle H \rangle \neq 0 \qquad \langle X^\mu \rangle = 0. \tag{3.92}$$

Since the transverse scalars X^μ are set to zero, this branch should be interpreted as describing the physics of the D1-branes when they coincide with the D5-brane. In this case the relevant scalar potential is V_3 which gives the equation

$$[Y^i, Y^j]_+ + H^a \sigma_{ab}^{ij} H^b = 0. \tag{3.93}$$

This is a rather famous equation: it is the so-called ADHM equation for $U(N)$ instantons of charge m on $\mathbf{R}^4$. That is, the Higgs modulus space is given by the instanton modulus space on $\mathbf{R}^4$

$$\mathcal{M}_H^{cl} = \mathcal{M}_{N,m}(\mathbf{R}^4). \tag{3.94}$$

Note that these are four-dimensional $U(N)$ instantons, even though we are dealing with a two-dimensional $U(m)$ gauge theory with N hypermultiplets. That is, we obtain instantons in the flavour group. There are no quantum corrections to the metric in this case because of the hyper-Kähler (HK) condition. For the Higgs branch the classical and quantum moduli spaces coincide

$$\mathcal{M}_H^{qu} = \mathcal{M}_H^{cl}. \tag{3.95}$$

This result confirms beautifully the picture that we derived from the alternative point of view of the D5-brane. The Higgs branch represents the 'dissolved' D-strings that can bind to the D5-brane as instantons. At the moment the D-strings touch at $x = 0$, they form point-like instantons. By turning on the vev of the H-fields, these instantons spread out to finite size. It is not straightforward to see from the perspective of the 1–5 strings why turning on an expectation value of them smooths out the pointlike instanton. Note, however, that in this derivation we are only able to treat the case of $\mathbf{R}^4$ since the non-trivial topology of the five-brane is hard to take into account.

So we obtain the following picture of the modulus space of the D1–D5-brane system. The modulus space consists of two branches: the Coulomb branch $\mathcal{M}_C$ with $H = 0$, $X \neq 0$ and the Higgs branch $\mathcal{M}_H$ with $H \neq 0$, $X = 0$ that meet at the points $H = X = 0$. The Coulomb branch describes the motion of the D-strings in the background of the D5-brane. The Higgs branch describes the degrees of freedom of the bound state. In the infrared limit the two-dimensional gauge theory will flow to a conformal field theory sigma model with target space $\mathcal{M}_C$ or $\mathcal{M}_H$.

Note that the two branches have very different properties. The Coulomb branch has dimension $4m$, whereas the Higgs branch has dimensions $4Nm$. The latter should be thought of as representing the degrees of freedom of the black hole that we will eventually make out of this system.

3.5.5 Six-dimensional strings

The case of a single five-brane deserves special attention. As we will show later in that case in the absence of any background fields (NS B-field or RR fields), the model reduces to a sigma-model on the orbifold

$$S^m X^4. \tag{3.96}$$

This was shown by Witten for the case $\mathbf{R}^4$ in analysing the ADHM equations [26].

As we will explain in great detail later any sigma-model on such a symmetric product $S^m X$ can be seen as a (sector of a) second-quantized string theory on X. In the case of N five-branes this would lead to a six-dimensional non-Abelian string theory, which is essentially a closed Green–Schwarz string with some kind of $U(N)$ gauge structure. Although at present no formulation of such a non-Abelian stringy theory is known, it is instructive to consider the Abelian theory, because for that model a Green–Schwarz light-cone formulation exists [19]. Such a model is not Lorentz invariant, but this point is believed to be rescued by the non-Abelian generalization.

Using the same notation for the $SO(4)_\parallel \times SO(4)_\perp$ symmetry group, such a world-sheet theory has scalar fields x^μ and the following fermion fields: a left-moving fermion $\theta_L^{a\alpha}$ in the $2^+ \otimes 2^+$ representation, and a right-moving fermion

$$\theta_R^{a\dot\alpha} 2^+ \otimes 2^- \text{ for the IIA 5-brane, 6d } \mathcal{N} = (2, 0) \text{ supersymmetry}$$

$$\theta_R^{\dot a\dot\alpha} 2^- \otimes 2^- \text{ for the IIB 5-brane, 6d } \mathcal{N} = (1, 1) \text{ supersymmetry.}$$

Quantizing the zero modes gives a representation of the Clifford algebra

$$\{\theta^{a\alpha}, \theta^{b\beta}\} = \epsilon^{ab}\epsilon^{\alpha\beta} \tag{3.97}$$

which gives the representation $2^+ \otimes 1 \oplus 1 \otimes 2^+$, or in more physical notation the ground states

$$|a\rangle + |\alpha\rangle. \tag{3.98}$$

This is a so-called 'half hypermultiplet' of the $\mathcal{N} = (1, 0)$ superalgebra. Tensoring the left-movers and right-movers gives the following bosonic multiplets. For the $(1, 1)$ theory we recover the Yang–Mills multiplet, that consists of a vector potential together with four scalar fields,

$$X^{\alpha\dot\beta}, A^{a\dot b} \Rightarrow X^\mu, A^i. \tag{3.99}$$

For the $(2, 0)$ theory we find the tensor multiplet,

$$X^{\alpha\dot\beta}, A^{ab} \Rightarrow X^\mu, Y, T_+^{ij}. \tag{3.100}$$

It is made up of a total of five scalar fields together with a two-form tensor field T that has a self-dual field strength $dT = *dT$.

3.6   Instanton strings

We have now obtained in two independent ways an effective description of the D1–D5-brane dynamics in terms of a two-dimensional sigma model with target space the instanton modulus space.

More precisely and more invariant: if we compactify the type IIB string on a four-manifold X we obtain a $(5 + 1)$-dimensional quantum gravity theory that includes a spectrum of strings, i.e. one-dimensional extended objects. These strings are labelled by a charge lattice Γ. They will be either fundamental strings, NS five-branes wrapped on X, or D-branes wrapped on the even-dimensional cycles of X. Let us for convenience work here with the case $X = K3$. In that case the lattice of strings is given by

$$\Gamma^{5,21} \tag{3.101}$$

and can be decomposed in terms of a lattice of NS-objects (strings and five-branes) and RR-objects (D-branes)

$$\Gamma^{1,1}_{NS} \oplus \Gamma^{4,20}_{RR}. \tag{3.102}$$

A D-brane state with charge $v \in H^*(X, \mathbf{Z}) \cong \Gamma^{4,20}$ can be described in terms of a vector bundle, or more generally a coherent sheaf $\mathcal{E}$ over X with Mukai vector

$$v = \mathrm{ch}(\mathcal{E})Td(X)^{1/2} = (r, c_1, r + \mathrm{ch}_2). \tag{3.103}$$

Here the Chern character is defined as $\mathrm{ch}(\mathcal{E}) = \mathrm{tr}\, \exp iF/2\pi$ and r is the rank of $\mathcal{E}$, $\mathrm{ch}_2 = -c_2 + \frac{1}{2}c_1^2$. Note that here we did not use the definition (3.76) of the generalized Mukai vector v', which includes the effect of the B-field. In fact the two are simply related as

$$v' = e^B \wedge v. \tag{3.104}$$

The point is that because of the B-field contribution the generalized Mukai vector is generically no longer an integer cohomology class, and we like to fix the charge lattice once and for all and identify it with $H^*(X, \mathbf{Z})$. We will instead account for the effect of the B-field in terms of the moduli of the $K3$ sigma-model.

It is natural to give $H^*(X, \mathbf{Z})$ the Mukai intersection product, defined as [28]

$$v \cdot v = \int_X (v^2 \wedge v^2 - 2v^0 \wedge v^4) \qquad v = (v^0, v^2, v^4), \ v^i \in H^i(X). \tag{3.105}$$

We will always identify $H^*(X, \mathbf{Z}) \cong \Gamma^{4,20}$ with Mukai's quadratic form. With this definition the modulus space $\mathcal{M}_v$ of simple sheaves with Mukai vector v has complex dimension

$$\dim \mathcal{M}_v = 2 + v \cdot v. \tag{3.106}$$

As we mentioned already, in the case of a $K3$ surface the total charge lattice is given by the unique even, self-dual lattice of signature $(5,21)$

$$\Gamma = \Gamma^{5,21}. \tag{3.107}$$

Given any charge vector $v \in \Gamma$ we can always apply a U-duality transformation in $O(5, 21, \mathbf{Z})$ to rotate the vector v to a configuration consisting of only $D5$- and possibly $D1$-branes. Quite generally if v consists of only D-branes, so

$$v \in \Gamma^{4,20} = H^*(X, \mathbf{Z}) \tag{3.108}$$

the above arguments would relate the BPS description of the D-brane bound state to the sigma-model on the instanton modulus space $\mathcal{M}_v$.

What can we say about this CFT? First of all, the modulus space $\mathcal{M}_v$ is a hyper-Kähler (HK) manifold of real dimension $4k$ with

$$\tfrac{1}{2}v^2 = k - 1. \tag{3.109}$$

If the vector v is primitive the moduli space is a smooth manifold. As required by U-duality all the spaces with equal v^2 are HK deformations of one particular differential manifold. This manifold is best known as the Hilbert scheme $X^{[k]}$ [29]. This is a canonical smooth resolution of the symmetric product

$$S^k X = X^k / S_k. \tag{3.110}$$

This implies in particular that many topological properties of the space $\mathcal{M}_v$ can be encoded in the orbifold $S^k X$.

The precise relation arises as follows [30], in close analogy with the familiar story about $K3$ surfaces. There is a modulus space of inequivalent HK structures that one can put on a Hilbert scheme. This modulus space takes the form

$$O(3, 20, \mathbf{Z}) \backslash O(3, 20) / O(3) \times O(20). \tag{3.111}$$

It arises in the following way. There is a canonical 'intersection product' on the second cohomology of $X^{[k]}$ of signature (3,20) given by

$$H^2(X^{[k]}, \mathbf{Z}) \cong \Gamma_k^{3,20} \tag{3.112}$$

where the lattice is defined as

$$\Gamma_k^{3,20} = \Gamma^{3,19} \oplus \mathbf{Z} \cdot u \qquad u^2 = 2 - 2k < 0. \tag{3.113}$$

Here $\Gamma^{3,19}$ denotes the lattice $H^2(K3, \mathbf{Z})$, the unique even self-dual lattice of that signature. The extra generator u has a beautiful interpretation. It arises because the Hilbert scheme desingularizes the symmetric product. In particular there is a $\mathbf{Z}_2$ orbifold singularity where at least two points in $S^k X$ coincide. Blowing up this singularity produces an extra divisor that is Poincaré dual to a two-form u. In CFT language this cohomology class is represented by a $\mathbf{Z}_2$ twist field.

The existence of the intersection form follows from a general result by Beauville, who showed that for any simple $4k$-dimensional HK manifold Y there exists a canonical quadratic form on $H^2(Y, \mathbf{Z})$ that generalizes the intersection form for $K3$ surfaces. It has rank $b = b_2$ and signature $(3, b - 3)$. Although the construction uses the complex structure and the associated holomorphic (2, 0) form η, the final result turns out to be independent of any choices and purely topological. Using the Hodge decomposition

$$H^2(Y, \mathbf{C}) = H^{2,0} \oplus H^{1,1} \oplus H^{0,2}. \tag{3.114}$$

Beauville's quadratic form is defined for any $w \in H^2(Y)$ as [31]

$$w \cdot w = \int_Y (\eta \bar{\eta})^{k-1} \left(2 w^{2,0} \wedge w^{0,2} + k w^{1,1} \wedge w^{1,1} \right). \tag{3.115}$$

Note that this reduces to the standard intersection form for $k = 1$. The normalization of η can be chosen such as to make the quadratic form on $H^2(Y, \mathbf{Z})$ integral and primitive. By construction the HK three-plane $U \subset H^2(Y, \mathbf{R})$ obtained by the period map is positive with respect to this quadratic form. In fact, one can prove a local Torelli theorem and use Yau's theorem to show that *locally* the moduli space of HK metrics of fixed volume on a general simple HK manifold Y is given by the Grassmannian

$$\frac{O(3, b - 3)}{O(3) \times O(b - 3)} \tag{3.116}$$

generalizing the results for $K3$ and T^4 with $b = 22$ and 6 respectively.

So also in our case a choice of HK structure gives a positive three-plane in $H^2(X^{[k]}, \mathbf{Z})$ spanned by the three HK forms $\omega_I, \omega_J, \omega_K$. One can now argue [30] that the Hilbert scheme generates the symmetric product orbifold exactly when the extra vector u is orthogonal to the HK three-plane. In this way we can describe the effective D1–D5-brane conformal field theory by the symmetric product and its deformations.

3.7   Conformal field theory of symmetric products

Given the fact that to a great extent the effective CFT that is responsible for the black hole degrees of freedom can be captured by a symmetric product sigma-model, it might be useful to discuss these particular orbifolds in greater detail.

Let X be a general Calabi–Yau manifold. We can consider the corresponding two-dimensional sigma-model. It will carry (at least) an $\mathcal{N} = (2, 2)$ superconformal symmetry with central charge $c = 3d$ with d the complex dimension of X. The Hilbert space $\mathcal{H}$ of such a sigma model can be considered as the space $\Omega^*(\mathcal{L}X)$ of semi-infinite differential forms on the loop space of X. We will always choose in the definition of $\mathcal{H}$ Ramond or periodic boundary conditions for the fermions. These boundary conditions respect the supersymmetry algebra; other boundary conditions can be obtained by spectral flow [32].

On this Hilbert space act two natural operators: the Hamiltonian H, roughly the generalized Laplacian on $\mathcal{L}X$, and the momentum operator P that generates the canonical circle action on the loop space corresponding to rotations of the loop,

$$e^{i\theta P} : x(\sigma) \mapsto x(\sigma + \theta). \tag{3.117}$$

In a conformal field theory the operators H and P are usually written in terms of left-moving and right-moving Virasoro generators L_0 and $\overline{L}_0$ as

$$H = L_0 + \overline{L}_0 - d/4 \qquad P = L_0 - \overline{L}_0. \tag{3.118}$$

(Note that we conveniently shifted the ground-state energy so that Ramond vacua will have $H = 0$.)

Any $\mathcal{N} = 2$ SCFT has a $U(1)_L \times U(1)_R$ R-symmetry which allows us to define separate left-moving and right-moving conserved fermion numbers F_L and F_R, that up to an infinite shift (that is naturally regularized in the QFT) represent the bidegrees in terms of the Dolbeault differential forms on $\Omega^*(\mathcal{L}X)$.

The most general partition function is written as

$$Z(X; q, y, \bar{q}, \bar{y}) = \text{Tr}_{\mathcal{H}}(-1)^F y^{F_L} \bar{y}^{F_R} q^{L_0 - \frac{d}{8}} \bar{q}^{\bar{L}_0 - \frac{d}{8}} \qquad (3.119)$$

with $F = F_L + F_R$ the total fermion number. The partition function Z represents the value of the path-integral on a torus or elliptic curve, and we can write $q = e^{2\pi i \tau}$, $y = e^{2\pi i z}$ with τ the modulus of the elliptic curve and z a point in its Jacobian that determines the line-bundle of which the fermions are sections. The spectrum of all four operators $L_0, \bar{L}_0, F_L, F_R$ is discrete with the further conditions

$$L_0, \bar{L}_0 \geq d/8 \qquad L_0 - \bar{L}_0 \in \mathbf{Z} \qquad F_L, F_R \in \mathbf{Z} + \tfrac{d}{2}. \qquad (3.120)$$

For a general Calabi–Yau manifold it is very difficult to compute the above partition function explicitly. Basically, only exact computations have been done for orbifolds and the so-called Gepner points, which are spaces with exceptional large quantum automorphism groups. This is not surprising, since even in the $\alpha' \to 0$ limit we would need to know the spectrum of the Laplacian, while for many Calabi–Yau spaces (even relatively simple ones such as $K3$) an explicit Ricci flat metric is not even known.

Just as for the quantum mechanics case, we learn a lot by considering the supersymmetric ground states $\psi \in V \subset \mathcal{H}$ that satisfy $H\psi = 0$. In the Ramond sector the ground states are canonically in one-to-one correspondence with the cohomology classes in the Dolbeault groups,

$$V \cong H^{*,*}(X). \qquad (3.121)$$

In fact, these states have special values for the conserved charges. Ramond ground states always have $L_0 = \bar{L}_0 = d/8$, and for a ground state that corresponds to a cohomology class $\psi \in H^{r,s}(X)$ the fermion numbers are shifted degrees

$$F_L = r - d/2 \qquad F_R = s - d/2. \qquad (3.122)$$

The shift in degrees by $d/2$ is a result from the fact that we had to 'fill up' the infinite Fermi sea. We see that there is an obvious reflection symmetry $F_{L,R} \to -F_{L,R}$ (Poincaré duality) around the middle-dimensional cohomology. If we take the limit $q, \bar{q} \to 0$, the partition function reduces essentially to the Poincaré–Hodge polynomial of X

$$\lim_{q,\bar{q} \to 0} Z(X; q, \bar{q}, y, \bar{y}) = h(X; y, \bar{y})$$

$$= \sum_{0 \leq r,s \leq d} (-1)^{r+s} y^{r - \frac{d}{2}} \bar{y}^{s - \frac{d}{2}} h^{r,s}(X). \qquad (3.123)$$

3.7.1 The elliptic genus

An interesting specialization of the sigma model partition function is the elliptic genus of X [33], defined as

$$\chi(X; q, y) = \text{Tr}_{\mathcal{H}}(-1)^F y^{F_L} q^{L_0 - \frac{d}{8}}. \qquad (3.124)$$

The elliptic genus is obtained as a specialization of the general partition function for $\bar{y} = 1$. Its proper definition is

$$\chi(X; q, y) = \mathrm{Tr}_{\mathcal{H}}(-1)^F y^{F_L} q^{L_0 - \frac{d}{8}} \bar{q}^{\bar{L}_0 - \frac{d}{8}}. \tag{3.125}$$

But, just as for the Witten index, because of the factor $(-1)^{F_R}$ there are no contributions of states with $\bar{L}_0 - d/8 > 0$. Only the right-moving Ramond ground states contribute. The genus is therefore holomorphic in q or τ. Since this fixes $L_0 - d/8$ to be an integer, the partition function becomes a topological index, with no dependence on the moduli of X.

Using general facts of modular invariance of conformal field theories, one deduces that for a Calabi–Yau d-fold the elliptic genus is a weak Jacobi form [34] of weight zero and index $d/2$. (For odd d one has to include multipliers or work with certain finite index subgroups, see [35, 36, 37].) The ring of Jacobi forms is finitely generated, and thus finite dimensional for fixed index[1]. It has a Fourier expansion of the form

$$\chi(X; q, y) = \sum_{m \geq 0, \, \ell} c(m, \ell) q^m y^\ell \tag{3.126}$$

with integer coefficients. The terminology 'weak' refers to the fact that the term $m = 0$ is included.

The elliptic genus has beautiful mathematical properties. In contrast with the full partition function, it does not depend on the moduli of the manifold X: it is a (differential) topological invariant. In fact, it is a genus in the sense of Hirzebruch—a ring-homomorphism from the complex cobordism ring $\Omega^*_U(pt)$ into the ring of weak Jacobi forms. That is, it satisfies the relations

$$\chi(X \cup X'; q, y) = \chi(X; q, y) + \chi(X'; q, y)$$

$$\chi(X \times X'; q, y) = \chi(X; q, y) \cdot \chi(X'; q, y) \tag{3.127}$$

$$\chi(X; q, y) = 0 \qquad \text{if } X = \partial Y$$

where the last relation is in the sense of complex bordism. The first two relations are obvious from the quantum field theory point of view; they are valid for all partition functions of sigma-models. The last condition follows basically from the definition in terms of classical differential topology, more precisely in terms of Chern classes of symmetrized products of the tangent bundle, that we will give in a moment. We already noted that in the limit $q \to 0$ the genus reduces to a weighted sum over the Hodge numbers, which is essentially the Hirzebruch χ_y-genus,

$$\chi(X; 0, y) = \sum_{r,s} (-1)^{r+s} h^{r,s}(X) y^{r - \frac{d}{2}} \tag{3.128}$$

[1] For example, in the case $d = 2$ it is one-dimensional and generated by the elliptic genus of $K3$.

and for $y = 1$ its equals the Witten index or Euler number of X

$$\chi(X; q, 1) = \text{Tr}_{\mathcal{H}}(-1)^F = \chi(X). \tag{3.129}$$

For smooth manifolds, the elliptic genus has an equivalent definition as

$$\chi(X; q, y) = \int_X \text{ch}(E_{q,y})td(X) \tag{3.130}$$

with the formal sum of vector bundles

$$E_{q,y} = y^{-\frac{d}{2}} \bigotimes_{n>0} \left(\Lambda_{-yq^{n-1}}T_X \otimes \Lambda_{-y^{-1}q^n}\overline{T}_X \otimes S_{q^n}T_X \otimes S_{q^n}\overline{T}_X \right) \tag{3.131}$$

where T_X denotes the holomorphic tangent bundle of X. If the bundle $E_{y,q}$ is expanded as

$$E_{q,y} = \bigoplus_{m,\ell} q^m y^\ell E_{m,\ell} \tag{3.132}$$

the coefficients $c(m, \ell)$ give the index of the Dirac operator on X twisted with the vector bundle $E_{m,\ell}$, and are therefore integers. This definition follows from the sigma-model by taking the large-volume limit, where curvature terms can be ignored and one essentially reduces to the free model, apart from the zero modes that give the integral over X.

Physically, the elliptic genus appears as a counting function of perturbative string BPS states. If one constrains the states of the string to be in a right-moving ground state, i.e., to satisfy $\overline{L}_0 = d/8$, the states are invariant under part of the space–time supersymmetry algebra and called BPS. The generating function of such states is naturally given by the elliptic genus. Because we weight the right-movers with the chiral Witten index $(-1)^{F_R}$, only the right-moving ground states contribute.

3.7.2 Symmetric products

The usual step of second quantization now consists of considering a system of N of these (super) particles. It is implemented by the taking the Nth symmetric product of the single particle Hilbert space $\mathcal{H}$

$$S^N \mathcal{H} = \mathcal{H}^{\otimes n}/S_N \tag{3.133}$$

or more properly the direct sum over all N

$$S^* \mathcal{H} = \bigoplus_{N \geq 0} S^N \mathcal{H}. \tag{3.134}$$

We now propose to reverse roles. Instead of taking the symmetric product of the Hilbert space of functions or differential forms on the manifold X (i.e., the symmetrization of the quantized manifold), we will take the Hilbert space of functions or differential forms on the symmetric product $S^N X$ (i.e., the quantization of the symmetrized manifold)

$$S^* X = \coprod_{N \geq 0} S^N X. \tag{3.135}$$

But first we have to address the issue that the symmetric space $S^N X$ is not a smooth manifold but an orbifold, namely the quotient by the symmetric group S_N on N elements,

$$S^N X = X^N / S_N. \tag{3.136}$$

We will first be interested in computing the ground states for this symmetric product, which are in general counted by the Euler number. Actually, the relevant concept will turn out to be the orbifold Euler number. Using this concept there is a beautiful formula that was first discovered by Göttsche [38] in the context of Hilbert schemes of algebraic surfaces, but which is much more generally valid in the context of orbifolds, as was pointed out by Hirzebruch and Höfer [39].

First some notation. It is well known that many formulas for symmetric products take a much more manageable form if we introduce generating functions. For a general graded vector space we will use the notation

$$S_p V = \bigoplus_{N \geq 0} p^N S^N V \tag{3.137}$$

for the weighted formal sum of symmetric products. Note that for graded vector spaces the symmetrization under the action of the symmetric group S_N is always to be understood in the graded sense, i.e., antisymmetrization for the odd-graded pieces. We recall that for an even vector space

$$\dim S_p V = \sum_{N \geq 0} p^N \dim S^N V = (1 - p)^{-\dim V} \tag{3.138}$$

whereas for an odd vector space

$$\text{sdim} \, S_p V = \sum_{N \geq 0} (-1)^N p^N \dim \bigwedge^N V = (1 - p)^{\dim V}. \tag{3.139}$$

These two formulas can be combined into the single formula valid for an arbitrary graded vector space that we will use often[2]

$$\text{sdim} \, S_p V = (1 - p)^{-\text{sdim} \, V}. \tag{3.140}$$

Similarly we introduce for a general space X the 'vertex operator'

$$S_p X = \text{`exp} \, pX' = \coprod_{N \geq 0} p^N S^N X. \tag{3.141}$$

Using this formal expression the formula we are interested in reads (see also [41])

---

[2] This can of course be generalized to traces of operators as $\text{sTr} \, (S_p A) = \text{sdet}(1 - pA)^{-1}$.

Theorem 1 [38, 39]—*The orbifold Euler numbers of the symmetric products $S^N X$ are given by*

$$\chi_{orb}(S_p X) = \prod_{n>0} (1 - p^n)^{-\chi(X)}. \tag{3.142}$$

3.7.3 The orbifold Euler character

The crucial ingredient in theorem 1 is the orbifold Euler character, a concept that is very nicely explained in [39]. Here we give a brief summary of its definition.

Suppose a finite group G acts on a manifold M. In general this action will not be free and the space M/G is not a smooth manifold but an orbifold instead. The *topological* Euler number of this singular space, defined as for any topological space, can be computed as the alternating sum of the dimensions of the invariant piece of the cohomology,

$$\chi_{top}(M/G) = \text{sdim } H^*(M)^G. \tag{3.143}$$

In the de Rham cohomology one can also simply take the complex of differential forms that are invariant under the G-action and compute the cohomology of the standard differential d. This expression can be computed by averaging over the group

$$\chi_{top}(M/G) = \frac{1}{|G|} \sum_{g \in G} \text{sTr}_{H^*(M)} g$$

$$= \frac{1}{|G|} \sum_{g \in G} g \, \boxed{}_{\mathbf{1}}. \tag{3.144}$$

Alternatively, using the Lifshitz fixed point formula, we can rewrite this Euler number as a sum of fixed point contributions. Let M^g denote the fixed point set of the element $g \in G$. (Note that for the identity $M^1 = M$.) Then we have

$$\chi_{top}(M/G) = \frac{1}{|G|} \sum_{g \in G} \text{sdim } H^*(M^g)$$

$$= \frac{1}{|G|} \sum_{g \in G} \mathbf{1} \, \boxed{}_{g}. \tag{3.145}$$

In the above two formulas we used the familiar string theory notation

$$h \, \boxed{}_{g} = \text{sTr}_{H^*(M^g)} h \tag{3.146}$$

for the trace of the group element h in the 'twisted sector' labelled by g. Note that the two expressions (3.144) and (3.145) for the topological Euler number are related by a 'modular S-transformation,' that acts as

$$g \boxed{} \rightarrow 1 \boxed{}. \qquad (3.147)$$
$$\quad 1 \qquad\quad g$$

The *orbifold* Euler number is the proper equivariant notion. We see in a moment how it naturally appears in string theory. In the orbifold definition we remember that on each fixed point set M^g there is still an action of the centralizer or stabilizer subgroup C_g that consists of all elements $h \in G$ that commute with g. The orbifold cohomology is defined by including the fixed point loci M^g, but now taking only the contributions of the C_g invariants. That is, we have a sum over the conjugacy classes $[g]$ of G of the topological Euler character of these strata

$$\chi_{orb}(M/G) = \sum_{[g]} \chi_{top}(M^g/C_g). \qquad (3.148)$$

Note that this definition always gives an integer, in contrast with other natural definitions of the Euler number of orbifolds. From this point of view the topological Euler number only takes into account the trivial class $g = 1$ (the 'untwisted sector'). If we use the elementary fact that $|[g]| = |G|/|C_g|$, we obtain in this way

$$\chi_{orb}(M/G) = \frac{1}{|G|} \sum_{g \in G} \text{sdim } H^*(M^g)^{C_g}$$

$$= \frac{1}{|G|} \sum_{g,h \in G,\ gh=hg} \text{sTr}_{H^*(M^g)} h$$

$$= \frac{1}{|G|} \sum_{g,h \in G,\ gh=hg} \cdot h \boxed{}. \qquad (3.149)$$
$$\qquad\qquad\qquad\qquad\qquad\qquad g$$

This definition is manifest invariant under the 'S-duality' that exchanges h and g. We see that, compared with the topological definition, the orbifold Euler number contains extra contribution of the 'twisted sectors' corresponding to the non-trivial fixed point loci M^g. Using again Lifshitz's formula, it can be written alternatively in terms of the cohomology of the subspaces $M^{g,h}$ that are left fixed by both g and h as

$$\chi_{orb}(M/G) = \frac{1}{|G|} \sum_{g,h \in G,\ gh=hg} \text{sdim } H^*(M^{g,h}). \qquad (3.150)$$

3.7.4   The orbifold Euler number of a symmetric product

We now apply the above formalism to the case of the quotient X^N/S_N. For the topological Euler number the result is elementary. We simply replace $H^*(X)$ by its symmetric product $S^N H^*(X)$. Since we take the sum over all symmetric products, graded by N, this is just the free symmetric algebra on the generators of $H^*(X)$, so that

$$\chi_{top}(S_p X) = \text{sdim } S_p H^*(X) = (1 - p)^{-\chi(X)}. \qquad (3.151)$$

In order to prove the orbifold formula (3.142) we need to include the contributions of the fixed point sets. Thereto we recall some elementary facts about the symmetric group. First, the conjugacy classes $[g]$ of S^N are labelled by partitions $\{N_n\}$ of N, since any group element can be written as a product of elementary cycles (n) of length n,

$$[g] = (1)^{N_1} (2)^{N_2} \ldots (k)^{N_k} \qquad \sum_{n>0} n N_n = N. \qquad (3.152)$$

The fixed point set of such an element g is easy to describe. The symmetric group acts on N-tuples $(x_1, \ldots, x_N) \in X^N$. A cycle of length n only leaves a point in X^N invariant if the n points on which it acts coincide. So the fixed point locus of a general g in the above conjugacy class is isomorphic to

$$(X^N)^g \cong \prod_{n>0} X^{N_n}. \qquad (3.153)$$

The centralizer of such an element is a semi-direct product of factors S_{N_n} and $\mathbf{Z}_n$,

$$C_g = S_{N_1} \times (S_{N_2} \ltimes \mathbf{Z}_2^{N_2}) \times \ldots (S_{N_k} \ltimes \mathbf{Z}_k^{N_k}). \qquad (3.154)$$

Here the factors S_{N_n} permute the N_n cycles (n), while the factors $\mathbf{Z}_n$ act within one particular cycle (n). The action of the centralizer C_g on the fixed point set $(X^N)^g$ is obvious: only the subfactors S_{N_n} act non-trivially giving

$$(X^N)^g / C_g \cong \prod_{n>0} S^{N_n} X. \qquad (3.155)$$

We now only have to assemble the various components to compute the orbifold Euler number of $S^N X$:

$$\begin{aligned}
\chi_{orb}(S_p X) &= \sum_{N \geq 0} p^N \chi_{orb}(S^N X) \\
&= \sum_{N \geq 0} p^N \sum_{\substack{\{N_n\} \\ \sum n N_n = N}} \prod_{n>0} \chi_{top}(S^{N_n} X) \\
&= \prod_{n>0} \sum_{N \geq 0} p^{nN} \chi_{top}(S^N X) \\
&= \prod_{n>0} (1 - p^n)^{-\chi(X)} \qquad (3.156)
\end{aligned}$$

which concludes the proof of (3.142).

3.7.5   Orbifold quantum mechanics on symmetric products

The above manipulation can be extended beyond the computation of the Euler number to the actual cohomology groups. We will only be able to fully justify these definitions (because that's what it is at this point) from the string theory considerations that we present in the next section. For the moment let us just state that in particular cases where the symmetric product allows for a natural smooth resolution (as for the algebraic surfaces studied in [38] where the Hilbert scheme provides such a resolution), we expect the orbifold definition to be compatible with the usual definition in terms of smooth resolution.

We easily define a second-quantized, infinite-dimensional graded Fock space whose graded superdimensions equal the Euler numbers that we just computed. Starting with the single-particle ground-state Hilbert space

$$\mathcal{V} = H^*(X) \tag{3.157}$$

we define it as the symmetric algebra of an infinite number of copies $\mathcal{V}^{(n)}$ graded by $n = 1, 2, \ldots$

$$\mathcal{F}_p = \bigotimes_{n>0} S_{p^n} \mathcal{V}^{(n)} = S\left(\bigoplus_{n>0} p^n \mathcal{V}^{(n)}\right). \tag{3.158}$$

Here $\mathcal{V}^{(n)}$ is a copy of $\mathcal{V}$ where the 'number operator' N is defined to have eigenvalue n, so that

$$\chi_{orb}(S_p X) = \mathrm{Tr}_{\mathcal{F}}(-1)^F p^N = \prod_{n>0} (1 - p^n)^{-\chi(X)}. \tag{3.159}$$

We will see later that the degrees in $\mathcal{V}^{(n)}$ are naturally shifted by $(n-1)\frac{d}{2}$ with d the dimension of X, so that

$$\mathcal{V}^{(n)} \cong H^{*-(n-1)\frac{d}{2}}(X) \qquad n > 0. \tag{3.160}$$

Of course, this definition makes only good sense for even d, which will be the case since we will always consider Kähler manifolds.

This result can be interpreted as follows. We have seen that the fixed point loci consist of copies of X. These copies $X^{(n)}$ appear as the big diagonal inside $S^n X$ where all n points come together. If we think in terms of middle-dimensional cohomology, which is particularly relevant for Kähler and hyper-Kähler manifolds, this result tells us that the middle-dimensional cohomology of X contributes through $X^{(n)}$ to the middle-dimensional cohomology of $S^n X$.

So, if we define the Poincaré polynomial as

$$P(X; y) = Z(X; 0, y) = \mathrm{Tr}_{\mathcal{V}}(-1)^F y^F = \sum_{0 \le k \le d} (-1)^k y^k b_k(X) \tag{3.161}$$

then we claim that the orbifold Poincaré polynomials of the symmetric products $S^N X$ are given by

$$P_{orb}(S_p X; y) = \prod_{\substack{n>0 \\ 0 \le k \le d}} \left(1 - y^{k+(n-1)\frac{d}{2}} p^n\right)^{-(-1)^k b_k}. \tag{3.162}$$

This is actually proved for the Hilbert scheme of an algebraic surface in [42, 43].

Although we will only be in a position to understand this well in the next section, we can also determine the full partition function that encodes the quantum mechanics on $S^N X$. Again the Hilbert space is a Fock space built on an infinite number of copies of the single-particle Hilbert space $\mathcal{H}(X)$

$$\mathcal{H}_{orb}(S_p X) = \bigotimes_{n>0} S_{p^n} \mathcal{H}^{(n)}(X). \tag{3.163}$$

The contribution to the total Hamiltonian of the states in the sector $\mathcal{H}^{(n)}$ turns out to be scaled by a factor of n relative to the first-quantized particle, whereas the fermion number are shifted as before, so that

$$\mathcal{H}^{(n)} \cong \Omega^{*-(n-1)\frac{d}{2}}(X) \qquad H^{(n)} = -\tfrac{1}{2}\Delta/n. \tag{3.164}$$

To be completely explicit, let $\{h(m, k)\}_{m \geq 0}$ be the spectrum of H on the subspace $\Omega^k(X)$ of k-forms with degeneracies[3] $c(m, k)$, so that the single-particle partition function reads

$$Z(X; q, y) = \mathrm{Tr}_{\mathcal{H}}(-1)^F y^F q^H = \sum_{\substack{n>0 \\ 0 \leq k \leq d}} c(m, k) y^k q^{h(m,k)}. \tag{3.165}$$

Then we have for the symmetric product (in the orbifold sense)

$$Z_{orb}(S_p X; q, y) = \mathrm{Tr}_{\mathcal{H}_{orb}(SX)}(-1)^F p^N q^H y^F$$

$$= \prod_{\substack{n>0,\, m \geq 0 \\ 0 \leq k \leq d}} \left(1 - p^n q^{h(m,k)/n} y^{k+(n-1)\frac{d}{2}}\right)^{-c(m,k)}. \tag{3.166}$$

3.7.6 Second-quantized elliptic genera

We now come to an important formula that expresses the elliptic genus of a orbifold symmetric product

Theorem 2 [44]—*The orbifold elliptic genus of the symmetric products $S^N X$ is given by*

$$\chi_{orb}(S_p X; q, y) = \prod_{n>0,\, m \geq 0,\, \ell} (1 - p^n q^m y^\ell)^{-c(nm,\ell)}. \tag{3.167}$$

---

[3] These degeneracies are consistently defined as superdimensions of the eigenspaces, so that $c(m, k) \leq 0$ for k odd, and $c(0, k) = (-1)^k b_k$.

In order to prove this result, we have to compute the elliptic genus or, more generally, the string partition function for the orbifold M/G with $M = S^N X$ and $G = S_N$. The computation follows closely the computation of the orbifold Euler character that was relevant for the point-particle case.

First of all, the decomposition of the Hilbert space in superselection sectors labelled by the conjugacy class of an element $g \in G$ follows naturally. The superconformal sigma model with target space M can be considered as a quantization of the loop space $\mathcal{L}M$. If we choose as our target space an orbifold M/G, the loop space $\mathcal{L}(M/G)$ will have disconnected components of loops in M satisfying the twisted boundary condition

$$x(\sigma + 2\pi) = g \cdot x(\sigma) \qquad g \in G \qquad (3.168)$$

and these components are labelled by the conjugacy classes $[g]$. In this way, we find that the Hilbert space of any orbifold conformal field theory decomposes naturally into twisted sectors. Furthermore, in the untwisted sector we have to take the states that are invariant under G. For the twisted sectors we can only take invariance under the centralizer C_g, which is the largest subgroup that commutes with g. If $\mathcal{H}_g$ indicates the sector twisted by g, the orbifold Hilbert space has therefore the general form [45]

$$\mathcal{H}(M/G) = \bigoplus_{[g]} \mathcal{H}_g^{C_g} . \qquad (3.169)$$

In the point-particle limit $\alpha' \to 0$ the size of all loops shrinks to zero. For the twisted boundary condition this means that the loop gets necessarily concentrated on the fixed point set M^g and we are in fact dealing with a point-particle on M^g/C_g. In this way the string computation automatically produces the prescription for the orbifold cohomology that we discussed before. Indeed, as we stress, the quantum mechanical model of the previous section can best be viewed as a low-energy limit of the string theory.

In the case of the symmetric product $S^N X$, the orbifold superselection sectors correspond to partitions $\{N_n\}$ of N. Furthermore, we have seen that for a given partition the fixed point locus is simply the product

$$\prod_n S^{N_n} X^{(n)} . \qquad (3.170)$$

Here we introduce the notation $X^{(n)}$, to indicate a copy of X obtained as the diagonal in X^N where n points coincide. In the case of point-particles this distinction was not very important but for strings it is absolutely crucial.

The intuition is best conveyed with the aid of figure 3.1, where we have depicted a generic twisted sector of the orbifold sigma-model. The crucial point is that such a configuration can be interpreted as describing long strings[4] whose

---

[4] The physical significance of this picture was developed in among others [1, 46, 19] and made precise in [44].

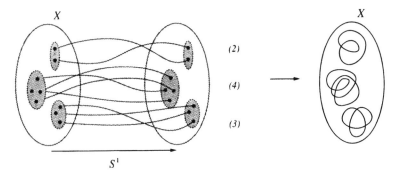

Figure 3.1. A twisted sector of a sigma model on $S^N X$ can describe fewer than N strings. (Here $N = 9$ and the sector contains three 'long strings.')

number can be smaller than N. Indeed, as we clearly see, a twisted boundary condition containing a elementary cycle of length n gives rise to a single string of 'length' n built out of n 'string bits.' If the cycle permutes the coordinates $(x_1, \ldots, x_n) \in X^n$ as

$$x_k(\sigma + 2\pi) = x_{k+1}(\sigma) \qquad k \in (1, \ldots, n) \qquad (3.171)$$

we can construct a new loop $x(\sigma)$ by gluing the n strings $x_1(\sigma), \ldots, x_n(\sigma)$ together:

$$x(\sigma) = x_k(\sigma') \quad \text{if} \quad \sigma = \tfrac{1}{n}\big(2\pi(k-1) + \sigma'\big) \qquad \sigma' \in [0, 2\pi]. \qquad (3.172)$$

If the twist element is the cycle $(N) \in S_N$, such a configuration describes one single long string of length N, instead of the N short strings that we would expect.

In this fashion we obtain from a cyclic twist (n) one single copy of the loop space $\mathcal{L}X$ that we denote as $\mathcal{L}X^{(n)}$. We use the notation $\mathcal{H}^{(n)}$ for its quantization. The twisted loop space $\mathcal{L}X^{(n)}$ is distinguished from the untwisted loop space $\mathcal{L}X$ in that the canonical circle action is differently normalized. We now have

$$e^{i\theta P} : x(\sigma) \rightarrow x(\sigma + \theta/n). \qquad (3.173)$$

So we find that only for $\theta = 2\pi n$ do we have a full rotation of the loop. This is obvious from the twisted boundary condition (3.171). It seems to imply that the eigenvalues for the operator $P = L_0 - \bar{L}_0$ in this sector are quantized in units of $1/n$. Together with the fact that in the elliptic genus only states with $\bar{L}_0 = 0$ contribute, this would suggest that the contribution of the sector $\mathcal{H}^{(n)}$ to the elliptic genus is[5]

$$\chi(\mathcal{H}^{(n)}; q, y) \stackrel{?}{=} \sum_{m, \ell} c(m, \ell) q^{m/n} y^\ell. \qquad (3.174)$$

---

[5] We use the more general notation $\chi(\mathcal{H}; q, y) = \mathrm{Tr}_{\mathcal{H}}(-1)^F q^{L_0 - \frac{d}{8}} y^{F_L}$ for any Hilbert space $\mathcal{H}$.

However, we must remember that the centralizer of a cycle of length n contains a factor $\mathbf{Z}_n$. This last factor did not play a role in the point-particle case, but here it does act non-trivially. In fact, it is precisely generated by $e^{2\pi i P}$. The orbifold definition includes a prescription to take the states that are invariant under the action of the centralizer. So only the states with integer eigenvalues of P survive. In this way only the states with m congruent to 0 modulo n survive and we obtain an integer q-expansion,

$$\chi(\mathcal{H}^{(n)}; y, q) = \sum_{m,\ell} c(nm, \ell) q^m y^\ell. \tag{3.175}$$

We now again assemble the various components to finish the proof of (3.167) (for more details see [44]).

$$\sum_{N \geq 0} p^N \chi_{orb}(S^N X; q, y) = \sum_{N \geq 0} p^N \sum_{\substack{N_n \\ \sum n N_n = N}} \prod_{n > 0} \chi(S^{N_n} \mathcal{H}^{(n)}; q, y)$$

$$= \prod_{n > 0} \sum_{N \geq 0} p^{nN} \chi(S^N \mathcal{H}^{(n)}; q, y)$$

$$= \prod_{n > 0,\, m,\, \ell} (1 - p^n q^m y^\ell)^{-c(nm,\ell)}. \tag{3.176}$$

The infinite product formula has strong associations to automorphic forms and denominator formulas of generalized Kac–Moody algebras [47] and string one-loop amplitudes [48], see also [49].

3.7.7 General partition function

It is not difficult to repeat the above manipulations in symmetric algebra for the full partition function. In fact, we can write a general formula for the second-quantized string Fock space, similarly as we did for the point-particle case in (3.158). This Fock space is again of the form

$$\mathcal{F}_p = \bigotimes_{n > 0} S_{p^n} \mathcal{H}^{(n)}. \tag{3.177}$$

Here $\mathcal{H}^{(n)}$ is the Hilbert space obtained by quantizing a single string that is wound n times. It is isomorphic to the subspace of the single-string Hilbert space $\mathcal{H} = \mathcal{H}^{(1)}$ with

$$L_0 - \bar{L}_0 = 0 \ (\mathrm{mod}\ n). \tag{3.178}$$

The action of the operators L_0 and $\bar{L}_0$ on $\mathcal{H}^{(n)}$ are then rescaled by a factor $1/n$ compared with the action on $\mathcal{H}$

$$L_0^{(n)} = L_0^{(1)}/n \qquad \bar{L}_0^{(n)} = \bar{L}_0^{(1)}/n. \tag{3.179}$$

As we explained already, this rescaling is due to the fact that the string has now length $2\pi n$ instead of 2π. Even though the world-sheet Hamiltonians $L_0^{(n)}$, $\overline{L}_0^{(n)}$ have fractional spectra compared to the single-string Hamiltonians, the momentum operator still has an integer spectrum,

$$L_0^{(n)} - \overline{L}_0^{(n)} = 0 \ (\text{mod } 1) \tag{3.180}$$

due to the restriction (3.178) that is implemented by the orbifold $\mathbf{Z}_n$ projection.

It is interesting to reconsider the ground states of $\mathcal{H}^{(n)}$, in particular their $U(1)_L \times U(1)_R$ charges, since this will teach us something about the orbifold cohomology of $S^N X$. A ground state $\psi^{(n)} \in V^{(n)}$ that corresponds to a cohomology class $\psi \in H^{r,s}(X)$ still has fermion charges F_L, F_R given by

$$F_L = r - d/2 \qquad F_R = s - d/2. \tag{3.181}$$

Making the string longer does not affect the $U(1)$ current algebra. However, since these states now appear as ground states of a conformal field theory with target space $S^n X$, which is of complex dimension nd, these fermion numbers have a different topological interpretation. The corresponding degrees $r^{(n)}$, $s^{(n)}$ of the same state now considered as a differential form in the orbifold cohomology of $S^n X \subset S^N X$ are therefore shifted as

$$r^{(n)} = r + (n-1)d/2 \qquad s^{(n)} = s + (n-1)d/2. \tag{3.182}$$

That is, we have

$$V^{(n)} \cong H^{*-(n-1)\frac{d}{2}, *-(n-1)\frac{d}{2}}(X). \tag{3.183}$$

In the quantum mechanics limit, the twisted loops that give rise to the contribution $\mathcal{H}^{(n)}$ in the Fock space become pointlike and produce another copy $X^{(n)}$ of the fixed point set X. However, this copy of X is the big diagonal in X^n. We see that this gives another copy of $H^*(X)$ however now shifted in degree. In the full Fock space we have an infinite number of copies, shifted by positive multiples of $(d/2, d/2)$.

We can encode this all in the generating function of Hodge numbers (3.123) as

$$h_{orb}(S_p X; y, \overline{y}) = \prod_{n>0, r, s} \left(1 - p^n y^r \overline{y}^s\right)^{-(-1)^{r+s} h^{r,s}(X)}. \tag{3.184}$$

For the full partition function we can write a similar expression

$$Z_{orb}(S_p X; q, \overline{q}, y, \overline{y}) = \prod_{n>0} \prod_{\substack{h, \overline{h}, r, s \\ h-\overline{h}=0 \ (\text{mod } n)}} \left(1 - p^n q^h \overline{q}^{\overline{h}} y^r \overline{y}^s\right)^{-c(h, \overline{h}, r, s)} \tag{3.185}$$

where

$$Z(X; q, \overline{q}, y, \overline{y}) = \sum_{h, \overline{h}, r, s} c(h, \overline{h}, r, s) q^h \overline{q}^{\overline{h}} y^r \overline{y}^s \tag{3.186}$$

is the single-string partition function.

3.7.8   Spectral flow

We have already mentioned that the Hilbert space of the CFT has two sectors corresponding to the possible boundary conditions of the fermions: Neveu–Schwarz (NS) with anti-periodic boundary conditions and Ramond (R) with periodic boundary conditions. Spectral flow relates the values of the conformal dimension $L_0 = h$ and $U(1)$ charge $F_L = q$ of the fields of these two sectors as follows [32]

$$h_{NS} = h_R + \tfrac{1}{2}q_r - d/8$$

$$q_{NS} = q_R + d/2 \tag{3.187}$$

and a similar formula for the right-moving charges. In particular the RR ground states (with $h_R = d/8$ and $q_R = r - d/2$) we obtain the chiral primary fields that are in one-to-one correspondence with the cohomology classes in $H^{r,s}(X)$ with

$$h_{NS} = r/2 \qquad q_{NS} = r. \tag{3.188}$$

3.8   Entropy computations

We have seen that the relevant degrees of freedom for the near-extremal black hole can be described in terms of a $(1 + 1)$-dimensional conformal field theory with central charge $6k$ where $k = Q_1 Q_5$. This conformal field theory carries an $\mathcal{N} = (4, 4)$ supersymmetry algebra. As we have explained, for many purposes we can think of this SCFT as a sigma-model with target space the kth symmetric product of the four-manifold X, which will be either a T^4 or a $K3$. Given this description of the quantum gravity solution with these particular charges, it is now straightforward to compute thermodynamic quantities such as the entropy from the CFT.

For this identification we should keep in mind that the momentum vector $p_5 = n/R_5$ in space–time can be identified with the CFT momentum

$$P = L_0 - \overline{L}_0 = n \tag{3.189}$$

and that the mass of the black hole can be identified with the CFT energy

$$H = L_0 + \overline{L}_0 - k/2. \tag{3.190}$$

Furthermore the transversal rotation group $SO(4)$ of the black hole can be decomposed as

$$SO(4)_\perp \cong SU(2)_L \times SU(2)_R \tag{3.191}$$

where $SU(2)_{L,R}$ are the global left-moving and right-moving R-symmetries of the $\mathcal{N} = (4, 4)$ SCFT. These symmetries are lifted to *affine* symmetries in the SCFT. The Cartan torus

$$U(1)_L \times U(1)_R \tag{3.192}$$

is generated by the fermion number operators F_L and F_R. The eigenvalues of these operators can be identified with the two spins J_+, J_- that label the representations of $SO(4)_\perp$.

3.8.1 Extremal solutions

We will first consider the extremal solution. This is the original computation of Strominger and Vafa [1]. In terms of the black hole supergravity solution all the quantities N_R^i now vanish. In that case the BPS condition will restrict us to pure left-moving states of the CFT, that are counted by the elliptic genus

$$\chi(S^k X; q, y) = \sum_{n,m} c(k, n, m) q^n y^m. \tag{3.193}$$

Here we can identify the eigenvalues of P with the quantum number n. But since we are imposing the BPS condition $\overline{L}_0 = 0$ we have

$$H = L_0 - k/4. \tag{3.194}$$

Furthermore the spin of the extremal black hole can be identified with the quantum number m.

We now want to know the (asymptotic) behaviour of the degeneracies $c(k, n, m)$, since the microscopic entropy for an extremal black hole with charges Q_1, Q_5, n and spin $J = J_+ = J_-$ is related to these degeneracies as

$$S_{BH} = \log c(Q_1 Q_5, n, J) \tag{3.195}$$

in the macroscopic limit $Q_1, Q_5, n \to \infty$, where we expect to make contact with black hole thermodynamics.

This can be done by using standard formulas about the growth of states in CFT. For the moment let us ignore the spin. Quite generally the number of states at level $L_0 - c/24 = n$ in a CFT of central charge c grows as

$$\exp 2\pi \sqrt{c\,n/6}. \tag{3.196}$$

It is not difficult to derive this: Consider the partition function

$$Z(\tau) = \mathrm{Tr}\, q^{L_0 - c/24} = \sum_n c(n) q^n \qquad q = e^{2\pi i \tau}. \tag{3.197}$$

The coefficient $c(n)$ is extracted as by a contour around $q = 0$ or $\tau = i\infty$,

$$c(n) = \int \frac{dq}{2\pi i q} q^{-n} Z(q). \tag{3.198}$$

Modular invariance $Z(\tau) = Z(-1/\tau)$ together with the above expansion imply the following leading behaviour of $Z(\tau)$ for small values of τ

$$Z(\tau) \sim e^{2\pi i c/24\tau}. \tag{3.199}$$

We can now make the saddle-point approximation in (3.198) for large n and find

$$\tau^2 = -c/24n. \tag{3.200}$$

Note that this saddle-point approximation is consistent since $n \to \infty$ implies $\tau \to 0$. Substituting this value of τ gives the promised result

$$c(n) \sim \exp 2\pi \sqrt{cn/6}. \tag{3.201}$$

Applying this result for our sigma-model we find with $c = 6k = 6Q_1Q_5$

$$S_{BH} = 2\pi \sqrt{Q_1 Q_5 n} \tag{3.202}$$

which is exactly the result predicted by the black hole geometry (3.59).

One can do slightly better and give a direct estimate for the degeneracies in

$$\chi(S_p X; q, y) = \sum_{k,n,m} c(k, n, m) p^k q^n y^m \tag{3.203}$$

using the explicit result (3.167) [49]. One then finds

$$c(k, n, m) \approx \exp 2\pi \sqrt{kn - m^2/4}. \tag{3.204}$$

Plugging in the physical variables this leads to the 'correct' entropy

$$S_{BH} = 2\pi \sqrt{6Q_1 Q_5 n - J^2/4}. \tag{3.205}$$

3.8.2   Non-extremal solutions

Non-extremal black holes that can still be confidently described in the CFT model have both left-moving and right-moving excitations. In terms of the macroscopic charges of the black hole, we allow left-moving and right-moving contributions n^L, n^R to the momentum giving a total mass and momentum given by

$$P = n^L - n^R = n$$

$$H = n^L + n^R. \tag{3.206}$$

In these models the D1-brane charge $Q_1 = N$ and D5-brane charge $Q_5 = m$ are still fixed, so we exclude anti-branes and have

$$N^L = N \qquad m^L = k \qquad N^R = m^R = 0. \tag{3.207}$$

In this case we can simply estimate the total number of states with $L_0 = n_L$ and $\bar{L}_0 = n_R$ by applying Cardy's formula separately for both excitations to find an entropy

$$S_{BH} = 2\pi \sqrt{Nm} \left(\sqrt{n_L} + \sqrt{n_R} \right). \tag{3.208}$$

This again neatly reproduces the predicted formula (3.51).

In the more general case we allow also anti-five-branes and anti-one-branes. Concentrating on the latter we interpret this as a gas of instantons and anti-instantons on a fixed number N of five-branes. There are of course solutions of the SYM gauge fields that look like such configurations. They simply break more supersymmetry then the case of pure (anti-)self-dual solutions. One interpretation [50] is to think of that as an interacting gas of instanton strings. In such a gas we have effective occupation numbers

$$m^L, \ m^R \qquad m^L - m^R = m. \tag{3.209}$$

Martinec and Li now argue convincingly that due to the strong interactions of these strings, the dominant contribution will be one 'long string' carrying all the degrees of freedom. Such a string then has a total energy

$$E = m^L + m^R + n^L + n^R \tag{3.210}$$

and a total momentum

$$P = n^L - n^R = n. \tag{3.211}$$

In addition the total winding number of the string (the total length of the string) is required to be

$$W = m^L - m^R = m. \tag{3.212}$$

Now in this situation we cannot simply identify $n^{L,R}$ with the excitation levels $L_0, \bar{L}_0$. These are determined by solving the usual Virasoro constraints

$$L_0 = N\left(E^2 - (P + W)^2\right)$$

$$\bar{L}_0 = N\left(E^2 - (P - W)^2\right). \tag{3.213}$$

Working out these formulas one is led to the expected result

$$S_{BH} = 2\pi \sqrt{N}\left(\sqrt{m^L} + \sqrt{m^R}\right)\left(\sqrt{n^L} + \sqrt{n^R}\right). \tag{3.214}$$

3.9 Matrix strings and black holes

Up to now we have worked from the perspective of type IIB string theory. We now want to turn to a non-perturbative description using M-theory. The essential starting point will be the beautiful *Ansatz* for a non-perturbative formulation of M-theory known as matrix theory [5]. See for example the reviews [51, 53, 52, 57] for more information about matrix theory.

3.9.1 Two-dimensional supersymmetric Yang–Mills theory

Matrix string theory gives a very simple *Ansatz* of what non-perturbative IIA string theory looks like in light-cone gauge [54, 55, 56]. It is simply given by the maximally supersymmetric two-dimensional Yang–Mills theory with gauge group $U(N)$ in the limit $N \to \infty$ (or with finite N in DLCQ).

To be more precise, let us consider two-dimensional $U(N)$ SYM theory with 16 supercharges. It can be obtained by dimensionally reducing the $\mathcal{N} = 1$ SYM theory in ten dimensions. Its field content consists of the following fields. First we pick a (necessarily trivial) $U(N)$ principle bundle P on the world-sheet $S^1 \times \mathbf{R}$. Let A be a connection on this bundle. We further have eight scalar fields X^i in the vector representation V of Spin(8), and eight left-moving fermions θ^a in the spinor representation S^+ and eight right-moving fermions $\bar{\theta}^{\dot{a}}$ in the conjugated spinor representation S^-. All these fields are Hermitian $N \times N$ matrices, or if one wishes sections of the adjoint bundle ad(P).

The action for the SYM theory reads

$$S_{SYM} = \int d^2\sigma \, \text{Tr} \left(\tfrac{1}{2} |DX^i|^2 + \theta^a \overline{D}\theta^a + \bar{\theta}^{\dot{a}} D\bar{\theta}^{\dot{a}} \right.$$

$$\left. + \frac{1}{2g^2} |F_A|^2 + g^2 \sum_{i<j} [X^i, X^j]^2 + g \, \bar{\theta}^{\dot{a}} \gamma^i_{a\dot{a}} [X^i, \theta^a] \right).$$

$$(3.215)$$

Here g is the SYM coupling constant—a dimensionful quantity with dimension 1/length in two dimensions. This means in particular that the SYM model is not conformally invariant. In fact, at large length scales (in the IR) the model becomes strongly interacting. So we have a one-parameter family of QFTs labelled by the coupling constant g or equivalently a length scale $\ell = 1/g$.

The relation with string theory is the following. First of all for finite N the Hilbert space of states of the SYM theory should be identified with the DLCQ *second-quantized* IIA string Hilbert space. The integer N that gives the rank of the gauge group is then related to the total longitudinal momentum in the usual way as

$$p^+ = N/R \qquad (3.216)$$

whereas the total light-cone energy is given by

$$p^- = \frac{N}{p^+} H_{SYM} \qquad (3.217)$$

with H_{SYM} the Hamiltonian of the SYM model. Note that in the decompactification of the null circle where we will take $N, R \to \infty$, keeping their ratio finite, only SYM states with energy

$$H_{SYM} \sim \frac{1}{N} \qquad (3.218)$$

will contribute a finite amount to p^-. Finally, the IIA string coupling constant g_s (a dimensionless constant) is identified as

$$g_s = (g\ell_s)^{-1} \qquad (3.219)$$

with ℓ_s the string length, $\alpha' = \ell_s^2$.

From this identification we see that free string theory ($g_s = 0$) is recovered at strong SYM coupling ($g = \infty$). This is equivalent to the statement that free string theory is obtained in the IR limit. In this scaling limit—the fixed point of the renormalization group flow—we expect on general grounds to recover a superconformal field theory with 16 supercharges. We will now argue that this SCFT is the supersymmetric sigma-model with target space $S^N \mathbf{R}^8$. We can then use our previous analysis of orbifold sigma-models to conclude that the point $g_s = 0$ indeed describes the second-quantized free IIA string.

The analysis proceeds in two steps. First we observe that because of the last two terms in the action (3.215), in the limit $g_s = 0$ which is equivalent to $g = \infty$, the fields X and θ necessarily have to commute. This means that we can write the matrix coordinates as

$$X^i(\sigma) = U(\sigma) \cdot x^i(\sigma) \cdot U^{-1}(\sigma) \tag{3.220}$$

with $U \in U(N)$ and x^i a diagonal matrix with eigenvalues $x_1^i, \ldots, x_N^i$. Now the matrix valued fields $X^i(\sigma)$ are single valued, being a section of the trivial bundle $U(N)$ vector ad(P). But this does not imply that the fields $U(\sigma)$ and $x^i(\sigma)$ are too. In fact, it is possible that after a shift $\sigma \to \sigma + 2\pi$ the individual eigenvalues are permuted due to a spectral flow. Only the set of eigenvalues (or more properly the set of common eigenstates) of the commuting matrices X^i is a gauge invariant quantity. So we should allow for configurations of the form

$$x^i(\sigma + 2\pi) = g \cdot x^i(\sigma) \cdot g^{-1} \tag{3.221}$$

with $g \in S_N$ the Weyl group of $U(N)$. Effectively this tells us that we are dealing with an orbifold with target space

$$\mathbf{R}^{8N}/S_N = S^N \mathbf{R}^8 \tag{3.222}$$

given Lie theoretically as t^8/W with t the Cartan Lie algebra and W the Weyl group of $U(N)$.

As we have analysed before this implies that the Hilbert space decomposes in superselection sectors labelled by the conjugacy classes $[g]$ of S_N, which in turn are given by partitions of N. This structure indicates that the Hilbert space is a Fock space of second-quantized IIA strings. A sector twisted by

$$g = (n_1) \ldots (n_k) \tag{3.223}$$

describes k strings of longitudinal momentum

$$p_i^+ = \frac{n_i}{R} = \frac{n_i}{N} p_{tot}^+ \qquad i = 1, \ldots, k. \tag{3.224}$$

We have also seen how for a string with a twist (n) of 'length' n the $\mathbf{Z}_n$ projection of the orbifold projects the Hilbert space to a subsector conditioned to

$$L_0 - \overline{L}_0 = 0 \ (\text{mod } n) \tag{3.225}$$

that we now interpret as the usual DLCQ level-matching condition. In the large-N limit, also the individual n_i go to infinity, effectively decompactifying the null circle.

The second step consists of analysing the behaviour of the gauge field. The possibly twisted configurations of $X^i(\sigma)$ break the gauge group $U(N)$ to an Abelian subgroup T that commutes with the configuration $X^i(\sigma)$. In fact, if the twist sector is labelled by a partition

$$n_1 + \cdots + n_k = N \qquad (3.226)$$

describing k strings of length $n_1, \ldots, n_k$, the unbroken gauge group is

$$T \cong U(1)^k. \qquad (3.227)$$

Because of the Higgs effect all the broken components of the gauge field acquire masses of the order g and thus decouple in the IR limit. This leaves us with a free Abelian gauge theory on $\mathbf{R} \times S^1$. This model has been analysed in great detail. Dividing by the gauge symmetries leaves us with the holonomy along the S^1

$$\mathrm{Hol}(A) = \exp \oint_{S^1} A \in T \qquad (3.228)$$

as the only physical degree of freedom. The gauge theory is therefore described by the quantum mechanics on the torus T with Hamiltonian given by

$$H = -g^2 \Delta \qquad (3.229)$$

with Δ the Laplacian on T. The eigenstates are given by the characters of the irreducible representations of T with eigenvalues (energies) g^2 times the second Casimir invariant of the representation. Clearly in the limit $g \to \infty$ only the vacuum state or trivial representation survives. This state has a constant wavefunction on T which has the interpretation that the Abelian gauge field is free to fluctuate, a result from the fact that in strong coupling the action $S = \frac{1}{g^2} \int |F|^2$ goes to zero. So, all in all, the gauge field sector only contributes a single vacuum state. This completes our heuristic derivation of the IR limit of SYM.

Since two-dimensional gauge theories are so well-behaved it would be interesting to make the above in a completely rigorous statement about the IR fixed point of SYM. One of the points of concern could be complications that emerge if some of the eigenvalues coincide. In that case unbroken non-Abelian symmetries appear. As we will show in the next section however, from the SCFT perspective such effects are always irrelevant and thus disappear in the IR limit. In fact, these effects are exactly responsible for the perturbative interactions at finite g.

3.9.2 Interactions

If the matrix string theory conjecture is correct, for finite coupling constant the SYM theory should reproduce the interacting string. A non-trivial check of this conjecture is to identify the correction for small g_s. This should be given by the joining and splitting interaction of the strings, producing surfaces with nontrivial topology.

This computation was done in [56] where the leading correction was computed. Let us try to summarize this computation. (It is also reviewed in [57].) The idea is to analyse the behaviour of the SYM theory in the neighbourhood of the IR fixed point. In leading order, a deformation to finite g, is given by the least irrelevant operator in the orbifold CFT. That is, we look for the operator $\mathcal{O}$ in the sigma model that preserves all the supersymmetries and the Spin(8) R-symmetry and that has the smallest scaling dimensions. The deformed QFT then has an action of the form

$$S = S_{SCFT} + (g_s)^{h-2} \int \mathcal{O} + \cdots \qquad (3.230)$$

with h the toal scaling dimension of $\mathcal{O}$. We would like to see that the power of g_s is one (so that $h = 3$) and that this deformation induces the usual joining and splitting interaction.

Note that the Hilbert space of the matrix string was defined with Ramond boundary conditions for the supercurrent $G^{\dot{a}} = \gamma_i^{a\dot{a}} \theta^a \partial x^i$. That is, we have

$$G^{\dot{a}}(\sigma + 2\pi) = G^{\dot{a}}(\sigma). \qquad (3.231)$$

We have seen that the ground-state space $\mathcal{V}^{(n)}$ of a $\mathbf{Z}_n$ twisted sector $\mathcal{H}^{(n)}$ is isomorphic to the ground-state space of a single string

$$\mathcal{V}^{(n)} \cong (V \oplus S^-) \otimes (V \oplus S^+). \qquad (3.232)$$

Only the conformal dimensions are rescaled and given by

$$L_0 = \overline{L}_0 = nd/8 \qquad (3.233)$$

since the central charge of the SCFT is n times as big. Here d was the complex dimension of the target space, so in our case $d = 4$.

One way to understand this vacuum degeneracy is that $\mathbf{Z}_n$ action on the n fermions $\theta_1, \ldots, \theta_n$ can be diagonalized with eigenvalues $e^{2\pi ik/n}$, $k = 0, \ldots, n-1$. That is, there are linear combinations of the θ_k, let us denote them by $\tilde{\theta}_k$, that have boundary conditions

$$\tilde{\theta}_k(\sigma + 2\pi) = e^{\frac{2\pi ik}{n}} \tilde{\theta}_k(\sigma). \qquad (3.234)$$

So the linear sum

$$\tilde{\theta}_0 = \theta_1 + \cdots + \theta_n \qquad (3.235)$$

is always periodic and its zero modes give the 16-fold vacuum degeneracy. A similar story holds for the right-moving fermions.

Since we want to keep Ramond boundary conditions in the interacting theory, the local operator $\mathcal{O}$ that describes the first-order deformation should be in the NS-sector. This just tells us that the OPE

$$G^{\dot{a}}(z)\mathcal{O}(w) \qquad (3.236)$$

is single valued in $z - w$. So, using the familiar operator-state correspondence of CFT we have to look in the NS-sector of the Hilbert space. These are of course again labelled by twist fields. The only difference is that the fermions now have an extra minus sign in their monodromy, and satisfy the boundary conditions

$$\tilde{\theta}_k(\sigma + 2\pi) = -e^{\frac{2\pi i k}{n}} \tilde{\theta}_k. \tag{3.237}$$

Now depending on whether n is even or odd there is a periodic fermion or not. So we expect to find only a degeneracy for even n. It is not difficult to compute the conformal dimension of the NS ground state in a Z_n twisted sector. First of all, both for the bosons and fermions the Z_n action can be diagonalized. The bosonic twist field that implements a twist with eigenvalue $e^{2\pi i k/n}$ has conformal dimensions $dk(n-k)/2n^2$, with d the complex dimension of the transversal space ($d = 4$ for the IIA string). For the corresponding fermionic twist field we find conformal dimension $dm^2/2n^2$, where $m = \min(k, N - k)$. Adding up all the possible eigenvalues we obtain total conformal dimension

$$h = \begin{cases} n & n \text{ even} \\ n - \frac{1}{n} & n \text{ odd}. \end{cases} \tag{3.238}$$

In particular the lowest dimension $h = 2$ is given by the Z_2 twist field σ. Since $n = 2$ is even, this ground state has the usual degeneracy

$$\sigma \in (V \oplus S^-) \otimes (V \oplus S^+). \tag{3.239}$$

Note that the zero modes of the superpartner of the twisted boson x^i give this degeneracy. However, the NS ground state is not supersymmetric or Spin(8) invariant, and is therefore not a suitable candidate for our operator $\mathcal{O}$.

There is, however, a small modification that does respect the supersymmetry algebra. In the Z_2 twisted sector the coordinate x^i has a mode expansion

$$\partial x^i = \sum_{n \in \mathbf{Z} + \frac{1}{2}} \alpha_n^i z^{-n-1}. \tag{3.240}$$

We now consider the first excited state

$$\mathcal{O} = \alpha^i_{-1/2} \bar{\alpha}^j_{-1/2} \sigma^{ij} \tag{3.241}$$

of conformal weights $2 + 1 = 3$ (here σ^{ij} indicates the components of σ in $V \otimes V$). This operator can be written as

$$\mathcal{O} = G^{\dot{a}}_{-1/2} \overline{G}^{b}_{-1/2} \sigma^{\dot{a}b} \tag{3.242}$$

and therefore satisfies

$$[G^{\dot{a}}_{-1/2}, \mathcal{O}] = \partial \overline{G}^{b}_{-1/2} \sigma^{\dot{a}b} \tag{3.243}$$

which is sufficient. Since $\mathcal{O}$ is both SUSY and Spin(8) invariant, it is the leading irrelevant operator that we were looking for.

What is the interpretation of the field $\mathcal{O}$ in string perturbation theory? It clearly maps superselection sectors with two strings into sectors with one string and vice versa. It is therefore exactly the usual joining and splitting interaction. In fact, the perturbation in the operator $\mathcal{O}$ reproduces the standard light-cone perturbation theory.

There is also a clear geometric interpretation of the twist field interaction $\mathcal{O}$. Consider the manifold $\mathbf{R}^8/\mathbf{Z}_2$ or if one wishes the compact version $T^8/\mathbf{Z}_2$. This is a Calabi–Yau orbifold and defines a perfectly well behaved superconformal sigma-model. One could now try to blow up the Z_2 singularity to obtain a smooth Calabi–Yau space. It is well known that this cannot be done without destroying the Calabi–Yau property; the orbifold $\mathbf{R}^8/\mathbf{Z}_2$ is rigid. In the SCFT language this is expressed by the fact that corresponding deformation does not respect the superconformal algebra. Algebraically, preserving the conformal invariance implies that the operator is marginal with scaling dimension 2. The fact that we found weight 3 is therefore in accordance with the fact that the two-dimensional field theory deforms to a massive field theory with a length scale—two-dimensional SYM.

However, we see that if the transverse target space had been four dimensional, the twist field interaction would have $L_0 = \overline{L}_0 = 1$ and would have represented a marginal operator. This is a simple reflection of the fact that the orbifold $\mathbf{R}^4/\mathbf{Z}_2$ or $T^4/\mathbf{Z}_2$ can be resolved to a smooth Calabi–Yau manifold, respectively a hyper-Kähler ALE space or a $K3$ surface.

3.9.3 Black holes in matrix theory

According to the matrix theory proposal of [5] non-perturbative type II string dynamics can be captured by means of an appropriate large-N limit of $U(N)$ supersymmetric quantum mechanics. This proposal originated in the description of D-particles and their bound states [4, 18]. The relation with the fundamental string degrees of freedom appears most manifestly in the representation of matrix theory as $(1 + 1)$-dimensional supersymmetric Yang–Mills theory, described by the Hamiltonian

$$H = \oint d\sigma \, \mathrm{tr} \left(\Pi_i^2 + DX_i^2 + \theta^T \gamma^9 D\theta \right.$$

$$\left. + \frac{1}{g_s} \theta^T \gamma^i [X_i, \theta] + \frac{1}{g_s^2} (E^2 + [X^i, X^j]^2) \right) \qquad (3.244)$$

defined on the circle $0 \le \sigma < 2\pi$. Here Π_i denotes the conjugate field to X^i, E is the electric field and D is the covariant derivative in the σ-direction.

The matrix string formalism quite naturally combines the perturbative string degrees of freedom with non-perturbative excitations such as D-branes and NS five-branes. The latter configurations can arise because after toroidal compactifications matrix theory becomes equivalent to higher-dimensional supersymmetric large-N gauge theory [5, 58]. This in particular opens up the possibility of adding new charged objects, essentially by considering gauge field configurations that carry non-trivial fluxes or other topological quantum numbers in the extra compactified directions.

To investigate this question, we will now consider the compactification of matrix theory to $4+1$ and $5+1$ dimensions. For the $(4+1)$-dimensional case we can make contact with the computation of Strominger and Vafa [1] of the D1–D5-brane bound state. In that model we have seen how effective six-dimensional strings provided the essential degrees of freedom. These strings arose from the D-string confined to a D5-brane. In matrix theory we will consider the dual BPS configuration with only NS charges made from bound states of NS five-branes with fundamental strings.

3.9.4   Five-branes and black strings

In matrix string theory, longitudinal NS five-branes that extend in the light-cone directions correspond to configurations with a non-vanishing topological quantum number [59, 60]

$$W_{ijkl}^+ = \oint d\sigma \; \text{Tr} \, X^{[i} X^j X^k X^{l]}. \tag{3.245}$$

This charge can naturally become non-vanishing when we compactify four of the transversal coordinates, so that W_{ijkl}^+ measures the winding number m of the five-brane around the corresponding four-torus. In the matrix string theory Hamiltonian (3.244), such a compactification is described via the identification of the compact X^i as the covariant derivatives in a $(5+1)$-dimensional super Yang–Mills theory on $S^1 \times T^4$. In gauge theory language, the longitudinal five-brane appears as a YM instanton configuration, and the wrapping number m translates into the instanton charge on the T^4. Note that in $5+1$ dimensions these instanton solutions represent string-like objects.

In the remaining six uncompactified dimensions, the wrapped five-brane looks like an infinitely long string. This becomes a black string with a finite horizon area per unit length, provided we adorn the brane with additional charges. In particular, we can consider the bound state with a longitudinal type IIA string. In the matrix string language, the winding number n of this string is identified with the total quantized world-sheet momentum $P = \oint T_{01}$ of the matrix string

$$P = \oint d\sigma \, \text{tr}\Big(\Pi_i D X^i + \theta^T D\theta\Big) = n. \tag{3.246}$$

In addition, we will assume that this bound state of NS five-branes and type IIA strings carries a non-zero longitudinal momentum p_9 per unit length.

In the standard large limit of matrix theory [5], the momentum p_+ is represented by the ratio $p_+ = N/R_9$ in a limit where both N and R_9 tend to infinity, keeping their ratio fixed. Here, however, we have to take a different limit that leads to a finite longitudinal momentum density

$$\frac{p_+}{R_9} = \frac{N}{R_9^2}. \tag{3.247}$$

Clearly, this implies that in the large-N limit the radius R_9 necessarily scales as $\sqrt{N}$.

In the presence of the black string configuration, the unbroken part of the supersymmetry algebra takes the form

$$\{Q^\alpha, Q^\beta\} = (p_- 1 + w^+ \gamma^9 + W^+_{ijkl} \gamma^{ijkl})^{\alpha\beta} \qquad (3.248)$$

with the finite central charge densities (in string units $\alpha' = 1$)

$$\frac{w^+}{R_9} = n \qquad \frac{W^+_{ijkl}}{R_9} = m \frac{V}{g_s^2} \epsilon_{ijkl} \qquad (3.249)$$

where V is the volume of the four-torus. The black string is annihilated by the right-hand side of (3.248). From this one derives the following expression for the black string tension

$$\frac{M_{ext}}{R_9} = \frac{N}{R_9^2} + m + \frac{kV}{g_s^2}. \qquad (3.250)$$

The last term represents the contribution of the five-brane with tension $1/g_s^2$ wrapped k times around the four-torus. This characteristic dependence on g_s of the mass of the solitonic NS five-brane can be read off immediately from the Hamiltonian (3.244), using the definitions (3.245) and (3.249).

3.9.5   The extremal 5D black hole

Upon compactification of the spatial light-cone direction x^9 on a circle of radius R_9, the six-dimensional black string described above becomes a five-dimensional black hole with quantized momentum

$$p_9 = \frac{N}{R_9}.$$

In the extremal case, this black hole has a mass given by

$$M_{ext} = \frac{N}{R_9} + n R_9 + \frac{m R_9 V}{g_s^2} \qquad (3.251)$$

and is directly related through a sequence of dualities to the five-dimensional type IIB black hole solution considered previously.

Let us briefly pause to recall the intermediate steps of this duality mapping. Starting on the type IIA side, we apply a T-duality along the compactified x^9-direction mapping the type IIA string on a type IIB string, and interchanging winding number and momentum. Under this T-duality the IIA NS five-brane becomes the IIB NS five-brane. Next, via the strong–weak coupling symmetry of the type IIB theory, we can turn this into a configuration of N D-strings in the presence of m D5-branes. Finally, by a complete T-duality along all four directions of T^4 we obtain N D5-branes and a D-string with winding number m.

Hence, through this duality sequence, the NS five-brane eventually becomes a D-string, while the original type IIA string has become the D5-brane. As a result, the matrix-valued fields X^i on the matrix string are identified with the covariant derivatives D_i in the $U(N)$ SYM theory on the world-volume of the D5-brane, while the D-strings represent the YM instanton configurations. Tracing this back to the type IIA language, one obtains the soliton description of the NS-five-brane given above.

For large values of the given charges (m, n, N) we have derived that this extremal black hole has a Bekenstein–Hawking entropy given by

$$S_{BH} = 2\pi \sqrt{Nmn}. \tag{3.252}$$

For BPS configurations, it turns out that this compactification of the ninth dimension (which turns the light-cone plane into a cylinder $S^1 \times \mathbf{R}$) can be achieved by considering the matrix string model with finite N, which becomes identified with the discrete nine-momentum.

The winding number n around the compactified circle R_9 is incorporated in the matrix string theory via an appropriate modification of the mass shell and level matching conditions. The level matching condition equates the total world sheet momentum of the matrix string to the integer winding number n around the R_9 direction, while the mass-shell relation takes the form

$$NH = p_0^2 - p_9^2 - w_9^2. \tag{3.253}$$

For BPS states, this exact structure indeed arises from the matrix string theory at finite N [56].

We would now like to use this construction to consider the BPS bound states of m NS five-branes and n IIA strings in the compactified light-cone situation. In this situation the central charges that appear in the unbroken supersymmetry algebra

$$\{Q^\alpha, Q^\beta\} = (p_- \mathbf{1} + w_9 \gamma^9 + W_{9ijkl}\gamma^{ijkl})^{\alpha\beta} \tag{3.254}$$

are finite quantities. The BPS condition

$$\epsilon^\alpha Q_\alpha |\text{BPS}\rangle = 0 \tag{3.255}$$

is compatible with this algebra, provided the spinor ϵ satisfies

$$(p_- + w_9\gamma^9 + W_{ijkl9}\gamma^{ijkl})\epsilon = 0. \tag{3.256}$$

This makes ϵ eigenspinor of both γ^9 and γ^{ijkl}. For a five-brane wrapped m times around the torus $T^4 \times S^1$ we have $W_{ijkl9} = m R_9 V \epsilon_{ijkl}$, where the epsilon symbol is short-hand for the unit volume element of the T^4.

The resulting BPS equation of motion is obtained by requiring that the super-symmetry variation of the fermions vanishes

$$\delta_\epsilon \theta = \left(\gamma^9 E + \gamma^i \Pi_i + \gamma^{i9} DX_i + \gamma^{ij}[X_i, X_j]\right)\epsilon = 0 \tag{3.257}$$

for the parameters ϵ determined above. This gives the equations

$$[X_i, X_j] = \epsilon_{ijkl} [X^k, X^l] \tag{3.258}$$

and

$$DX_i = \Pi_i \qquad\qquad [X_i, X_I] = 0 \tag{3.259}$$

$$E = 0 \qquad\qquad [X_I, X_J] = 0. \tag{3.260}$$

Here i, j, k, l denote only the internal T^4-directions, while I, J denote the transversal uncompactified dimensions.

Via the identification of X^i with a covariant derivative on the dual T^4, we recognize in equation (3.258) the self-duality equation of the corresponding gauge field configuration. The solutions of the five-brane equation (3.259) are thus in one-to-one correspondence with maps from $S^1 \times \mathbf{R}$ (with $\mathbf{R}$ the time direction) into the space of self-dual Yang–Mills instantons on T^4. The equation (3.259) requires that these maps are holomorphic. Finally, the last two equations (3.260) require that the four scalar fields X^I, that describe the four transverse coordinates to the five-brane, mutually commute and be covariantly constant on the T^4. For generic instantons, this condition leaves only the constant matrices $X^I = x_{cm}^I \mathbf{1}$ that define the centre of mass of the five-brane.

3.10 Near-horizon limits and *AdS* string theory

We now turn to the final approach to the black hole: string theory close to the horizon [61].

3.10.1 The near-horizon limit

Let us return to the geometry of a collection of Q Dp-branes. We have seen that the classical supergravity solution behaves as

$$ds^2 = f(r)^{1/2} dx_\perp^2 + f(r)^{-1/2} dx_\parallel^2 \qquad e^{2\Phi} = f(r)^{(3-p)/2} g_s^2 \tag{3.261}$$

with the function $f(r)$ is given by (in string units and up to an irrelevant numerical constant)

$$f(r) = 1 + \frac{g_s Q}{r^{7-p}} \qquad r^2 = x_\perp^2. \tag{3.262}$$

The D-brane geometry can be considered as a tunnelling between two geometries: the flat space at large r where the function $f(r)$ can be approximated by

$$f(r) \approx 1 \tag{3.263}$$

and a regime close to the origin $r \ll 1$, very much smaller than the string scale, where one can make the approximation

$$f(r) \approx \frac{g_s Q}{r^{7-p}}. \tag{3.264}$$

It is the latter region that can be described by a gauge theory.

Recall that the world-volume theory of N D-branes contains as massless fields a super-Yang–Mills theory with 16 supercharges. The bosonic fields consist of a $U(N)$ gauge field A_μ^{IJ} together with a set of adjoint-valued scalar fields U_i^{IJ} that describe the separations of the branes. If we ignore α' and g_s corrections the action reads

$$S = \frac{1}{g_s^2} \int d^{p+1}x \, \mathrm{Tr} \left(F_{\mu\nu}^2 + (D_\mu U_i)^2 + [U_i, U_j]^2 \right). \tag{3.265}$$

The Yang–Mills coupling is given by

$$g_{YM}^2 = g_s \ell_s^{p-3}. \tag{3.266}$$

The scalar fields U_i can be considered as compactified components of a ten-dimensional $\mathcal{N} = 1$ SYM gauge field, and have dimension of [mass]. They are related to the transverse oscillations x^i by

$$U^i = \frac{x^i}{\ell_s^2}. \tag{3.267}$$

In fact, the expectation value $\langle U^{II} - U^{JJ} \rangle$ can be thought of as the mass of the string stretched between the Ith and Jth brane. So the radial distance r between the branes translates into a mass-scale

$$m = \frac{r}{\ell_s^2} \tag{3.268}$$

on the D-brane world-volume. In the case that $r \to 0$ and we probe the shortest distances in space–time, this mass goes to zero and we reach the infrared long-distance dynamics of the gauge theory. This phenomenon that short-distance/UV physics in space–time corresponds to low energy/IR on the world-volume of the D-brane is one of the most important recent insights in string theory.

Of course, we still have to worry about stringy corrections to the SYM theory. These can be turned off by taking the limit $\alpha' \to 0$, which effectively is a low-energy limit on the brane. In order to preserve some interesting dynamics, one should at the same time tune the coupling constant g_s so that the gauge coupling g_{YM}^2 remains finite. We see from (3.266) that this implies that for $p \le 3$ in this limit $g_s \to 0$ so that the bulk physics decouples. We will return to the case $p > 3$.

Furthermore one should tune the Higgs expectation values such that finite energies on the brane are attained, that is we should also keep

$$\langle U \rangle = r/\alpha' \tag{3.269}$$

finite. This implies we are in the substringy regime $r \ll \ell_s$.

3.10.2 The D3-brane

The most famous example where this near-horizon limit gives interesting result is the case of the D3-brane.

From the point of view of the gauge theory we are working with a $(3 + 1)$-dimensional $\mathcal{N} = 4$ SYM with gauge group $U(N)$. This is a remarkable theory because it is finite and conformally invariant. The Yang–Mills coupling constant $g_{YM}^2 = g_s$ is dimensionless only in $3 + 1$ dimensions. The global symmetry group is

$$SO(4, 2) \times SO(6) \tag{3.270}$$

the product of the conformal group in $3 + 1$ dimensions $SO(4, 2)$ and a $SU(4) \cong SO(6)$ R-symmetry that can be thought of as rotations transversal to the three-brane. This SYM theory is strongly believed to be invariant under the Montonen–Olive S-duality

$$g_{YM}^2 \rightarrow 1/g_{YM}^2 \tag{3.271}$$

which is a reflection of the $g_s \rightarrow 1/g_s$ S-duality of the type IIB string.

From the supergravity point of view, a D3-brane is a solution of the general form (3.261) with function $f(r)$ given by (in SYM notation)

$$f(r) = 1 + \frac{g_{YM}^2 N}{U^4 \alpha'^2} \qquad e^{\Phi} = g_s. \tag{3.272}$$

In the decoupling or near-horizon limit $\alpha' \rightarrow 0$ we can make the approximation

$$f(r) \approx \frac{g_{YM}^2 N}{U^4 \alpha'^2}. \tag{3.273}$$

Now the geometry takes the simpler form

$$ds^2 = \alpha' \left(\frac{U^2}{g_{YM}\sqrt{N}} dx_{\parallel}^2 + \frac{g_{YM}\sqrt{N}}{U^2} \left(dU^2 + U^2 d\Omega_5^2 \right) \right) \tag{3.274}$$

which can be written as

$$ds^2 = \alpha' \left(\frac{U^2}{R^2} dx_{\parallel}^2 + \frac{R^2}{U^2} \left(dU^2 + U^2 d\Omega_5^2 \right) \right) \tag{3.275}$$

with $R^2 = g_{YM}\sqrt{N}$. Here we recognize the metric on the manifold

$$AdS^5 \times S^5 \tag{3.276}$$

where the radius of the metric of constant negative and positive curvature on the AdS^5 and S^5 is equal and given by R. This metric can be trusted if curvatures are small, which is for

$$g_{YM}^2 N \gg 1. \tag{3.277}$$

This near-horizon space–time geometry has isometry group

$$SO(4, 2) \times SO(6) \tag{3.278}$$

which coincides with the symmetries of the $\mathcal{N} = 4$ SYM theory.

From a Kaluza–Klein perspective supergravity on this background might be considered as five-dimensional gauged supergravity with gauge group $SO(6)$ on an anti-de Sitter space–time. There are various massless fields, among others the metric, the $SO(6)$ gauge fields and various scalar fields.

All this led Maldacena in [61] to conjecture that for any value of g_{YM} and N the $U(N)$ SYM theory is equivalent to the full ten-dimensional type IIB string theory on the $AdS^5 \times S^5$ background. This corresponds relates the two free parameters of the SYM model, the coupling constant (complexified by adding the theta-angle) and the rank N, to the two free parameters of the type IIB string, again the complexified coupling constant and the string scale (or more properly the dimensionless curvature $\mathcal{R}$ of the space–time in string units) as

$$g_{YM}^2 = g_s \tag{3.279}$$

$$g_{YM}^2 N = \mathcal{R}^2. \tag{3.280}$$

The correspondence can be seen as a holographic one, in the sense that a $(4+1)$-dimensional quantum gravity theory is related to a $(3+1)$-dimensional gauge theory, that can be thought to live on the boundary of the five-dimensional space–time. Indeed, the boundary of AdS^5 is (conformally compactified) Minkowski space–time. With the geometric picture in mind, the correspondence can be formulated as follows: to every field Φ_i in the supergravity (or string) theory is related a local operator $\mathcal{O}_i$ in the SYM theory. The partition function $Z_{grav}(\Phi_i)$ of the gravity theory with boundary values $\Phi_i(x)$ is related to the SYM partition function by [62, 63]

$$Z_{grav}(\Phi_i) = \left\langle e^{\int d^4x \, \Phi_i(x)\mathcal{O}_i(x)} \right\rangle_{SYM}. \tag{3.281}$$

In particular the metric is related in this fashion to the stress tensor

$$G_{\mu\nu} \sim T_{\mu\nu} = \partial_\mu U^i \partial_\nu U_i + \cdots. \tag{3.282}$$

The $SO(6)$ gauge fields correspond to the R-symmetry currents of the SYM theory

$$A_\mu^{ij} \sim J_\mu^{ij} = U^i \partial_m U^j + \cdots. \tag{3.283}$$

We notice that the massless fields of the supergravity theory are all composed (in leading order) of bilinears in the SYM fields. This is a familar fact of the representation theory of the conformal or AdS group $SO(4, 2)$. The gauge-theory multiplet exists, but this so-called singleton is not local in the five-dimensional sense. It is essentially four dimensional. Furthermore the standard massless representations can be made out a tensor product of two singletons.

3.10.3   D1–D5-branes and string theory on AdS^3

The case of the D3-brane is special because for that solution the dilaton was constant. In particular it did not blow up in the near-horizon limit. This guaranteed the appearance of the superconformal group. We now claim that a similar limit can be made for the D5–D1-brane system.

Working through the limit of the black hole geometry in a similar fashion as for the D3-brane, one finds that the geometry of a D1–D5-brane system compactified on $X^4 \times S^1$ simplifies to the metric (with $k = Q_1 Q_5$ and up to irrelevant numerical factors) [61]

$$ds^2 = \sqrt{k}\left(r^2\, dx^+ dx^- + r^{-2}\, dr^2 + d\Omega_3^2\right) + ds_X^2. \tag{3.284}$$

This describes (locally) the geometry of the space

$$AdS^3 \times S^3 \times X^4. \tag{3.285}$$

In this fashion Maldacena has conjectured that type IIB string theory on this geometry should be equivalent to the near-horizon limit of the D1–D5-brane system. But we have already seen how this system is described by the $(1 + 1)$-dimensional superconformal sigma-model on the instanton modulus space.

This remarkable equivalence has sparked a sizable literature, see e.g. [64, 65, 66, 10, 67], so we will here restrict ourselves to pointing out some of the more remarkable features.

What is the evidence that these two descriptions are equivalent? First there are the global symmetries groups. On the space $AdS^3 \times S^3$ the global symmetry is

$$SO(2, 2) \times SO(4). \tag{3.286}$$

This symmetry can be identified with the global superconformal transformations of the $\mathcal{N} = (4, 4)$ SCFT which are

$$SL(2, \mathbf{R})_L \times SU(2)_L \times SL(2, \mathbf{R})_R \times SU(2)_R. \tag{3.287}$$

Here the global conformal group $SL(2, \mathbf{R})_{L,R}$ are generated by the Virasoro generators L_1, L_0, L_{-1} and $\overline{L}_1$, $\overline{L}_0$, $\overline{L}_{-1}$. But in a two-dimensional SCFT we have of course a much bigger symmetry group generated by all modes of the (super-) currents, in particular all Virasoro generators L_n. How are these realized in the AdS^3 geometry? Quite remarkably this Virasoro algebra can be found in the supergravity theory and even in the full type IIB string theory.

The AdS^3 geometry has a boundary that looks like a cylinder $S^1 \times \mathbf{R}$ parametrized by the coordinates $x^\pm$. It was shown by Brown and Henneaux [68] that the group of diffeomorphisms of AdS^3 that respect the asymptotic form of the AdS metric act like conformal transformations on the boundary and the algebra reproduces the Virasoro algebra of central charge $c = 6k$.

One way to understand this Virasoro algebra is by realizing that $(2 + 1)$-dimensional gravity with a negative cosmological constant can be formulated as a Chern–Simons gauge theory with gauge group

$$SL(2, \mathbf{R}) \times SL(2, \mathbf{R}). \tag{3.288}$$

As explained at the beginning of these notes such a topological gauge theory will induce a CFT at the boundary, that in this case can be argued to be Liuoville theory.

It is a beautiful result that this Virasoro algebra can also be recovered by studying strings moving in the AdS geometry [69]. Roughly the relation is obtained as follows. First one rewrites the AdS^3 metric as

$$ds^2 = d\varphi^2 + e^{2\varphi}\, d\gamma\, d\overline{\gamma}. \tag{3.289}$$

In the sigma-model metric one can introduce auxiliary spin one fields $\beta, \overline{\beta}$, so that the action takes the form

$$S = \int \partial\varphi\overline{\partial}\varphi + \beta\overline{\partial}\gamma + \overline{\beta}\partial\overline{\gamma} - e^{-2\varphi}\beta\overline{\beta}. \tag{3.290}$$

Now the essential observation is that this CFT contains an infinite set of conserved currents that for large values of φ (that is, close to the boundary of AdS^3) take the form

$$L_n(z) = \gamma^{n+1}\beta. \tag{3.291}$$

To these currents we can associate global conserved charges $L_n = \oint L_n(z)$ and one can then show that these charges give rise to the Virasoro algebra

$$[L_n, L_m] = (n-m)L_{n+m} + \tfrac{1}{2}k(n^3-n)\partial_{n+m} \tag{3.292}$$

where the central charge is defined by the winding number

$$k = \oint \frac{\partial\gamma}{\gamma}. \tag{3.293}$$

So here we have a case where a local affine symmetry of the CFT gives rise to a *space–time* conformal algebra.

Finally, one of the most striking results is the computation of the elliptic genus (3.167) from the supergravity point of view [70].

References

[1] Strominger A and Vafa C 1996 Microscopic origin of the Bekenstein–Hawking entropy *Phys. Lett.* B **379** 99–104

[2] Peet A W 1998 The Bekenstein formula and string theory (N-brane theory) *Class. Quantum Grav.* **15** 3291–338

[3] Polchinski J 1998 *String Theory* vols 1, 2 (Cambridge: Cambridge University Press)

[4] Polchinski J 1995 Dirichlet-branes and Ramond–Ramond charges *Phys. Rev. Lett.* **75** 4724–7

[5] Banks T, Fischler W, Shenker S H and Susskind L 1997 M theory as a matrix model: a conjecture *Phys. Rev.* D **55** 5112–28

[6] 't Hooft G 1993 Dimensional reduction in quantum gravity, gr-qc/9310026

[7] Susskind L 1995 The world as a hologram *J. Math. Phys.* **36** 6377

[8] Witten E 1989 Quantum field theory and the Jones polynomial *Commun. Math. Phys.* **121** 351

[9] Elitzur S, Moore G, Schwimmer A and Seiberg N 1989 Remarks on the canonical quantization of Chern–Simons–Witten theory *Nucl. Phys.* B **326** 108

[10] Martinec E 1998 Conformal field theory, geometry, and entropy, hep-th/9809021

[11] Douglas M R, Kabat D, Pouliot P and Shenker S H 1997 D-branes and short distances in string theory *Nucl. Phys.* B **485** 85–127

[12] Connes A, Douglas M R and Schwarz A 1998 Noncommutative geometry and matrix theory: compactification on tori *J. High Energy Phys.* 9802(1998)003

[13] Callan C, Harvey J and Strominger A 1991 World-sheet approach to heterotic solitons and instantons *Nucl. Phys.* **359** 611

[14] Cvetic M and Youm D 1996 General rotating five dimensional black holes of toroidally compactified heterotic string *Nucl. Phys.* B **476** 118–32

[15] Horowitz G, Maldacena J and Strominger A 1996 Nonextremal black hole microstates and U-duality *Phys. Lett.* B **383** 151–9

[16] Hull C and Townsend P 1995 Unity of superstring dualities *Nucl. Phys.* B **438** 109

[17] Cremmer E 1981 Supergravities in 5 dimensions *Superspace and Supergravities* ed S W Hawking and M Rocek (Cambridge: Cambridge University Press) p 267

[18] Witten E 1996 Bound states of strings and p-branes *Nucl. Phys.* B **460** 335–50

[19] Dijkgraaf R, Verlinde E and Verlinde H 1997 BPS spectrum of the five-brane and black bole entropy *Nucl. Phys.* B **486** 77–88

Dijkgraaf R, Verlinde E and Verlinde H 1997 BPS quantization of the five-brane *Nucl. Phys.* B **486** 89–113

[20] Seiberg N 1997 Matrix description of M-theory on T^5 and T^5/Z_2 *Phys. Lett.* B **408** 98–104

[21] Li M 1996 Boundary states of D-branes and Dy-strings *Nucl. Phys.* B **460** 351

[22] Douglas M 1995 Branes within branes, hep-th/9512077

[23] Green M, Harvey J and Moore G 1997 I-brane inflow and anomalous couplings on D-branes *Class. Quantum Grav.* **14** 47–52

[24] Harvey J A and Moore G 1998 On the algebra of BPS states *Commun. Math. Phys.* **197** 489–519

[25] Berkooz M and Douglas M 1997 Fivebranes in M(atrix) theory *Phys. Rev.* D **55** 5112–28

[26] Witten E 1997 On the conformal field theory of the Higgs branch *J. High Energy Phys.* 07(1997)003

[27] Douglas M, Polchinski J and Strominger A 1997 Probing five-dimensional black holes with D-branes *J. High Energy Phys.* 971(1997)003

[28] Mukai S 1987 On the moduli space of bundles on $K3$ surfaces I *Vector Bundles on Algebraic Varieties* ed M F Atiyah *et al* (Oxford: Oxford University Press)

[29] Huybrechts D 1997 Compact hyperkähler manifolds: basic results alg-geom/9705025

[30] Dijkgraaf R 1998 Instanton strings and hyper-Kähler geometry hep-th/9810210

[31] Beauville A 1983 Variétés Kähleriennes dont la première classe de Chern est nulle *J. Diff. Geom.* **18** 755–82

[32] Schwimmer A and Seiberg N 1987 Comments on the $N = 2, N = 3, N = 4$ superconformal algebras in two-dimensions *Phys. Lett.* B **184** 191

[33] Landweber P S (ed) 1988 *Elliptic Curves and Modular Forms in Algebraic Topology* (Berlin: Springer)

Witten E 1987 Elliptic genera and quantum field theory *Commun. Math. Phys.* **109** 525

Schellekens A and Warner N 1986 Anomalies and modular invariance in string theory *Phys. Lett* B **177** 317

Schellekens A and Warner N 1986 Anomalies, characters and strings *Nucl. Phys.* B **287** 317

Alvarez O, Killingback T P, Mangano M and Windey P 1987 The Dirac–Ramond operator in string theory and loop space index theorems *Nucl. Phys.* B **1** 89

Alvarez O, Killingback T P, Mangano M and Windey P 1987 String theory and loop space index theorems *Commun. Math. Phys.* **111** 1

Eguchi T, Ooguri H, Taormina A and Yang S-K 1989 Superconformal algebras and string compactification on manifolds with $SU(N)$ holonomy *Nucl. Phys.* B **315** 193

Kawai T, Yamada Y and Yang S-K 1994 Elliptic genera and $N = 2$ superconformal field theory *Nucl. Phys.* B **414** 191–212

[34] Eichler M and Zagier D 1985 *The Theory of Jacobi Forms* (Basle: Birkhäuser)

[35] Neumann C D D 1996 The elliptic genus of Calabi–Yau 3- and 4-folds, product formulae and generalized Kac–Moody algebras, hep-th/9607029

[36] Kawai T 1996 $N = 2$ heterotic string threshold correction, $K3$ surface and generalized Kac–Moody superalgebra *Phys. Lett.* B **372** 59–64

[37] Kawai T 1997 K3 surfaces, Igusÿ cusp form and string theory, hep-th/9710016

[38] Göttsche L 1990 The Betti numbers of the Hilbert scheme of points on a smooth projective surface *Math. Ann.* **286** 193–207

Göttsche L 1990 *Hilbert Schemes of Zero-dimensional Subschemes of Smooth Varieties (Lecture Notes in Mathematics vol 1572)* (Berlin: Springer) p 1572

[39] Hirzebruch F and Höfer T 1990 On the Euler number of an orbifold *Math. Ann.* **286** 255

[40] Witten E 1992 Mirror manifolds and topological field theory *Essays on Mirror manifolds* ed S-T Yau (Hong Kong: International Press)

[41] Vafa C and Witten E 1994 A strong coupling test of S-duality *Nucl. Phys.* B **431** 3–77

[42] Göttsche L and Soergel W 1993 Perverse sheaves and the cohomology of Hilbert schemes of smooth algebraic surfaces *Math. Ann.* **296** 235–45

[43] Cheah J 1996 On the cohomology of Hilbert schemes of points *J. Alg. Geom.* **5** 479–511

[44] Dijkgraaf R, Moore G, Verlinde E and Verlinde H 1997 Elliptic genera of symmetric products and second quantized strings *Commun. Math. Phys.* **185** 197–209

[45] Dixon L, Harvey J, Vafa C and Witten E 1985 Strings on orbifolds *Nucl. Phys.* B **261** 620
Dixon L, Harvey J, Vafa C and Witten E 1985 Strings on orbifolds 2 *Nucl. Phys.* B **274** 285

[46] Maldacena J M and Susskind L 1996 D-branes and fat black holes *Nucl. Phys.* B **475** 679

[47] Borcherds R E 1995 Automorphic forms on $O_{s+2,2}(R)$ and infinite products *Invent. Math.* **120** 161

[48] Harvey J and Moore G 1996 Algebras, BPS states, and strings *Nucl. Phys.* B **463** 315–68

[49] Dijkgraaf R, Verlinde E and Verlinde H 1997 Counting dyons in $N = 4$ string theory *Nucl. Phys.* B **484** 543

[50] Li M and Martinec E 1997 Matrix black holes, hep-th/9703211

[51] Bilal A 1997 M(atrix) theory: a pedagogical introduction, hep-th/9710136

[52] Banks T 1997 Matrix theory, hep-th/9710231

[53] Bigatti D and Susskind L 1997 Review of matrix theory, hep-th/9712072

[54] Motl L 1997 Proposals on non-perturbative superstring interactions, hep-th/ 9701025

[55] Banks T and Seiberg N 1997 Strings from matrices *Nucl. Phys.* B **497** 41–55

[56] Dijkgraaf R, Verlinde E and Verlinde H 1997 Matrix string theory *Nucl. Phys.* B **500** 43–61

[57] Dijkgraaf R, Verlinde E and Verlinde H 1997 Notes on matrix and micro strings, hep-th/9709107

[58] Taylor W 1997 D-brane field theory on compact spaces *Phys. Lett.* B **394** 283–7

[59] Ganor O J, Ramgoolam S and Taylor W 1997 Branes, fluxes and duality in M(atrix)-theory *Nucl. Phys.* B **492** 191–204

[60] Banks T, Seiberg N and Shenker S 1996 Branes from matrices, hep-th/9612157

[61] Maldacena J 1997 The large N limit of superconformal field theories and super-gravity, hep-th/9711200

[62] Gubser S S, Klebanov I R and Polyakov A M 1998 Gauge theory correlators from noncritical string theory *Phys. Lett.* B **428** 105–14

[63] Witten E 1998 Anti-de Sitter space and holography *Adv. Theor. Math. Phys.* **2** 253–91

[64] Strominger A 1997 Black hole entropy from near-horizon microstates, hep-th/ 9712251

[65] Maldacena J and Strominger A 1998 AdS^3 black holes and a stringy exclusion principle, hep-th/9804085

[66] Martinec E 1998 Matrix models of AdS gravity, hep-th/9804111

[67] Banks T, Douglas M R, Horowitz G T and Martinec E 1998 AdS dynamics from conformal field theory, hep-th/9808016

[68] Brown J D and Henneaux M 1986 Central charges in the canonical realization of asymptotic symmetries: an example from three-dimensional gravity *Commun. Math. Phys.* **104** 207–26

[69] Giveon A, Kutasov D and Seiberg N 1998 Comments on string theory on AdS_3, hep-th/9806194

[70] de Boer J 1998 Six-dimensional supergravity on $S^3 \times AdS_3$ and 2d conformal field theory, hep-th/9806104

de Boer J 1998 Large N elliptic genus and AdS/CFT correspondence, hep-th/9812240

PART 3

BPS BLACK HOLES IN SUPERGRAVITY:
DUALITY GROUPS, P-BRANES, CENTRAL CHARGES AND ENTROPY

Riccardo D'Auria[1] *and Pietro Fré*[2]

[1] *Dipartimento di Fisica del Politecnico di Torino,*
Corso Duca degli Abruzzi 24, I-10129 Torino, Italy
[2] *Dipartimento di Fisica Teorica dell'Università di Torino,*
Via P Giuria 1, I-10125 Torino, Italy

Chapter 4

Introduction

4.1 Extremal black holes from classical general relativity to string theory

Black hole physics has many aspects of great interest to physicists with very different cultural backgrounds. These range from astrophysics to classical general relativity, to quantum field theory in curved space–times, particle physics and finally string theory and supergravity. This is not surprising since black holes are one of the basic consequences of a fundamental theory, namely Einstein general relativity. Furthermore black holes have fascinating thermodynamical properties that seem to encode the deepest properties of the so far unestablished fundamental theory of quantum gravity. Central in this context is the Bekenstein–Hawking entropy:

$$S_{BH} = \frac{k_B}{G\hbar} \frac{1}{4} \text{area}_H \qquad (4.1.1)$$

where k_B is the Boltzmann constant, G is Newton's constant, $\hbar$ is Planck's constant and area_H denotes the area of the horizon surface.

This very precise relation between a thermodynamical quantity and a geometrical quantity such as the horizon area has been for more than 20 years the source of unextinguishable interest and meditation. Indeed a microscopic statistical explanation of the area law for the black hole entropy has been correctly regarded as possible only within a solid formulation of quantum gravity. Superstring theory is the most serious candidate for a theory of quantum gravity and as such should eventually provide such a microscopic explanation of the area law. Although superstrings have been around for more than 20 years significant progress in this direction came only recently [1], after the so called second string revolution (1995). Indeed black holes are a typical non-perturbative phenomenon and perturbative string theory could say very little about their entropy: only non-perturbative string theory can have a handle on it and our handle on non-perturbative string theory came after 1995 through the recognition of the role of string dualities. These dualities allow us to relate the strong coupling regime of one superstring model to the weak coupling regime of another one and are all encoded in the symmetry group (the U-duality group) of the low energy *supergravity effective action*. Paradoxically this low energy action is precisely the handle on the non-perturbative aspects of superstrings.

Since these lectures are addressed to an audience assumed to be mostly unfamiliar with supergravity and superstrings, we have briefly summarized this recent conceptual revolution in order to put what follows in the right perspective. What we want to emphasize is that the first instance of a microscopic explanation of the area law within string theory has been limited to what in the language of general relativity would be an *extremal black hole*. This is not a capricious choice but has a deep reason. Indeed the extremality condition, namely the coincidence of two horizons, obtains, in the context of a supersymmetric theory, a profound reinterpretation that makes extremal black holes the most interesting objects to study. To introduce the concept consider the usual Reissner–Nordstrom metric describing a black hole of mass m and electric (or magnetic) charge q:

$$ds^2 = -dt^2 \left(1 - \frac{2m}{\rho} + \frac{q}{\rho^2}\right) + d\rho^2 \left(1 - \frac{2m}{\rho} + \frac{q}{\rho^2}\right)^{-1} + \rho^2 d\Omega^2 \quad (4.1.2)$$

where $d\Omega^2 = (d\theta^2 + \sin^2\theta \, d\phi^2)$ is the metric on a 2-sphere. As is well known the metric (4.1.2) admits two Killing horizons where the norm of the Killing vector $\frac{\partial}{\partial t}$ changes sign. The horizons are at the two roots of the quadratic form $\Delta \equiv -2m\rho + q^2 + \rho^2$ namely at:

$$\rho_\pm = m \pm \sqrt{m^2 - q^2}. \quad (4.1.3)$$

If $m < |q|$ the two horizons disappear and we have a naked singularity. For this reason in the context of classical general relativity the *cosmic censorship* conjecture was advanced that singularities should always be hidden inside horizons and this conjecture was formulated as the bound:

$$m \geq |q|. \quad (4.1.4)$$

As we shall see, in the context of a supersymmetric theory the bound (4.1.4) is always guaranteed by supersymmetry. As anticipated, of particular interest are the states that saturate the bound (4.1.4). If $m = |q|$ the two horizons coincide and the metric (4.1.2) can be rewritten in a new interesting way. Setting:

$$m = |q| \qquad \rho = r + m \qquad r^2 = \vec{x} \cdot \vec{x} \quad (4.1.5)$$

equation (4.1.2) becomes:

$$ds^2 = -dt^2 \left(1 + \frac{q}{r}\right)^{-2} + \left(1 + \frac{q}{r}\right)^2 \left(dr^2 + r^2 d\Omega^2\right)$$
$$= -H^{-2}(\vec{x}) dt^2 + H^2(\vec{x}) d\vec{x} \cdot d\vec{x} \quad (4.1.6)$$

where by:

$$H(\vec{x}) = \left(1 + \frac{q}{\sqrt{\vec{x} \cdot \vec{x}}}\right) \quad (4.1.7)$$

we have denoted a harmonic function in a three-dimensional space spanned by the three Cartesian coordinates $\vec{x}$ with the boundary condition that $H(\vec{x})$ goes to 1 at infinity.

The metric (4.1.6) already contains all the features of the black hole metrics we shall consider in these lectures.

(i)   It is of the form:

$$ds^2 = -e^{2U(r)}\, dt^2 + e^{-2U(r)}\, d\vec{x}^2 \tag{4.1.8}$$

where the radial function $U(r)$ is expressed as a linear combination of harmonic functions of the *transverse coordinates* $\vec{x}$:

$$U(r) = \sum_{i=1}^{p} \alpha_i \, \log \, H_i(\vec{x}). \tag{4.1.9}$$

(ii) It is asymptotically flat $\Longrightarrow U(\infty) = 0$.
(iii) It is a Maxwell–Einstein metric in the sense that it satisfies Einstein equations with the stress–energy tensor $T_{\mu\nu}$ contributed by a suitable collection of Abelian gauge fields A_μ^Λ ($\Lambda = 1, \ldots, \bar{n}$).
(iv) It saturates the cosmic censorship bound (4.1.4) in the sense that its ADM mass is expressed as an algebraic function of the electric and magnetic charges hidden in the harmonic functions $H_i(\vec{x})$.

In the next section we shall interpret black holes of this form as BPS saturated states namely as quantum states filling special irreducible representations of the supersymmetry algebra, the so called *short supermultiplets*, the shortening condition being precisely the saturation of the cosmic censorship bound (4.1.4). Indeed such a bound can be restated as the equality of the mass with the central charge which occurs when a certain fraction of the supersymmetry charges identically annihilate the state. The remaining supercharges applied to the BPS state build up a unitary irreducible representation of supersymmetry that is shorter than the typical one since it contains fewer states. As we stress in the next section it is precisely this interpretation that makes extremal black holes relevant to the string theorist. Indeed these classical solutions behave as *the other half* of the particle spectrum of superstring theory, the half not accessible to perturbative string theory.

When in the above sentence we said *classical solutions* we should have specified of which theory. The answer to this question is *supergravity*, that is the low energy effective field theory of superstrings. The *other half* of the quantum particle spectrum, the non-perturbative one, is actually mostly composed of classical solutions of supergravity. This miracle, as we shall further explain in the next section 4.2, is once again due to the BPS condition which protects these classical solutions from quantum corrections.

Furthermore we said *particles*. In modern parlance particles are just 0-branes, namely a particular instance of p-dimensional extended objects (the p-branes) that occur as quantum states in the non-perturbative superstring spectrum and can be retrieved as classical solutions of effective supergravity in D space–time dimensions. If we confine our attention to space–time dimensions $D = 4$ we have only 0-branes, but these are in fact extended p-branes that have *wrapped* along homology cycles of the compactified dimensions. We shall elaborate on this point in chapter 5. In the p-brane picture one divides the space–time coordinates in two subsets, those belonging to the *world volume* spanned by the moving brane and those *transverse* to it. The p-brane metric depends on space–time coordinates only through harmonic functions of the transverse coordinates. For 0-branes the world volume consists only of the time t and the above general feature is precisely that displayed by the extremal Reissner–Nordstrom metric (4.1.6).

Hence to our reader whose background is classical general relativity or quantum gravity and who approaches the notion of supergravity p-branes for the first time we can just point out the following.

Statement 4.1.1. The extremal Reissner–Nordstrom metric (4.1.2) where $m = |q|$, once rewritten as in equation (4.1.6), is the prototype of the BPS saturated black holes treated in these lectures.

4.2   Extremal black holes as quantum BPS states

In the previous section we have reviewed the idea of extremal black holes as it arises in classical general relativity. Extremal black holes have become objects of utmost relevance in the context of superstrings since *the second string revolution* took place in 1995. Indeed supersymmetric extremal black holes have been studied in depth in a vast recent literature [2, 3, 4]. This interest is just part of a more general interest in the p-brane classical solutions of supergravity theories in all dimensions $4 \leq D \leq 11$ [14, 5]. This interest streams from the interpretation of the classical solutions of supergravity that preserve a fraction of the original supersymmetries as the BPS non-perturbative states necessary to complete the perturbative string spectrum and make it invariant under the many conjectured duality symmetries [6, 7, 8, 9, 10]. Extremal black holes and their parent p-branes in higher dimensions are therefore viewed as additional *particle-like* states that compose the spectrum of a fundamental quantum theory. The reader of these notes who has a background in classical general relativity and astrophysics should be advised that the holes we are discussing here are neither stellar-mass, nor mini black holes: their mass is typically of the order of the Planck mass:

$$M_{black\ hole} \sim M_{Planck}. \tag{4.2.1}$$

The Schwarzschild radius is therefore microscopic.

Yet, as the monopoles in gauge theories, these non-perturbative quantum states originate from regular solutions of the classical field equations, the same Einstein equations one deals with in classical general relativity and astrophysics. The essential new ingredient, in this respect, is supersymmetry that requires the presence of *vector fields* and *scalar fields* in appropriate proportions. Hence the black holes we are going to discuss are solutions of generalized Einstein–Maxwell-dilaton equations.

The above mentioned identification between classical p-brane solutions (black holes are instances of 0-branes) and the non-perturbative quantum states of string theory required by duality has become quite circumstantial with the advent of D-branes [11] and the possibility raised by them of a direct construction of the BPS states within the language of perturbative string theory extended by the choice of Dirichlet boundary conditions [12].

A basic feature of the non-perturbative states of the string spectrum is that they can carry Ramond–Ramond charges forbidden at the perturbative level. On the other hand, the important observation by Hull and Townsend [8, 9] is that at the level of the low energy supergravity Lagrangians all fields of both the Neveu–Schwarz–Neveu–Schwarz (NS–NS) and the Ramond–Ramond (R–R) sector are unified by the group of duality transformations U which is also the isometry group of the homogenous scalar manifold $\mathcal{M}_{scalar} = U/H$. At least this is true in theories with sufficiently large number of supersymmetries that is with $N \geq 3$ in $D = 4$ or, in a dimensional reduction invariant language with no of supercharges ≥ 12. This points out that the distinction between R–R and NS–NS sectors is just an artifact of perturbative string theory. It also points out the fact that the unifying symmetry between the perturbative and non-perturbative sectors is already known from supergravity, namely it is the U-duality group. Indeed the basic conjecture of Hull and Townsend is that the restriction to integers $U(\mathbb{Z})$ of the U Lie group determined by supergravity should be an exact symmetry of non-perturbative string theory.

From an abstract viewpoint BPS saturated states are characterized by the fact that they preserve, in modern parlance, $1/2$ (or $1/4$, or $1/8$) of the original supersymmetries. What this actually means is that there is a suitable projection operator $\mathbb{P}_{BPS}^2 = \mathbb{P}_{BPS}$ acting on the supersymmetry charge Q_{SUSY}, such that:

$$(\mathbb{P}_{BPS} \, Q_{SUSY}) \, |\, \text{BPS state} \rangle \, = \, 0. \qquad (4.2.2)$$

Since the supersymmetry transformation rules of any supersymmetric field theory are linear in the first derivatives of the fields equation (4.2.2) is actually a *system of first order differential equations*. This system has to be combined with the second order field equations of supergravity and the common solution to both systems of equations is a classical BPS saturated state. That it is actually an exact state of non-perturbative string theory follows from supersymmetry representation theory. The classical BPS state is by definition an element of a *short supermultiplet* and, if supersymmetry is unbroken, it cannot be renormalized to a *long supermultiplet*.

Translating equation (4.2.2) into an explicit first order differential system requires knowledge of the supersymmetry transformation rules of supergravity. These latter have a rich geometrical structure that is the purpose of the present lectures to illustrate to an audience assumed to be unfamiliar with supergravity theory. Indeed the geometrical structure of supergravity which originates in its scalar sector is transferred into the physics of extremal black holes by the BPS saturation condition.

In order to grasp the significance of the above statement let us first rapidly review, as an example, the algebraic definition of $D = 4$ BPS states in a theory with an even number of supercharges $N = 2v$. The case with $N =$ odd can be similarly treated but needs some minor modifications due to the fact the eigenvalues of an antisymmetric matrix in odd dimensions are $\{\pm i\lambda_i, 0\}$.

4.2.1 General definition of BPS states in a 4D theory with $N = 2 \times p$ supersymmetries

The $D = 4$ supersymmetry algebra with $N = 2 \times p$ supersymmetry charges is given by

$$\{\overline{Q}_{A\alpha}, \overline{Q}_{B\beta}\} = i(C\gamma^\mu)_{\alpha\beta} P_\mu \delta_{AB} - C_{\alpha\beta} Z_{AB}$$
$$(A, B = 1, \ldots, 2p) \qquad (4.2.3)$$

where the SUSY charges $\overline{Q}_A \equiv Q_A^\dagger \gamma_0 = Q_A^T C$ are Majorana spinors, C is the charge conjugation matrix, P_μ is the 4-momentum operator and the antisymmetric tensor $Z_{AB} = -Z_{BA}$ is the central charge operator. It can always be reduced to normal form

$$Z_{AB} = \begin{pmatrix} \epsilon Z_1 & 0 & \cdots & 0 \\ 0 & \epsilon Z_2 & \cdots & 0 \\ \cdots & \cdots & \cdots & \cdots \\ 0 & 0 & \cdots & \epsilon Z_p \end{pmatrix} \qquad (4.2.4)$$

where ϵ is the 2×2 antisymmetric matrix, (every zero is a 2×2 zero matrix) and the p skew eigenvalues Z_I of Z_{AB} are the central charges.

If we identify each index $A, B, \ldots$ with a pair of indices

$$A = (a, I) \qquad a, b, \ldots = 1, 2 \qquad I, J, \ldots = 1, \ldots, p \qquad (4.2.5)$$

then the superalgebra (4.2.3) can be rewritten as:

$$\{\overline{Q}_{aI|\alpha}, \overline{Q}_{bJ|\beta}\} = i(C\gamma^\mu)_{\alpha\beta} P_\mu \delta_{ab} \delta_{IJ} - C_{\alpha\beta} \epsilon_{ab} \times Z_{IJ} \qquad (4.2.6)$$

where the SUSY charges $\overline{Q}_{aI} \equiv Q_{aI}^\dagger \gamma_0 = Q_{aI}^T C$ are Majorana spinors, C is the charge conjugation matrix, P_μ is the 4-momentum operator, ϵ_{ab} is the two-dimensional Levi–Civita symbol and the central charge operator is now represented by the *symmetric tensor* $Z_{IJ} = Z_{JI}$ which can always be diagonalized $Z_{IJ} = \delta_{IJ} Z_J$. The p eigenvalues Z_J are the skew eigenvalues introduced in equation (4.2.4).

The Bogomolny bound on the mass of a generalized monopole state:

$$M \geq |Z_I| \qquad \forall Z_I, I = 1, \ldots, p \qquad (4.2.7)$$

is an elementary consequence of the supersymmetry algebra and of the identification between *central charges* and *topological charges*. To see this it is convenient to introduce the following reduced supercharges:

$$\overline{S}_{aI|\alpha}^\pm = \tfrac{1}{2}(\overline{Q}_{aI}\gamma_0 \pm i \epsilon_{ab} \overline{Q}_{bI})_\alpha. \qquad (4.2.8)$$

They can be regarded as the result of applying a projection operator to the super-symmetry charges:

$$\overline{S}_{aI}^{\pm} = \overline{Q}_{bI} \, \mathbb{P}_{ba}^{\pm}$$

$$\mathbb{P}_{ba}^{\pm} = \tfrac{1}{2}(1\delta_{ba} \pm i\epsilon_{ba}\gamma_0).$$

(4.2.9)

Combining equation (4.2.6) with the definition (4.2.8) and choosing the rest frame where the 4-momentum is $P_\mu =(M, 0, 0, 0)$, we obtain the algebra:

$$\{\overline{S}_{aI}^{\pm}, \overline{S}_{bJ}^{\pm}\} = \pm\epsilon_{ac} \, C \, \mathbb{P}_{cb}^{\pm} \, (M \mp Z_I) \, \delta_{IJ}.$$

(4.2.10)

By positivity of the operator $\{\overline{S}_{aI}^{\pm}, \overline{S}_{bJ}^{\pm}\}$ it follows that on a generic state the Bogomolny bound (4.2.7) is fulfilled. Furthermore it also follows that the states which saturate the bounds:

$$(M \pm Z_I) \, |\text{BPS state}, i\rangle = 0$$

(4.2.11)

are those which are annihilated by the corresponding reduced supercharges:

$$\overline{S}_{aI}^{\pm} \, |\text{BPS state}, i\rangle = 0.$$

(4.2.12)

On one hand equation (4.2.12) defines *short multiplet representations* of the original algebra (4.2.6) in the following sense: one constructs a linear representation of (4.2.6) where all states are identically annihilated by the operators $\overline{S}_{aI}^{\pm}$ for $I = 1, \ldots, n_{max}$. If $n_{max} = 1$ we have the minimum shortening; if $n_{max} = p$ we have the maximum shortening. On the other hand equation (4.2.12) can be translated into a first order differential equation on the bosonic fields of supergravity.

Indeed, let us consider a configuration where all the fermionic fields are zero. Setting the fermionic SUSY rules appropriate to such a background equal to zero we find the following Killing spinor equation:

$$0 = \delta\text{fermions} = \text{SUSY rule (bosons}, \epsilon_{AI})$$

(4.2.13)

where the SUSY parameter satisfies the following conditions[1]:

$$
\begin{aligned}
\xi^\mu \, \gamma_\mu \, \epsilon_{aI} &= i \, \varepsilon_{ab} \, \epsilon^{bI} & I &= 1, \ldots, n_{max} \\
\epsilon_{aI} &= 0 & I &> n_{max}.
\end{aligned}
$$

(4.2.14)

Hence equations (4.2.13) with a parameter satisfying the condition (4.2.14) will be our operative definition of BPS states.

---

[1] ξ^μ denotes a timelike Killing vector.

4.3   The horizon area and central charges

In the main body of these lectures we are going to see how equations (4.2.13) can be translated into explicit differential equations. Solving such first order equations, together with the second order Einstein–Maxwell equations, we can obtain BPS saturated black hole solutions of the various versions of supergravity theory. This programme requires the use of the rich and complex structure of supergravity Lagrangians. In these lectures, however, we do not dwell on the technicalities of supersymmetry theory and except for the supersymmetry transformations rules needed to write (4.2.13) we almost nowhere mention the fermionic fields. What we rather explain in detail the geometric symplectic structure of the supergravity Lagrangians related to the simultaneous presence of *vector* and *scalar* fields, the latter interpreted à la σ-model as coordinates of a suitable scalar manifold $\mathcal{M}_{scalar}$. This symplectic structure which is enforced by supersymmetry and which allows the definition of generalized *duality rotations* fits into a rather general pattern and it is responsible for the most fascinating and most intriguing result in the analysis of supergravity BPS black holes: the interpretation of the horizon area appearing in the Bekenstein–Hawking formula (4.1.1) as a *topological U-duality invariant* depending only on the magnetic and electric charges of the hole.

If we go back to our prototype metric (4.1.2) and we calculate the area of the horizon, we find:

$$\text{area}_H = \int_{\rho=\rho_+} \sqrt{g_{\theta\theta}\, g_{\phi\phi}}\, d\theta\, d\phi = 4\pi\, \rho_+^2 = 4\pi \left(m + \sqrt{m^2 - |q|^2} \right)^2.$$
(4.3.1)

In the case of an extremal black hole ($m = |q|$) equation (4.3.1) becomes:

$$\frac{\text{area}_H}{4\pi} = |q|^2.$$
(4.3.2)

The rather innocent looking formula (4.3.2) contains a message of the utmost relevance. For BPS black holes the horizon area and hence the entropy is a function solely of the electric and magnetic charges of the hole. It is also a very specific function, since it is an algebraic invariant of a duality group. But this is difficult to see in a too simple theory where there is just one electromagnetic field. Equation (4.3.2) reveals its hidden surprising structure when it is generalized to a theory containing several vector fields interacting with gravity and scalars within the symplectic scheme enforced by supersymmetry and proper to the supergravity Lagrangians.

Explaining the generalization of equation (4.3.2) both at a general level and in explicit examples taken from the maximally extended supergravity theory ($N = 8$) is the main purpose of this series of lectures.

Chapter 5

Supergravity p-branes in higher dimensions

As we already stressed, in line with the topics of this school, our main goal is to classify BPS *black hole solutions of* 4D *supergravity* and to unravel the fascinating group theoretical structure of their Bekenstein–Hawking entropy.

Black holes are instances of 0-branes, namely objects with zero-dimensional spacelike extension that can evolve in time. They carry *quantized electric and magnetic charges* $\{q_\Sigma, p^\Lambda\}$ under the host of gauge fields A_μ^Λ ($\Lambda = 1, \ldots, \overline{n}$) appearing in the spectrum of supergravity, that is the low energy effective theory of strings, or M-theory. As we shall see, the entropy

$$S_{BH} = S(q, p) \tag{5.0.1}$$

is a topological invariant that depends solely on such quantized charges. Actually it is not only a topological invariant insensitive to continuous deformations of the black hole solution with respect to the parameters it depends on (*the moduli*) but also an invariant in the group-theoretical sense. Indeed $S(q, p)$ is an invariant of the U-duality group that unifies the various perturbative and non-perturbative duality of string theory. Such a complicated structure of the black hole arises from the compactification of too many of the actual dimensions of space–time. The black hole appears to us as a 0-brane only because its spatial extension has been hidden in the six- (or seven-) dimensional compact manifold $\mathcal{M}_{comp}$ that is not directly accessible at low energies. The 0-brane black hole is actually a p-brane (or intersection of many p-branes) that are *wrapped* on the homology p-cycles of the internal manifold $\mathcal{M}_{comp}$.

So although our main focus will be on the $D = 4$ theory, in order to obtain a better insight into the meaning of such complicated objects as we propose to study, it is appropriate to start from the vantage point of higher dimensions $D > 4$ where p-branes become much simpler by freely unfolding into non-compact directions.

5.1 Definition and general features of dilatonic p-brane solutions in dimension D

The basic idea of a p-brane solution can be illustrated by considering a very simple action functional in D space–time dimensions that contains just only three fields:

(i)   the metric $g_{\mu\nu}$, namely the *graviton*
(ii)   a scalar field $\phi(X)$, namely the *dilaton*
(iii)   a $(p + 1)$-form gauge field $A_{[p+1]} \equiv A_{M_1 \ldots M_{p+1}} \, dX^{M_1} \wedge \cdots \wedge dX^M_{p+1}$.

Explicitly we write :

$$I_D = \int d^D x \sqrt{-g} \left[R - \frac{1}{2} \nabla_M \phi \nabla^M \phi - \frac{1}{2n!} e^{a\phi} F^2_{[p+2]} \right] \qquad (5.1.1)$$

where $F_{[p+2]} \equiv dA_{[p+1]}$ is the field strength of the $(p + 1)$-form gauge potential and a is some real number whose profound meaning will become clear in the later discussion of the solutions. For various values of

$$n = p + 2 \quad \text{and} \quad a \qquad (5.1.2)$$

the functional I_D is a consistent truncation of some supergravity bosonic action S_D^{SUGRA} in dimension D. By consistent truncation we mean that a subset of the bosonic fields have been put equal to zero but in such a way that all solutions of the truncated action are also solutions of the complete one. For instance if we choose:

$$a = 1 \qquad n = \begin{cases} 3 \\ 7 \end{cases} \qquad (5.1.3)$$

equation (5.1.1) corresponds to the bosonic low energy action of the $D = 10$ heterotic superstring ($N = 1$, supergravity) where the $E_8 \times E_8$ gauge fields have been deleted. The two choices 3 or 7 in equation (5.1.3) correspond to the two formulations (electric/magnetic) of the theory. Other choices correspond to truncations of the type IIA or type IIB action in the various intermediate dimensions $4 \le D \le 10$. Since the $(n - 1)$-form $A_{[n-1]}$ couples to the world volume of an extended object of dimension:

$$p = n - 2 \qquad (5.1.4)$$

namely a p-brane, the choice of the truncated action (5.1.1) is motivated by the search for p-brane solutions of supergravity. According to the interpretation (5.1.4) we set:

$$n = p + 2 \qquad d = p + 1 \qquad \tilde{d} = D - p - 3 \qquad (5.1.5)$$

where d is the world-volume dimension of an electrically charged *elementary* p-brane solution, while $\tilde{d}$ is the world-volume dimension of a magnetically charged *solitonic $\tilde{p}$-brane* with $\tilde{p} = D - p - 4$. The distinction between elementary and solitonic is the following. In the elementary case the field configuration we shall discuss is a true vacuum solution of the field equations following from the action (5.1.1) everywhere in D-dimensional space–time except for a singular locus of dimension d. This locus can be interpreted as the location of an elementary p-brane source that is coupled to supergravity via an electric charge spread over its own world volume. In the solitonic case, the field configuration we shall consider is instead a bona fide solution of the supergravity field equations everywhere in space–time without the need to postulate external elementary sources. The field energy is, however, concentrated around a locus of dimension $\tilde{p}$. These solutions have been derived and discussed thoroughly in the literature [13]. Good reviews of such results are [14, 15]. Defining:

$$\Delta = a^2 + 2\,\frac{d\tilde{d}}{D-2} \tag{5.1.6}$$

it was shown in [13] that the action (5.1.1) admits the following elementary p-brane solution

$$
\begin{aligned}
ds^2 &= \left(1 + \frac{k}{r^{\tilde{d}}}\right)^{-\frac{4\tilde{d}}{\Delta(D-2)}} dx^\mu \otimes dx^\nu\, \eta_{\mu\nu} \\
&\quad - \left(1 + \frac{k}{r^{\tilde{d}}}\right)^{\frac{4d}{\Delta(D-2)}} dy^m \otimes dy^n\, \delta_{mn} \\
F &= \lambda(-)^{p+1} \epsilon_{\mu_1 \dots \mu_{p+1}}\, dx^{\mu_1} \wedge \cdots \\
&\quad \cdots \wedge dx^{\mu_{p+1}} \wedge \frac{y^m\, dy^m}{r} \left(1 + \frac{k}{r^{\tilde{d}}}\right)^{-2} \frac{1}{r^{\tilde{d}+1}} \\
e^{\phi(r)} &= \left(1 + \frac{k}{r^{\tilde{d}}}\right)^{-\frac{2a}{\Delta}}
\end{aligned}
\tag{5.1.7}
$$

where the coordinates X^M $(M = 0, 1 \dots, D-1)$ have been split into two subsets:

- x^μ $(\mu = 0, \dots, p)$ are the coordinates on the p-brane world volume,
- y^m $(m = D - d + 1, \dots, D)$ are the coordinates transverse to the brane.

By $r \equiv \sqrt{y^m y_m}$ we denote the radial distance from the brane and by k the value of its electric charge. Finally, in equation (5.1.8) we have set:

$$\lambda = 2\,\frac{\tilde{d}\,k}{\sqrt{\Delta}}. \tag{5.1.8}$$

The same authors of [13] show that that the action (5.1.1) admits also the following solitonic $\tilde{p}$-brane solution:

$$
\begin{aligned}
ds^2 &= \left(1 + \frac{k}{r^d}\right)^{-\frac{4d}{\Delta(D-2)}} dx^\mu \otimes dx^\nu\, \eta_{\mu\nu} \\
&\quad - \left(1 + \frac{k}{r^d}\right)^{\frac{4\tilde{d}}{\Delta(D-2)}} dy^m \otimes dy^n\, \delta_{mn}
\end{aligned}
$$

$$\tilde{F}_{[D-n]} = \lambda \epsilon_{\mu_1 \dots \mu_{\tilde{d}} p} \, dx^{\mu_1} \wedge \dots \wedge dx^{\mu_{\tilde{d}}} \wedge \frac{y^p}{r^{d+2}} \tag{5.1.9}$$

$$e^{\phi(r)} = \left(1 + \frac{k}{r^d}\right)^{\frac{2a}{\Delta}}$$

where the $(D - p - 2)$-form $\tilde{F}_{[D-n]}$ is the dual of $F_{[n]}$, k is now the magnetic charge and:

$$\lambda = -2 \frac{\tilde{d} k}{\sqrt{\Delta}}. \tag{5.1.10}$$

These p-brane configurations are solutions of the second order field equations obtained by varying the action (5.1.1). However, when (5.1.1) is the truncation of a supergravity action both (5.1.8) and (5.1.10) are also the solutions of a *first order differential system of equations*. This happens because they are BPS-extremal p-branes which preserve a fraction of the original supersymmetries. As an example we consider the ten-dimensional case.

5.1.1 The elementary string solution of heterotic supergravity in $D = 10$

Here we have

$$D = 10 \qquad d = 2 \qquad \tilde{d} = 6 \qquad a = 1 \qquad \Delta = 4 \qquad \lambda = \pm 6k \tag{5.1.11}$$

so that the elementary string solution reduces to:

$$ds^2 = \exp[2U(r)] \, dx^\mu \otimes dx^\nu - \exp[-\tfrac{2}{3}U(r)] \, dy^m \otimes dy^m$$

$$\exp[2U(r)] = \left(1 + \frac{k}{r^6}\right)^{-3/4}$$

$$F = 6k \, \epsilon_{\mu\nu} \, dx^\mu \wedge dx^\nu \wedge \frac{y^m \, dy^m}{r} \left(1 + \frac{k}{r^6}\right)^{-2} \frac{1}{r^7}$$

$$\exp[\phi(r)] = \left(1 + \frac{k}{r^6}\right)^{-1/2}. \tag{5.1.12}$$

As already pointed out, with the values (5.1.11), the action (5.1.1) is just the truncation of heterotic supergravity where, besides the fermions, also the $E_8 \times E_8$ gauge fields have been set to zero. In this theory the supersymmetry transformation rules we have to consider are those of the gravitino and of the dilatino. They read:

$$\delta \psi_\mu = \nabla_\mu \epsilon + \tfrac{1}{96} \exp[\tfrac{1}{2}\phi] \left(\Gamma_{\lambda\rho\sigma\mu} + 9\Gamma_{\lambda\rho} g_{\sigma\mu}\right) F^{\lambda\rho\sigma} \epsilon$$

$$\delta \chi = i \frac{\sqrt{2}}{4} \partial^\mu \phi \, \Gamma_\mu \epsilon - i \frac{\sqrt{2}}{24} \exp[-\tfrac{1}{2}\phi] \, \Gamma_{\mu\nu\rho} \epsilon \, F^{\mu\nu\rho}. \tag{5.1.13}$$

Expressing the ten-dimensional gamma-matrices as tensor products of the two-dimensional gamma-matrices γ_μ ($\mu = 0, 1$) on the 1-brane world sheet with the eight-dimensional gamma-matrices Σ_m ($m = 2, \ldots, 9$) on the transverse space it is easy to check that in the background (5.1.12) the SUSY variations (5.1.13) vanish for the following choice of the parameter:

$$\epsilon = \left(1 + \frac{k}{r^6}\right)^{-3/16} \epsilon_0 \otimes \eta_0 \qquad (5.1.14)$$

where the constant spinors ϵ_0 and η_0 are respectively the 2-component and 16-component and both have positive chirality:

$$\gamma_3 \epsilon_0 = \epsilon_0 \qquad \Sigma_{10} \eta_0 = \eta_0 . \qquad (5.1.15)$$

Hence we conclude that the extremal p-brane solutions of all maximal (and non-maximal) supergravities can be obtained by imposing the supersymmetry invariance of the background with respect to a projected SUSY parameter of the type (5.1.14).

5.2 *M2-branes in D = 11 and the issue of the horizon geometry*

To illustrate a very crucial feature of p-brane solutions, namely the structure of their horizon geometry, we choose another example in $D = 11$. The bosonic spectrum of $D = 11$ supergravity [16] is very simple since besides the metric g_{MN} it contains only a three-form gauge field $A_{[3]}$ and no scalar field ϕ. This means that $a = 0$ and that there is an elementary electric 2-brane solution and a magnetically charged five-brane, the $M2$-brane and the $M5$-brane, respectively. This universally adopted nomenclature follows from $D = 11$ supergravity being the low energy effective action of M-theory. Accordingly we have:

$$a = 0 \qquad d = 3 \qquad \tilde{d} = 6 \qquad \Delta = 4 \qquad D = 11 \qquad (5.2.1)$$

and following equation (5.1.8) the metric corresponding to the elementary $M2$-brane can be written as:

$$ds_{11}^2 = \left(1 + \frac{k}{r^{\tilde{d}}}\right)^{-\frac{\tilde{d}}{9}} dx^\mu \, dx^\nu \, \eta_{\mu\nu} + \left(1 + \frac{k}{r^{\tilde{d}}}\right)^{\frac{d}{9}} dX^I \, dX^J \, \delta_{IJ} \qquad (5.2.2)$$

where

$$d \equiv 3 \; ; \; \tilde{d} \equiv 11 - 3 - 2 = 6. \qquad (5.2.3)$$

What we would like to note is that in the *bulk* of this solution, namely for generic values of r, the isometry of the 11-dimensional metric (5.2.2) is:

$$\mathcal{I}_{2\text{-brane}} = ISO(1, 2) \otimes SO(8) \qquad (5.2.4)$$

yet if we take the limiting value of this metric in the two limits

(i) infinity: $r \to \infty$
(ii) horizon: $r \to 0$

then the isometry is enhanced and we respectively obtain:

$$ISO(1, 10) \text{ at infinity}$$
$$SO(2, 3) \times SO(8) \text{ at the horizon.} \tag{5.2.5}$$

The reason for this is simple. As one can realize by inspection, at $r = \infty$ the metric (5.2.2) becomes the flat 11-dimensional metric which obviously admits the 11D Poincaré group as isometry, while in the vicinity of $r = 0$, the metric (5.2.2) is approximated by the following metric:

$$ds^2 = \rho^2(-dt^2 + d\vec{z} \cdot d\vec{z}) + \rho^{-2} d\rho^2 + d\Omega_7^2 \tag{5.2.6}$$

where the last term $d\Omega_7^2$ is the $SO(8)$ invariant metric on a 7-sphere S^7, while the previous ones correspond to a particular parametrization of the anti de Sitter metric on AdS_4, which, by definition, admits $SO(2, 3)$ as isometry. To see that this is indeed the case it suffices to use polar coordinates for the eight transverse directions to the brane:

$$dy^I dy^J \delta_{IJ} = dr^2 + r^2 d\Omega_7^2 \tag{5.2.7}$$

set $\rho = r^2$ and take the limit $\rho \to 0$ in the metric (5.2.2).

The conclusion therefore is that in the vicinity of the horizon the geometry of the $M2$-brane (5.2.2) is:

$$M_2^{hor} = AdS_4 \times S^7 \tag{5.2.8}$$

which, by itself, would be an exact solution of $D = 11$ supergravity. It follows that the $M2$-brane can be seen as an M-theory *soliton*, in the usual sense of soliton theory: it is an exact solution of the field equations that interpolates smoothly between two vacua of the theory, 11D Minkowski space at infinity and the space $AdS_4 \times S^7$ at the horizon.

This feature of the $M2$-brane (5.2.2) corresponds to a very general property of p-branes and can be generalized in two ways. First we can make the following.

Statement 5.2.1. Whenever $a = 0$ the limiting horizon geometry of an elementary or solitonic p-brane as defined in equation (5.1.8) or equation (5.1.10) is

$$M_p^{hor} = AdS_{p+2} \times S^{D-p-2}. \tag{5.2.9}$$

Secondly we can pursue the following.

Generalization 5.2.1. Rather than equations (5.1.8) or (5.1.10), as candidate solutions for the action (5.1.1) consider the following *ansatz*:

$$ds^2 = e^{2A(r)} dx^\mu dx^\nu \eta_{\mu\nu} + e^{2B(r)}[dr^2 + r^2 \lambda^{-2} ds_{G/H}^2] \tag{5.2.10}$$
$$A_{\mu_1 \dots \mu_d} = \epsilon_{\mu_1 \dots \mu_d} e^{C(r)} \tag{5.2.11}$$
$$\phi = \phi(r) \tag{5.2.12}$$

where the following apply.

(i) λ is a constant parameter with the dimensions of length.

(ii) The D coordinates X^M are split as follows: $X^M = (x^\mu, r, y^m)$, $\eta^{MN} = \mathrm{diag}(-,+++\cdots)$.

(iii) $\mu = 0, \ldots, d-1$ runs on the p-brane world-volume $(d = p+1)$.

(iv) $\bullet$ labels the r coordinate.

(v) $m = d+1, \ldots, D-1$ runs on some $(D-d-1)$-dimensional compact coset manifold G/H, G being a compact Lie group and $H \subset G$ a closed Lie subgroup.

(vi) $ds^2_{G/H}$ denotes a G-invariant metric on the above mentioned coset manifold.

The basic difference with respect to the previous case is that we have replaced the invariant metric $ds^2_{S^{D-d-1}}$ on a sphere S^{D-d-1} by the more general coset manifold metric $ds^2_{G/H}$. With such an *ansatz* exact solutions can be found as shown in [17]. If $a = 0$, namely if there is no dilaton, such G/H p-branes have as horizon geometry:

$$M_p^{hor} = AdS_{p+2} \times \left(\frac{G}{H}\right)_{D-p-2} \tag{5.2.13}$$

and their horizon symmetry is:

$$SO(2, p+1) \times G. \tag{5.2.14}$$

Applied to the case of $D = 11$ supergravity, the electric *ansatz* in equations (5.2.10), (5.2.11) produces G/H M2-brane solutions. For pedagogical purposes we describe the derivation of such solutions in a little detail.

5.2.1 Derivation of the G/H M2-brane solution

The field equations derived from (5.1.1) have the following form:

$$R_{MN} = \tfrac{1}{2}\partial_M\phi\partial_N\phi + S_{MN} \tag{5.2.15}$$

$$\nabla_{M_1}(e^{a\phi} F^{M_1\ldots M_n}) = 0 \tag{5.2.16}$$

$$\Box\phi = \frac{a}{2n!}F^2 \tag{5.2.17}$$

where S_{MN} is the energy–momentum tensor of the n-form F:

$$S_{MN} = \frac{1}{2(n-1)!} e^{a\phi}\left[F_{M\ldots}F_N^{\cdots} - \frac{n-1}{n(D-2)}F^2 g_{MN}\right]. \tag{5.2.18}$$

5.2.1.1 *The vielbein*

In order to prove that the *ansatz* (5.2.10)–(5.2.12) is a solution of the field equations it is necessary to calculate the corresponding vielbein, spin-connection and curvature tensors. We use the convention that tangent space indices are underlined. Then the vielbein components relative to the *ansatz* (5.2.10) are:

$$E^{\underline{\mu}} = e^A dx^\mu \qquad E^{\underline{\bullet}} = e^B d\,dr \qquad E^{\underline{m}} = e^B r\lambda^{-1} E^m \quad (5.2.19)$$
$$g_{\mu\nu} = e^{2A}\eta_{\mu\nu} \qquad g_{\bullet\bullet} = e^{2B} \qquad g_{mn} = e^{2B}r^2\lambda^{-2}g_{mn} \quad (5.2.20)$$

with $E^{\underline{m}} \equiv G/H$ vielbein and $g_{mn} \equiv G/H$ metric.

5.2.1.2   The spin connection

The Levi–Civita spin connection on our D-dimensional manifold is defined as the solution of the vanishing torsion equation:

$$dE^{\underline{M}} + \omega^{\underline{M}}_{\underline{N}} \wedge E^{\underline{N}} = 0. \qquad (5.2.21)$$

Solving equation (5.2.21) explicitly we obtain the spin-connection components:

$$\omega^{\underline{\mu\nu}} = 0 \qquad \omega^{\underline{\mu\bullet}} = e^{-B}A'E^{\underline{\mu}} \qquad \omega^{\underline{\mu n}} = 0$$

$$\omega^{\underline{mn}} = \omega^{\underline{mn}} \qquad \omega^{\underline{m\bullet}} = \exp[-B]\,(B'+r^{-1})E^{\underline{m}} \qquad (5.2.22)$$

where $A' \equiv \partial_\bullet A$ etc and $\omega^{\underline{mn}}$ is the spin connection of the G/H manifold.

5.2.1.3   The Ricci tensor

From the definition of the curvature 2-form :

$$R^{\underline{MN}} = d\omega^{\underline{MN}} + \omega^{\underline{M}}_{\underline{S}} \wedge \omega^{\underline{SN}} \qquad (5.2.23)$$

we find the Ricci tensor components:

$$R_{\mu\nu} = -\tfrac{1}{2}\eta_{\mu\nu}\,e^{2(A-B)}[A'' + d(A')^2 + \tilde{d}A'B' + (\tilde{d}+1)r^{-1}A'] \qquad (5.2.24)$$
$$R_{\bullet\bullet} = -\tfrac{1}{2}[d(A'' + (A')^2 - A'B') + (\tilde{d}+1)(B''+r^{-1}B')] \qquad (5.2.25)$$
$$R_{mn} = -\tfrac{1}{2}g_{mn}\frac{r^2}{\lambda^2}[dA'(B'+r^{-1}) + r^{-1}B' + B'' + \tilde{d}(B'+r^{-1})^2] + \mathsf{R}_{mn}$$
$$\qquad (5.2.26)$$

where R_{mn} is the Ricci tensor of the G/H manifold, and $\tilde{d} \equiv D - d - 2$.

5.2.1.4   The field equations

Inserting the electric *ansatz* into the field equations (5.2.15) yields:

$$A'' + d(A')^2 + \tilde{d}A'B' + (\tilde{d}+1)A'r^{-1} = \frac{\tilde{d}}{2(D-2)}S^2 \qquad (5.2.27)$$

$$d[A'' + (A')^2 - A'B'] + (\tilde{d} + 1)[B'' + r^{-1}B'] = \frac{\tilde{d}}{2(D-2)}S^2 - \frac{(\phi')^2}{2}$$

(5.2.28)

$$g_{mn}[dA'(B' + r^{-1}) + r^{-1}B' + B'' + \tilde{d}(B' + r^{-1})^2] - 2R_{mn}$$

$$= -\frac{d}{2(D-2)}g_{mn}S^2$$

(5.2.29)

while equations (5.2.16) and (5.2.17) become:

$$C'' + (\tilde{d} + 1)r^{-1}C' + (\tilde{d}B' - dA' + C' + a\phi')C' = 0 \qquad (5.2.30)$$

$$\phi'' + (\tilde{d} + 1)r^{-1}\phi' + [dA' + \tilde{d}B']\phi' = -\frac{a}{2}S^2 \qquad (5.2.31)$$

with

$$S \equiv e^{\frac{1}{2}a\phi + C - dA}C'. \qquad (5.2.32)$$

5.2.2 Construction of the BPS Killing spinors in the case of $D = 11$ supergravity

At this point we specialize our analysis to the case of $D = 11$ supergravity, whose action in the bosonic sector reads:

$$I_{11} = \int d^{11}x\sqrt{-g}\ (R - \frac{1}{48}F_{[4]}^2) + \frac{1}{6}\int F_{[4]} \wedge F_{[4]} \wedge A_{[3]} \qquad (5.2.33)$$

and we look for the further restrictions imposed on the electric *ansatz* by the requirement that the solutions should preserve a certain amount of supersymmetry. This is essential for our goal since we are interested in G/H M-branes that are BPS saturated states and the BPS condition requires the existence of Killing spinors.

As discussed in [14], the above action does not fall exactly in the general class of actions of type (5.1.1). Nevertheless, the results of sections 2.1 and 2.2 still apply: indeed it is straightforward to verify that the FFA term in the action (5.2.33) gives no contribution to the field equations once the electric or magnetic *ansatz* is implemented. Moreover no scalar fields are present in (5.2.33): this we handle by simply setting to zero the scalar coupling parameter a.

Imposing that the *ansatz* solution admits Killing spinors allows us to simplify the field equations drastically.

We recall the supersymmetry transformation for the gravitino:

$$\delta\psi_M = \tilde{D}_M\epsilon \qquad (5.2.34)$$

with

$$\tilde{D}_M = \partial_M + \frac{1}{4}\omega_M{}^{AB}\Gamma_{AB} - \frac{1}{288}[\Gamma^{PQRS}{}_M + 8\Gamma^{PQR}\delta^S_M]F_{PQRS}. \qquad (5.2.35)$$

Requiring that setting $\psi_M = 0$ be consistent with the existence of residual supersymmetry yields:

$$\delta\psi_{M|\psi=0} = \tilde{D}_M\epsilon = 0. \qquad (5.2.36)$$

Solutions $\epsilon(x, r, y)$ of the above equation are *Killing spinor fields* on the bosonic background described by our *ansatz*.

In order to discuss the solutions of (5.2.36) we adopt the following tensor product realization of the (32 × 32) $SO(1, 10)$ gamma-matrices:

$$\Gamma_A = [\gamma_\mu \otimes \mathbb{1}_8, \gamma_3 \otimes \mathbb{1}_8, \gamma_5 \otimes \Gamma_m]. \tag{5.2.37}$$

The above basis (5.2.37) is well adapted to our (3+1+7)-*ansatz*. The γ_μ ($\mu = 0, 1, 2, 3$) are usual $SO(1, 3)$ gamma-matrices, $\gamma_5 = i\gamma_0\gamma_1\gamma_2\gamma_3$, while Γ_m are 8×8 gamma-matrices realizing the Clifford algebra of $SO(7)$. Thus for example $\Gamma_\bullet = \gamma_3 \otimes \mathbb{1}_8$.

Correspondingly, we split the $D = 11$ spinor ϵ as follows

$$\epsilon = \varepsilon \otimes \eta(r, y) \tag{5.2.38}$$

where ε is an $SO(1, 3)$ constant spinor, while the $SO(7)$ spinor η, besides the dependence on the internal G/H coordinates y^m, is assumed to depend also on the radial coordinate r. Note the difference with respect to Kaluza–Klein supersymmetric compactifications where η depends only on y^m. Computing $\tilde{D}$ in the *ansatz* background yields:

$$\tilde{D}_\mu = \partial_\mu + \frac{1}{2}e^{-B-2A}\gamma_\mu\gamma_3\left[e^{3A}A' - \frac{i}{3}e^C C'\gamma_3\gamma_5\right] \otimes \mathbb{1}_8$$

$$\tilde{D}_\bullet = \partial_r + \frac{i}{6}e^{-3A}C'e^C\gamma_3\gamma_5 \otimes \mathbb{1}_8 \tag{5.2.39}$$

$$\tilde{D}_m = \mathcal{D}_m^{G/H} + \frac{r}{2\lambda}[(B' + r^{-1})i\gamma_3\gamma_5 + \frac{1}{6}e^{C-3A}C'] \otimes \Gamma_m$$

where all γ_μ, Γ_m have tangent space indices. The Killing spinor equation $\tilde{D}_\mu\epsilon = 0$ becomes equivalent to:

$$(1_4 - i\gamma_3\gamma_5)\varepsilon = 0 \qquad 3e^{3A}A' = e^C C'. \tag{5.2.40}$$

Thus half of the components of the 4-dim spinor ε are projected out. Moreover the second equation is solved by $C = 3A$. Considering next $\tilde{D}_\bullet\epsilon = 0$ leads to the equation (where we have used $C = 3A$):

$$\partial_r\eta + \frac{1}{6}C'\eta = 0 \tag{5.2.41}$$

whose solution is

$$\eta(r, y) = e^{-C(r)/6}\eta_\circ(y). \tag{5.2.42}$$

Finally, $\tilde{D}_m\epsilon = 0$ implies

$$B = -\frac{1}{6}C + \text{const} \tag{5.2.43}$$

$$\left[\mathcal{D}_m^{G/H} + \frac{1}{2\lambda}\Gamma_m\right]\eta_\circ = 0. \tag{5.2.44}$$

Equation (5.2.44) deserves attentive consideration. If we identify the Freund–Rubin parameter as:

$$e \equiv \frac{1}{2\lambda} \qquad (5.2.45)$$

then equation (5.2.44) is nothing else but the Killing spinor equation for a G/H spinor that one encounters while discussing the residual supersymmetry of Freund–Rubin vacua. Freund–Rubin vacua [18] are exact solutions of $D = 11$ supergravity where the 11-dimensional space is:

$$\mathcal{M}_{11} = AdS_4 \times \left(\frac{G}{H}\right)_7 \qquad (5.2.46)$$

having denoted by $AdS_D = SO(2, D-1)/SO(1, D-1)$ anti de Sitter space–time in dimension D and by $\left(\frac{G}{H}\right)_7$ a seven-dimensional coset manifold equipped with a G-invariant Einstein metric. Such a metric solves the field equations under the condition that the four-form field strength take a constant $SO(1, 3)$-invariant vev:

$$F_{\mu_1\mu_2\mu_3\mu_4} = e\,\epsilon_{\mu_1\mu_2\mu_3\mu_4} \qquad (5.2.47)$$

on AdS_4. The Freund–Rubin vacua have been exhaustively studied in the old literature on Kaluza–Klein supergravity (see [16] for a comprehensive review) and are all known. Furthermore it is also known how many Killing spinors each of them admits: such a number is denoted $N_{G/H}$.

In this way we have explicitly verified that the number of BPS Killing spinors admitted by the G/H M-brane solution is $N_{G/H}$, i.e. the number of Killing spinors admitted by the corresponding Freund–Rubin vacuum.

5.2.3 M-brane solution

To be precise the Killing spinors of the previous section are admitted by a configuration that has still to be shown to be a complete solution of the field equations. To prove this is immediate. Setting $D = 11$, $d = 3$, $\tilde{d} = 6$, the scalar coupling parameter $a = 0$, and using the relations $C = 3A$, $B = -C/6 + \text{const} = -A/2 + \text{const}$ we have just deduced, the field equations (5.2.27), (5.2.28), (5.2.29) become:

$$A'' + 7r^{-1}A' = \tfrac{1}{3}S^2 \qquad (5.2.48)$$

$$(A')^2 = \tfrac{1}{6}S^2 \qquad (5.2.49)$$

$$R_{mn} = \frac{3}{\lambda^2}g_{mn}. \qquad (5.2.50)$$

Combining the first two equations to eliminate S^2 yields:

$$\nabla^2 A - 3(A')^2 \equiv A'' + \frac{7}{r}A' - 3(A')^2 = 0 \qquad (5.2.51)$$

or:

$$\nabla^2 e^{-3A} = 0 \qquad (5.2.52)$$

whose solution is:

$$e^{-3A(r)} = H(r) = 1 + \frac{k}{r^6}. \tag{5.2.53}$$

We have chosen the integration constant such that $A(\infty) = 0$. The functions $B(r)$ and $C(r)$ are then given by $B = -A/2$ (so that $B(\infty) = 0$) and $C = 3A$. Finally, after use of $C = 3A$, the F-field equation (5.2.30) becomes equivalent to (5.2.51). The equation (5.2.52) determining the radial dependence of the function $A(r)$ (and consequently of $B(r)$ and $C(r)$) is the same here as in the case of ordinary branes, while to solve equation (5.2.50) it suffices to choose for the manifold G/H the G-invariant Einstein metric. Each of the Freund–Rubin cosets admits such an Einstein metric which was also constructed in the old Kaluza–Klein supergravity literature (see [19, 16]).

Summarizing: for $D = 11$ supergravity the field equations are solved by the *ansatz* (5.2.10), (5.2.11) where the A, B, C functions are

$$A(r) = -\frac{\tilde{d}}{18} \ln\left(1 + \frac{k}{r^{\tilde{d}}}\right) = -\frac{1}{3} \ln\left(1 + \frac{k}{r^6}\right)$$

$$B(r) = \frac{d}{18} \ln\left(1 + \frac{k}{r^{\tilde{d}}}\right) = \frac{1}{6} \ln\left(1 + \frac{k}{r^6}\right)$$

$$C(r) = 3\,A(r) \tag{5.2.54}$$

displaying the same r-dependence as the ordinary M-brane solution (5.2.2).

In this way we have illustrated the existence of G/H M-brane solutions (see [20]). Table 5.1 displays the Freund–Rubin cosets with non vanishing $N_{G/H}$. Each of them is associated with a BPS saturated M-brane. The notations are as in [19], [16].

Table 5.1. Supersymmetric Freund–Rubin cosets with Killing spinors.

G/H	G	H	$N_{G/H}$
S^7	$SO(8)$	$SO(7)$	8
squashed S^7	$SO(5) \times SO(3)$	$SO(3) \times SO(3)$	1
M^{ppr}	$SU(3) \times SU(2) \times U(1)$	$SU(2) \times U(1)^2$	2
N^{010}	$SU(3) \times SU(2)$	$SU(2) \times U(1)$	3
N^{pqr}	$SU(3) \times U(1)$	$U(1)^2$	1
Q^{ppp}	$SU(2)^3$	$U(1)^3$	2
B^7_{irred}	$SO(5)$	$SO(3)_{max}$	1
$V_{5,2}$	$SO(5)) \times U(1)$	$SO(3) \times U(1)$	2

Chapter 6

The symplectic structure of 4D supergravity and $N = 2$ BPS black holes

6.1 Introduction to tetradimensional BPS black holes and the general form of the supergravity action

In this chapter we begin the study of BPS black hole solutions in four space–time dimensions. We mainly focus on the case of $N = 2$ supergravity, leaving the higher N-cases, in particular that of $N = 8$ supergravity, to later chapters. Apart from its intrinsic interest we want to use the $N = 2$ example to illustrate the profound relation between the black hole structure, in particular its entropy, and the scalar geometry of supergravity, together with its symplectic embedding that realizes covariance with respect to electric–magnetic duality rotations. We approach this intricate and fascinating relation starting from a general description of the nature of BPS black hole solutions. These are field configurations satisfying the equations of motions derived from the bosonic part of the supergravity action that are characterized by the following defining properties:

(i) The metric has the form of a 0-brane solution admitting $\mathbb{R} \times SO(3)$ as isometry group:

$$ds^2 = e^{2U(r)} dt^2 - e^{-2U(r)} d\vec{x}^2 \qquad (r^2 = \vec{x}^2). \qquad (6.1.1)$$

(ii) The $\bar{n}$ gauge fields of the theory A^Λ carry both electric and magnetic charges and are in a 0-brane configuration, namely their field strengths F^Λ are of the form:

$$F^\Lambda = \frac{p^\Lambda}{2r^3} \epsilon_{abc} x^a \, dx^b \wedge dx^c - \frac{\ell^\Lambda(r)}{r^3} e^{2U} dt \wedge \vec{x} \cdot d\vec{x}. \qquad (6.1.2)$$

(iii) The m scalar fields of the theory ϕ^I have only radial dependence:

$$\phi^I = \phi^I(r) \qquad (I = 1, \ldots, m). \qquad (6.1.3)$$

(iv) The field configuration preserves a certain fraction of the original N-extended supersymmetry:

$$f_{SUSY} = \frac{N}{2 n_{max}} \qquad (f_{SUSY}) = \frac{1}{2} \text{ or } \frac{1}{4} \text{ or } \frac{1}{8} \qquad (6.1.4)$$

in the sense that there exists a BPS Killing spinor $\xi_A(x)$ subject to the condition:

$$\gamma^0 \xi_A = i \mathbb{C}_{AB} \xi^B \qquad A, B = 1, \ldots, n_{max} \qquad 2 \le n_{max} \le N$$

$$\xi_A = 0 \qquad A = n_{max} + 1, \ldots, N$$

$$\mathbb{C}_{AB} = -\mathbb{C}_{AB} = \text{antisymmetric matrix of rank } n_{max} \qquad (6.1.5)$$

such that in the considered bosonic background the supersymmetry transformation rule of all the fermions of the theory vanishes along the Killing spinor (6.1.5):

$$\delta_\epsilon^{SUSY} \text{ fermions } = 0 \quad \text{if SUSY parameter } \epsilon_A = \xi_A = \text{Killing spinor.}$$
$$(6.1.6)$$

Let us illustrate the above definition of BPS black holes by comparison with the results of chapter 4 on higher dimensional p-branes. To make such a comparison we have to assume that only one (say $A_{[1]} \equiv A^0$) of the $\bar{n}$ gauge fields A^Λ and only one (say $\phi = \phi^1$) of the m scalar fields ϕ^I are switched on in our solution. Furthermore we have to assume that upon truncation to such fields the supergravity Lagrangian takes the form (5.1.1). Under these conditions a comparison is possible and it reveals the meaning of the definition of BPS black holes we have adopted. It will become clear that the much richer structure of 4D black holes is due to the presence of many scalar and many vector fields and to the geometric structure of their mutual interactions completely fixed by supersymmetry.

So we begin by comparing the *ansatz* (6.1.1) for the 4D-metric with the general form of the metric in either an elementary or solitonic p-brane solution as given in equations (5.1.8) or (5.1.10). In this comparison we have also to recall equations (5.1.4), (5.1.5) and (5.1.6). Since the black hole is a 0-brane in $D = 4$, we have:

$$\begin{array}{llll} d & = & 1 & \tilde{d} & = & 1 \\ p & = & 0 & \Delta & = & a^2 + 1 \end{array} \qquad (6.1.7)$$

and for a solution coupled to a single vector field we would expect a metric of the form:

$$ds^2 = \left(1 + \frac{k}{r} \right)^{-\frac{2}{a^2+1}} dt^2 - \left(1 + \frac{k}{r} \right)^{\frac{2}{a^2+1}} d\vec{x}^2 \qquad (6.1.8)$$

that is precisely of the form (6.1.1) with a function U given by:

$$U(r) = -\frac{1}{a^2 + 1} \log H(r) \qquad (6.1.9)$$

where:

$$H(r) \equiv \left(1 + \frac{k}{r}\right)$$

$$\Delta_3 H(r) = \sum_{i=1}^{3} \frac{\partial^2}{\partial x_i^2} H(r) = 0 \qquad (6.1.10)$$

is a harmonic function depending on the charge k carried by the single vector field we have considered in this comparison.

Similarly for the single scalar field we would expect:

$$\phi(r) = \begin{cases} -\frac{2a}{a^2+1} \log H(r) & \text{electric 0-brane} \\ \frac{2a}{a^2+1} \log H(r) & \text{magnetic solitonic 0-brane} \end{cases} \qquad (6.1.11)$$

and for the field strength of the single vector field we would expect:

$$F = -2 \frac{k}{\sqrt{a^2+1}} \frac{1}{r^3} \left(1 + \frac{k}{r}\right)^{-2} dt \wedge \vec{x} \cdot \vec{x} \qquad \text{electric 0-brane}$$

$$F = -2 \frac{k}{\sqrt{a^2+1}} \frac{1}{r^3} \epsilon_{abc} x^a dx^b \wedge dx^c \qquad \text{magnetic solitonic 0-brane.}$$
$$(6.1.12)$$

Comparison of equations (6.1.12) with equations (6.1.2) shows that the *ansatz* we adopted is indeed consistent with the 0-brane interpretation of the black hole upon the identifications:

$$\begin{aligned} \ell(r) &= r^3 \frac{2\sqrt{a^2+1}}{a^2-1} \frac{d}{dr} [H(r)]^{\frac{1-a^2}{1+a^2}} & \text{electric 0-brane} \\ p &= -\frac{4}{\sqrt{a^2+1}} k & \text{magnetic solitonic 0-brane.} \end{aligned}$$
$$(6.1.13)$$

At this point we must observe that for this one-vector, one-scalar coupling there are critical values of the a-parameter, namely:

$$a = \begin{cases} \sqrt{3} & \implies \Delta = 4 & U = -\frac{1}{4} \log H(r) \\ 1 & \implies \Delta = 2 & U = -\frac{1}{2} \log H(r) \\ 0 & \implies \Delta = 1 & U = -\log H(r) \end{cases}$$

$$\phi = \mp\frac{\sqrt{3}}{2} \log H \qquad \begin{cases} \ell = 2r^3 \frac{d}{dr} H^{-\frac{1}{2}} \\ p = -k \end{cases}$$

$$\phi = \mp \log H \qquad \begin{cases} \ell = -2r^3 \frac{d}{dr} H^{-1} \\ p = -2\sqrt{2} k \end{cases} \qquad (6.1.14)$$

$$\phi = 0 \qquad \begin{cases} \ell = -2r^3 \frac{d}{dr} H \\ p = -4k. \end{cases}$$

The special property of these critical values of a is that for all of them we can write:

$$\Delta = 4 \times 2 \times f_{SUSY} \qquad \begin{cases} f_{SUSY} = \frac{1}{2} \leftrightarrow a = \sqrt{3} \\ f_{SUSY} = \frac{1}{4} \leftrightarrow a = 1 \\ f_{SUSY} = \frac{1}{8} \leftrightarrow a = 0. \end{cases} \qquad (6.1.15)$$

This numerical coincidence hints at a relation with the number of preserved supersymmetries. Indeed such a relation will be verified in the context of $N = 8$ BPS black hole solutions where the three cases of supersymmetry fractions make sense. The other important observation is that it is only in the case $a = 0$ that our black hole can be *dyonic*, namely that we can simultaneously assign both an electric and a magnetic part to our field strength. This is evident from equation (6.1.14). In the other cases where $\phi(r)$ is non-vanishing it cannot be simultaneously equal to const $\times \log H(r)$ and to minus the same expression. For the value $a = 0$ the 1-vector BPS black hole simply coincides with the extremum Reissner-Nordstrom black hole

$$ds^2_{RN} = \left(1 + \frac{k}{r}\right)^{-2} dt^2 - \left(1 + \frac{k}{r}\right)^2 d\vec{x}^2 \qquad (6.1.16)$$

and for small values of r it is approximated by the horizon geometry:

$$AdS_2 \times S^2 \qquad (6.1.17)$$

that corresponds to the Bertotti–Robinson metric [30]:

$$ds^2_{BR} = \frac{1}{m^2_{BR}} r^2 dt^2 - m^2_{BR} \frac{dr^2}{r^2} - m^2_{BR} \left(\sin^2(\theta)\, d\phi^2 + d\theta^2\right) \qquad (6.1.18)$$

the parameter:

$$m_{BR} = |k| \qquad (6.1.19)$$

being named the Bertotti–Robinson mass.

Our comparison between the higher dimensional p-brane solutions and the *ansatz* defining BPS black holes in $D = 4$ is now complete. By means of this comparison we have understood the form (6.1.1) assigned to the metric. It must reduce to equations (6.1.8), (6.1.9) when there is only one vector and only one scalar switched on. However, in the general case we are not allowed to identify directly the function $U(r)$ with the logarithm of the harmonic function $H(r)$ as in equation (6.1.9). This is so because there are many vector fields in the theory and with each of them we can associate a different charge and hence a different harmonic function. Similarly the presence of many scalar fields does not allow us to know, *a priori*, whether a given field strength F^Λ is electric, magnetic or dyonic. Correspondingly the general *ansatz* for F^Λ is that of equation (6.1.2), the electric and magnetic parts having the form expected for 0-branes.

Finally we have to discuss why the condition defining the BPS Killing spinor is given by equation (6.1.5). This is nothing else but the transcription, on the supersymmetry parameters of the projection operator:

$$\mathbb{P}^{\pm}_{AB} = \tfrac{1}{2}(\mathbb{1}\,\delta^B_A \pm i\mathbb{C}_{AB}\,\gamma_0) \qquad (6.1.20)$$

that acts on the supersymmetry generators in the abstract description (4.2.10).

Having illustrated the general form of the *ansatz* for BPS black holes in $D = 4$ space–time dimensions, in order to proceed one needs to insert such an *ansatz* into

- the second order bosonic field equations of supergravity
- the Killing spinor equations (6.1.6) that follow from the supersymmetry transformations rules of the fermions and are linear in the derivatives of the bosonic fields.

In this way one arrives at a coupled system of first and second order differential equations that depends crucially on two ingredients:

- the self-interaction of the scalar fields, that is on the metric $g_{IJ}(\phi)$ of the scalar manifold of which the ϕ^I are interpreted as coordinates
- the non-minimal coupling between the scalars ϕ^I and the vector fields A^Λ of which the exponential coupling $\exp[a\,\phi]\,F^2_{[p+2]}$ in the action (5.1.1) is just the simplest example.

Indeed for all tetradimensional supergravities theories the bosonic action takes the following general form:

$$\mathcal{L} = \int \sqrt{-g}\,\mathrm{d}^4x \left(2R + \operatorname{Im}\mathcal{N}_{\Lambda\Gamma}\,F_{\mu\nu}{}^\Lambda F^{\Gamma|\mu\nu} + \frac{1}{6}g_{IJ}(\phi)\partial_\mu\phi^I\partial^\mu\phi^J \right.$$
$$\left. + \frac{1}{2}\operatorname{Re}\mathcal{N}_{\Lambda\Gamma}\frac{\epsilon^{\mu\nu\rho\sigma}}{\sqrt{-g}}F_{\mu\nu}{}^\Lambda F^\Gamma{}_{\rho\sigma} \right) \qquad (6.1.21)$$

where, as already stated, $g_{IJ}(\phi)$ is the scalar metric on the m-dimensional scalar manifold $\mathcal{M}_{scalar}$ and

$$\mathcal{N}_{\Lambda\Sigma}(\phi) \qquad (6.1.22)$$

is a complex, symmetric, $\bar{n} \times \bar{n}$ matrix depending on the scalar fields which we name the *period matrix*. What varies, depending on the number N of supersymmetries is

(i) the relative and total number of vectors and scalars, that is $\bar{n}$ and m and
(ii) the geometry $g_{IJ}(\phi)$ and the isometry group G of the scalar manifold $\mathcal{M}_{scalar}$.

Yet the relation between this scalar geometry and the period matrix $\mathcal{N}$ has a very general and universal form. Indeed it is related to the solution of a general problem, namely *how to lift the action of the scalar manifold isometries from the scalar to the vector fields*. Such a lift is necessary because of supersymmetry since scalars and vectors generically belong to the same supermultiplet and must rotate coherently under symmetry operations. Such a problem is solved by considering a deep property inherent to Lagrangians of type (6.1.21): their possible covariance with respect to generalized *electromagnetic duality rotations*. In the next section we address this general property and we show how enforcing covariance with respect to such duality rotations leads to a determination of the period matrix $\overline{N}$. It goes without saying that the structure of $\overline{N}$ enters the black hole equations in a crucial way so that, not too surprisingly, at the end of the day, the topological invariant associated with the hole, that is its *entropy*, is an invariant of the group of electromagnetic duality rotations, the U-duality group. In table 6.1 we summarize the geometries of the scalar manifolds for the various values of N. In such a table the symplectic embedding of these geometries is already mentioned. What this means will become clear by working through the next section.

Table 6.1. Scalar manifolds of extended supergravities. (a) Number of scalars in scalar manifolds. (b) Number of scalars in vector manifolds. (c) Number of scalars in gravity manifolds. (d) Number of vectors in vector manifolds. (e) Number of vectors in gravity manifolds.

N	(a)	(b)	(c)	(d)	(e)	Γ_{cont}	$\mathcal{M}_{scalar}$
1	$2m$			n		$\mathcal{I}$ $\subset Sp(2n, \mathbb{R})$	Kähler
2	$4m$	$2n$		n	1	$\mathcal{I}$ $\subset Sp(2n+2, \mathbb{R})$	quaternionic $\otimes$ special Kähler
3		$6n$		n	3	$SU(3, n)$ $\subset Sp(2n+6, \mathbb{R})$	$\dfrac{SU(3, n)}{S(U(3) \times U(n))}$
4		$6n$	2	n	6	$SU(1, 1) \otimes SO(6, n)$ $\subset Sp(2n+12, \mathbb{R})$	$\dfrac{SU(1, 1)}{U(1)} \otimes \dfrac{SO(6, n)}{SO(6) \times SO(n)}$
5			10		10	$SU(1, 5)$ $\subset Sp(20, \mathbb{R})$	$\dfrac{SU(1, 5)}{S(U(1) \times U(5))}$
6			30		16	$SO^\star(12)$ $\subset Sp(32, \mathbb{R})$	$\dfrac{SO^\star(12)}{U(1) \times SU(6)}$
7, 8			70		56	$E_{7(-7)}$ $\subset Sp(128, \mathbb{R})$	$\dfrac{E_{7(-7)}}{SU(8)}$

6.2 Duality rotations and symplectic covariance

In this section, relying on the motivations given above, we review the general structure of an Abelian theory of vectors and scalars displaying covariance under a group of duality rotations. The basic reference is the 1981 paper by Gaillard and Zumino [25]. A general presentation in $D = 2p$ dimensions was recently given in [26]. Here we fix $D = 4$.

We consider a theory of $\bar{n}$ gauge fields A_μ^Λ, in a $D = 4$ space–time with Lorentz signature. They correspond to a set of $\bar{n}$ differential 1-forms

$$A^\Lambda \equiv A_\mu^\Lambda \, dx^\mu \qquad (\Lambda = 1, \ldots, \bar{n}). \qquad (6.2.1)$$

The corresponding field strengths and their Hodge duals are defined by

$$
\begin{aligned}
F^\Lambda &\equiv dA^\Lambda \equiv \mathcal{F}_{\mu\nu}^\Lambda \, dx^\mu \wedge dx^\nu \\
\mathcal{F}_{\mu\nu}^\Lambda &\equiv \tfrac{1}{2}(\partial_\mu A_\nu^\Lambda - \partial_\nu A_\mu^\Lambda) \\
{}^\star F^\Lambda &\equiv \widetilde{\mathcal{F}}_{\mu\nu}^\Lambda \, dx^\mu \wedge dx^\nu \\
\widetilde{\mathcal{F}}_{\mu\nu}^\Lambda &\equiv \tfrac{1}{2}\varepsilon_{\mu\nu\rho\sigma} \mathcal{F}^{\Lambda|\rho\sigma}.
\end{aligned} \qquad (6.2.2)
$$

Defining the space–time integration volume as

$$d^4 x \equiv -\frac{1}{4!} \varepsilon_{\mu_1 \ldots \mu_4} \, dx^{\mu_1} \wedge \cdots \wedge dx^{\mu_4} \qquad (6.2.3)$$

we obtain

$$F^\Lambda \wedge F^\Sigma = \varepsilon^{\mu\nu\rho\sigma} \mathcal{F}_{\mu\nu}^\Lambda \mathcal{F}_{\rho\sigma}^\Sigma \, d^4 x \qquad F^\Lambda \wedge {}^\star F^\Sigma = -2 \mathcal{F}_{\mu\nu}^\Lambda \mathcal{F}^{\Sigma|\mu\nu} \, d^4 x. \qquad (6.2.4)$$

In addition to the gauge fields let us also introduce a set of real scalar fields ϕ^I ($I = 1, \ldots, m$) spanning an $\bar{m}$-dimensional manifold $\mathcal{M}_{scalar}$[1] endowed with a metric $g_{IJ}(\phi)$. Utilizing the above field content we can write the following action functional:

$$
\begin{aligned}
S = \frac{1}{2} \int &\{[\gamma_{\Lambda\Sigma}(\phi) \, F^\Lambda \wedge {}^\star F^\Sigma + \theta_{\Lambda\Sigma}(\phi) \, F^\Lambda \wedge F^\Sigma] \\
&+ g_{IJ}(\phi) \, \partial_\mu \phi^I \, \partial^\mu \phi^J \, d^4 x\}
\end{aligned} \qquad (6.2.5)
$$

where the scalar field dependent $\bar{n} \times \bar{n}$ matrix $\gamma_{\Lambda\Sigma}(\phi)$ generalizes the inverse of the squared coupling constant $\frac{1}{g^2}$ appearing in ordinary gauge theories. The field dependent matrix $\theta_{\Lambda\Sigma}(\phi)$ is instead a generalization of the *theta*-angle of quantum chromodynamics. Both γ and θ are symmetric matrices. Introducing a formal operator j that maps a field strength into its Hodge dual

$$\left(j \, \mathcal{F}^\Lambda \right)_{\mu\nu} \equiv \tfrac{1}{2} \epsilon_{\mu\nu\rho\sigma} \mathcal{F}^{\Lambda|\rho\sigma} \qquad (6.2.6)$$

---

[1] Whether the ϕ^I can be arranged into complex fields is not relevant at this level of the discussion.

and a formal scalar product

$$(G, K) \equiv G^T K \equiv \sum_{\Lambda=1}^{\bar{n}} G^\Lambda_{\mu\nu} K^{\Lambda|\mu\nu} \tag{6.2.7}$$

the total Lagrangian of equation (6.2.5) can be rewritten as

$$\mathcal{L}^{(tot)} = \mathcal{F}^T \left(-\gamma \otimes \mathbb{1} + \theta \otimes j\right)\mathcal{F} + \tfrac{1}{2} g_{IJ}(\phi)\, \partial_\mu \phi^I\, \partial^\mu \phi^J. \tag{6.2.8}$$

The operator j satisfies $j^2 = -\mathbb{1}$ so that its eigenvalues are $\pm i$. Introducing self-dual and antiself-dual combinations

$$\mathcal{F}^\pm = \tfrac{1}{2}\left(\mathcal{F} \pm i j \mathcal{F}\right)$$
$$j\mathcal{F}^\pm = \mp i \mathcal{F}^\pm \tag{6.2.9}$$

and the field-dependent symmetric matrices

$$\mathcal{N} = \theta - i\gamma$$
$$\overline{\mathcal{N}} = \theta + i\gamma \tag{6.2.10}$$

the vector part of the Lagrangian (6.2.8) can be rewritten as

$$\mathcal{L}_{vec} = i[\mathcal{F}^{-T}\overline{\mathcal{N}}\mathcal{F}^- - \mathcal{F}^{+T}\mathcal{N}\mathcal{F}^+]. \tag{6.2.11}$$

Introducing the new tensors

$$\widetilde{\mathcal{G}}^\Lambda_{\mu\nu} \equiv \frac{1}{2}\frac{\partial \mathcal{L}}{\partial \mathcal{F}^\Lambda_{\mu\nu}} \leftrightarrow \mathcal{G}^{\mp\Lambda}_{\mu\nu} \equiv \mp\frac{i}{2}\frac{\partial \mathcal{L}}{\partial \mathcal{F}^{\mp\Lambda}_{\mu\nu}} \tag{6.2.12}$$

which, in matrix notation, corresponds to

$$j\mathcal{G} \equiv \frac{1}{2}\frac{\partial \mathcal{L}}{\partial \mathcal{F}^T} = -(\gamma \otimes \mathbb{1} - \theta \otimes j)\mathcal{F} \tag{6.2.13}$$

the Bianchi identities and field equations associated with the Lagrangian (6.2.5) can be written as

$$\partial^\mu \widetilde{\mathcal{F}}^\Lambda_{\mu\nu} = 0 \tag{6.2.14}$$
$$\partial^\mu \widetilde{\mathcal{G}}^\Lambda_{\mu\nu} = 0 \tag{6.2.15}$$

or equivalently

$$\partial^\mu \operatorname{Im} \mathcal{F}^{\pm\Lambda}_{\mu\nu} = 0 \tag{6.2.16}$$
$$\partial^\mu \operatorname{Im} \mathcal{G}^{\pm\Lambda}_{\mu\nu} = 0. \tag{6.2.17}$$

This suggests that we introduce the $2\bar{n}$ column vector

$$V \equiv \begin{pmatrix} j\mathcal{F} \\ j\mathcal{G} \end{pmatrix} \tag{6.2.18}$$

and that we consider general linear transformations on such a vector

$$
\begin{pmatrix} j\,\mathcal{F} \\ j\,\mathcal{G} \end{pmatrix}' = \begin{pmatrix} A & B \\ C & D \end{pmatrix} \begin{pmatrix} j\,\mathcal{F} \\ j\,\mathcal{G} \end{pmatrix}.
\tag{6.2.19}
$$

For any matrix $\begin{pmatrix} A & B \\ C & D \end{pmatrix} \in GL(2\bar{n}, \mathbb{R})$ the new vector V' of *magnetic and electric* field-strengths satisfies the same equations (6.2.15) as the old one. In a condensed notation we can write

$$
\partial V = 0 \quad \Longleftrightarrow \quad \partial V' = 0.
\tag{6.2.20}
$$

Separating the self-dual and anti-self-dual parts

$$
\mathcal{F} = \left(\mathcal{F}^+ + \mathcal{F}^- \right) \qquad \mathcal{G} = \left(\mathcal{G}^+ + \mathcal{G}^- \right)
\tag{6.2.21}
$$

and taking into account that we have

$$
\mathcal{G}^+ = \mathcal{N}\mathcal{F}^+ \qquad \mathcal{G}^- = \overline{\mathcal{N}}\mathcal{F}^-
\tag{6.2.22}
$$

the duality rotation of equation (6.2.19) can be rewritten as

$$
\begin{pmatrix} \mathcal{F}^+ \\ \mathcal{G}^+ \end{pmatrix}' = \begin{pmatrix} A & B \\ C & D \end{pmatrix} \begin{pmatrix} \mathcal{F}^+ \\ \mathcal{N}\mathcal{F}^+ \end{pmatrix} \qquad \begin{pmatrix} \mathcal{F}^- \\ \mathcal{G}^- \end{pmatrix}' = \begin{pmatrix} A & B \\ C & D \end{pmatrix} \begin{pmatrix} \mathcal{F}^- \\ \overline{\mathcal{N}}\mathcal{F}^- \end{pmatrix}.
\tag{6.2.23}
$$

The problem is that the transformation rule (6.2.23) of $\mathcal{G}^\pm$ must be consistent with the definition of the latter as variation of the Lagrangian with respect to $\mathcal{F}^\pm$ (see equation (6.2.12)). This request restricts the form of the matrix $\Lambda = \begin{pmatrix} A & B \\ C & D \end{pmatrix}$. As we are going to show, Λ must belong to the symplectic subgroup of the general linear group

$$
\Lambda \equiv \begin{pmatrix} A & B \\ C & D \end{pmatrix} \in Sp(2\bar{n}, \mathbb{R}) \subset GL(2\bar{n}, \mathbb{R})
\tag{6.2.24}
$$

the subgroup $Sp(2\bar{n}, \mathbb{R})$ being defined as the set of $2\bar{n} \times 2\bar{n}$ matrices that satisfy the condition

$$
\Lambda \in Sp(2\bar{n}, \mathbb{R}) \quad \longrightarrow \quad \Lambda^T \begin{pmatrix} 0 & \mathbb{1} \\ -\mathbb{1} & 0 \end{pmatrix} \Lambda = \begin{pmatrix} 0 & \mathbb{1} \\ -\mathbb{1} & 0 \end{pmatrix}
\tag{6.2.25}
$$

that is, using $n \otimes n$ block components

$$
A^T C - C^T A = B^T D - D^T B = 0 \qquad A^T D - C^T B = 1.
\tag{6.2.26}
$$

To prove the statement we just made, we calculate the transformed Lagrangian $\mathcal{L}'$ and then we compare its variation $\frac{\partial \mathcal{L}'}{\partial \mathcal{F}^T}$ with $\mathcal{G}^{\pm\prime}$ as follows from the postulated transformation rule (6.2.23). To perform such a calculation we rely on the following basic idea. While the duality rotation (6.2.23) is performed on the field strengths and on their duals, also the scalar fields are transformed by the action of some diffeomorphism $\xi \in \text{Diff}(\mathcal{M}_{scalar})$ of the scalar manifold and, as a consequence of that, also the matrix $\mathcal{N}$ changes. In other words given the scalar manifold $\mathcal{M}_{scalar}$ we assume that there exists a homomorphism of the form

$$\iota_\delta : \text{Diff}(\mathcal{M}_{scalar}) \longrightarrow GL(2\bar{n}, \mathbb{R}) \qquad (6.2.27)$$

so that

$$\forall \; \xi \; \in \text{Diff}(\mathcal{M}_{scalar}) : \phi^I \overset{\xi}{\longrightarrow} \phi^{I'}$$

$$\exists \iota_\delta(\xi) = \begin{pmatrix} A_\xi & B_\xi \\ C_\xi & D_\xi \end{pmatrix} \in GL(2\bar{n}, \mathbb{R}). \qquad (6.2.28)$$

(In the following the suffix ξ will be omitted when no confusion can arise and be reinstalled when necessary for clarity.)

Using such a homomorphism we can define the simultaneous action of ξ on all the fields of our theory by setting

$$\xi : \begin{cases} \phi \longrightarrow \xi(\phi) \\ V \longrightarrow \iota_\delta(\xi)\,V \\ \mathcal{N}(\phi) \longrightarrow \mathcal{N}'(\xi(\phi)) \end{cases} \qquad (6.2.29)$$

where the notation (6.2.18) has been utilized. In the gauge sector the transformed Lagrangian is

$$\mathcal{L}'_{vec} = i[\mathcal{F}^{-T}\left(A+B\overline{\mathcal{N}}\right)^T \overline{\mathcal{N}}'(A+B\overline{\mathcal{N}})\mathcal{F}^- - \mathcal{F}^{+T}\left(A+B\mathcal{N}\right)^T \mathcal{N}'(A+B\mathcal{N})\mathcal{F}^+].$$
$$(6.2.30)$$

Consistency with the definition of $\mathcal{G}^+$ requires that

$$\mathcal{N}' \equiv \mathcal{N}'(\xi(\phi)) = (C + D\mathcal{N}(\phi))\,(A + B\mathcal{N}(\phi))^{-1} \qquad (6.2.31)$$

while consistency with the definition of $\mathcal{G}^-$ imposes the transformation rule

$$\overline{\mathcal{N}}' \equiv \overline{\mathcal{N}}'(\xi(\phi)) = \left(C + D\overline{\mathcal{N}}(\phi)\right)\left(A + B\overline{\mathcal{N}}(\phi)\right)^{-1}. \qquad (6.2.32)$$

It is from the transformation rules (6.2.31) and (6.2.32) that we derive a restriction on the form of the duality rotation matrix $\Lambda \equiv \iota_\delta(\xi)$. Indeed by requiring that the transformed matrix $\mathcal{N}'$ be again symmetric one easily finds that Λ must obey equation (6.2.25), namely $\Lambda \in Sp(2\bar{n}, \mathbb{R})$. Consequently the homomorphism of equation (6.2.27) specializes as

$$\iota_\delta : \text{Diff}(\mathcal{M}_{scalar}) \longrightarrow Sp(2\bar{n}, \mathbb{R}). \qquad (6.2.33)$$

Clearly, since $Sp(2\bar{n}, \mathbb{R})$ is a finite dimensional Lie group, while $\text{Diff}(\mathcal{M}_{scalar})$ is infinite dimensional, the homomorphism ι_δ can never be an isomorphism. Defining the Torelli group of the scalar manifold as

$$\text{Diff}(\mathcal{M}_{scalar}) \supset \text{Tor}(\mathcal{M}_{scalar}) \equiv \ker \iota_\delta \qquad (6.2.34)$$

we always have

$$\dim \text{Tor}(\mathcal{M}_{scalar}) = \infty. \tag{6.2.35}$$

The reason why we have given the name of Torelli to the group defined by equation (6.2.34) is because of its similarity with the Torelli group that occurs in algebraic geometry.

What should be clear from the above discussion is that a family of Lagrangians as in equation (6.2.5) will admit a group of duality-rotations/field-redefinitions that will map elements of the family into each other, as long as a *kinetic matrix* $\mathcal{N}_{\Lambda\Sigma}$ can be constructed that transforms as in equation (6.2.31). A way to obtain such an object is to identify it with the *period matrix* occurring in problems of algebraic geometry. At the level of the present discussion, however, this identification is by no means essential: any construction of $\mathcal{N}_{\Lambda\Sigma}$ with the appropriate transformation properties is acceptable. Note also that so far we have used the words *duality-rotations/field-redefinitions* and not the word duality symmetry. Indeed the diffeomorphisms of the scalar manifold we have considered were quite general and, as such had no pretension to be symmetries of the action, or of the theory. Indeed the question we have answered is the following: what are the appropriate transformation properties of the tensor gauge fields and of the generalized coupling constants under diffeomorphisms of the scalar manifold? The next question is obviously that of duality symmetries.

As is the case with the difference between general covariance and isometries in the context of general relativity, duality symmetries correspond to the subset of duality transformations for which we obtain an invariance in form of the theory. In this respect, however, we have to stress that what is invariant in form cannot be the Lagrangian but only the set of field equations plus Bianchi identities. Indeed, while any $\Lambda \in Sp(2\bar{n}, \mathbb{R})$ can, in principle, be an invariance in form of equations (6.2.17), the same is not true for the Lagrangian. One can easily find that the vector kinetic part of this latter transforms as follows:

$$\begin{aligned}
\text{Im}\mathcal{F}^{-\Lambda}\overline{\mathcal{N}}_{\Lambda\Sigma}\mathcal{F}^{-\Sigma} &\to \text{Im}\,\widetilde{\mathcal{F}}^{-\Lambda}\,\widetilde{\mathcal{G}}_{\Sigma}^{-} \\
&= \text{Im}\,(\mathcal{F}^{-\Lambda}\mathcal{G}_{\Lambda}^{-} + 2\mathcal{F}^{-\Lambda}\,(C^{T}B)_{\Lambda}{}^{\Sigma}\,\mathcal{G}_{\Sigma}^{-} \\
&\quad + \mathcal{F}^{-\Lambda}\,(C^{T}A)_{\Lambda\Sigma}\mathcal{F}^{-\Sigma} + \mathcal{G}_{\Lambda}^{-}\,(D^{T}B)^{\Lambda\Sigma}\,\mathcal{G}_{\Sigma}^{-})
\end{aligned} \tag{6.2.36}$$

whence we conclude that proper symmetries of the Lagrangian are to be looked for only among matrices with $C = B = 0$. If $C \neq 0$ and $B = 0$, the Lagrangian varies through the addition of a topological density. Elements of $Sp(2\bar{n}, \mathbb{R})$ with $B \neq 0$, cannot be symmetries of the classical action under any circumstance.

The scalar part of the Lagrangian, on the other hand, is invariant under all those diffeomorphisms of the scalar manifolds that are *isometries* of the scalar metric g_{IJ}. Naming $\xi^{*} : T\mathcal{M}_{scalar} \to T\mathcal{M}_{scalar}$ the push-forward of ξ, this means that

$$\forall X, Y \in T\mathcal{M}_{scalar}$$
$$g(X, Y) = g(\xi^{*}X, \xi^{*}Y) \tag{6.2.37}$$

and ξ is an exact global symmetry of the scalar part of the Lagrangian in equation (6.2.5). In view of our previous discussion these symmetries of the scalar sector are not guaranteed to admit an extension to symmetries of the complete action. Yet we can insist that they extend to symmetries of the field equations plus Bianchi identities, namely to duality symmetries in the sense defined above. This requires that the group of isometries of the scalar metric $\mathcal{I}(\mathcal{M}_{scalar})$ be suitably embedded into the duality group $Sp(2\bar{n}, \mathbb{R})$ and that the kinetic matrix $\mathcal{N}_{\Lambda\Sigma}$ satisfies the covariance law:

$$\mathcal{N}(\xi(\phi)) = (C_\xi + D_\xi \mathcal{N}(\phi))(A_\xi + B_\xi \mathcal{N}(\phi))^{-1}. \qquad (6.2.38)$$

6.3 Symplectic embeddings of homogenous spaces

A general construction of the kinetic coupling matrix $\mathcal{N}$ can be derived in the case where the scalar manifold is taken to be a homogeneous space $\mathcal{G}/\mathcal{H}$. This is what happens in all extended supergravities for $N \geq 3$ and also in specific instances of $N = 2$ theories. For this reason we shortly review the construction of the *kinetic period matrix* $\mathcal{N}$ in the case of homogeneous spaces.

The relevant homomorphism ι_δ (see equation (6.2.33)) becomes:

$$\iota_\delta : \text{Diff}\left(\frac{\mathcal{G}}{\mathcal{H}}\right) \longrightarrow Sp(2\bar{n}, \mathbb{R}). \qquad (6.3.1)$$

In particular, focusing on the isometry group of the canonical metric defined on $\frac{\mathcal{G}}{\mathcal{H}}$[2]: $\mathcal{I}\left(\frac{\mathcal{G}}{\mathcal{H}}\right) = \mathcal{G}$ we must consider the embedding:

$$\iota_\delta : \mathcal{G} \longrightarrow Sp(2\bar{n}, \mathbb{R}). \qquad (6.3.2)$$

That in equation (6.3.1) is a homomorphism of finite dimensional Lie groups and as such it constitutes a problem that can be solved in explicit form. What we just need to know is the dimension of the symplectic group, namely the number $\bar{n}$ of gauge fields appearing in the theory. Without supersymmetry the dimension m of the scalar manifold (namely the possible choices of $\frac{\mathcal{G}}{\mathcal{H}}$) and the number of vectors $\bar{n}$ are unrelated so that the possibilities covered by equation (6.3.2) are infinitely many. In supersymmetric theories, instead, the two numbers m and $\bar{n}$ are related, so that there are finitely many cases to be studied corresponding to the possible embeddings of given groups $\mathcal{G}$ into a symplectic group $Sp(2\bar{n}, \mathbb{R})$ of fixed dimension $\bar{n}$. Actually taking into account further conditions on the holonomy of the scalar manifold that are also imposed by supersymmetry, the solution for the symplectic embedding problem is unique for all extended supergravities with $N \geq 3$ (see for instance [16]).

Apart from the details of the specific case considered once a symplectic embedding is given there is a general formula one can write down for the *period*

---

[2] Actually, in order to be true, the equation $\mathcal{I}(\frac{\mathcal{G}}{\mathcal{H}}) = \mathcal{G}$ requires that that the normalizer of $\mathcal{H}$ in $\mathcal{G}$ be the identity group, a condition that is verified in all the relevant examples.

matrix $\mathcal{N}$ that guarantees symmetry ($\mathcal{N}^T = \mathcal{N}$) and the required transformation property (6.2.38). This is the result we want to review.

The real symplectic group $Sp(2\bar{n}, \mathbb{R})$ is defined as the set of all *real* $2\bar{n} \times 2\bar{n}$ matrices $\Lambda = \begin{pmatrix} A & B \\ C & D \end{pmatrix}$ satisfying equation (6.2.25), namely

$$\Lambda^T \mathbb{C} \Lambda = \mathbb{C} \tag{6.3.3}$$

where $\mathbb{C} \equiv \begin{pmatrix} 0 & \mathbb{1} \\ -\mathbb{1} & 0 \end{pmatrix}$. If we relax the condition that the matrix should be real but we still impose equation (6.3.3) we obtain the definition of the complex symplectic group $Sp(2\bar{n}, \mathbb{C})$. It is a well known fact that the following isomorphism is true:

$$Sp(2\bar{n}, \mathbb{R}) \sim Usp(\bar{n}, \bar{n}) \equiv Sp(2\bar{n}, \mathbb{C}) \cap U(\bar{n}, \bar{n}). \tag{6.3.4}$$

By definition an element $\mathcal{S} \in Usp(\bar{n}, \bar{n})$ is a complex matrix that satisfies simultaneously equation (6.3.3) and a pseudo-unitarity condition, that is:

$$\mathcal{S}^T \mathbb{C} \mathcal{S} = \mathbb{C} \qquad \mathcal{S}^\dagger \mathbb{H} \mathcal{S} = \mathbb{H} \tag{6.3.5}$$

where $\mathbb{H} \equiv \begin{pmatrix} \mathbb{1} & 0 \\ 0 & -\mathbb{1} \end{pmatrix}$. The general block form of the matrix $\mathcal{S}$ is:

$$\mathcal{S} = \begin{pmatrix} T & V^\star \\ V & T^\star \end{pmatrix} \tag{6.3.6}$$

and equations (6.3.5) are equivalent to:

$$T^\dagger T - V^\dagger V = \mathbb{1} \qquad T^\dagger V^\star - V^\dagger T^\star = 0. \tag{6.3.7}$$

The isomorphism of equation (6.3.4) is explicitly realized by the so called Cayley matrix:

$$\mathcal{C} \equiv \frac{1}{\sqrt{2}} \begin{pmatrix} \mathbb{1} & i\mathbb{1} \\ \mathbb{1} & -i\mathbb{1} \end{pmatrix} \tag{6.3.8}$$

via the relation:

$$\mathcal{S} = \mathcal{C} \Lambda \mathcal{C}^{-1} \tag{6.3.9}$$

which yields:

$$T = \frac{1}{2}(A + D) - \frac{i}{2}(B - C) \qquad V = \frac{1}{2}(A - D) - \frac{i}{2}(B + C). \tag{6.3.10}$$

When we set $V = 0$ we obtain the subgroup $U(\bar{n}) \subset Usp(\bar{n}, \bar{n})$, that in the real basis is given by the subset of symplectic matrices of the form $\begin{pmatrix} A & B \\ -B & A \end{pmatrix}$. The basic idea, to obtain the general formula for the period matrix, is that the symplectic embedding of the isometry group $\mathcal{G}$ will be such that the isotropy subgroup $\mathcal{H} \subset \mathcal{G}$ is embedded into the maximal compact subgroup $U(\overline{n})$, namely:

$$\mathcal{G} \xrightarrow{\iota_\delta} Usp(\overline{n}, \overline{n}) \qquad \mathcal{G} \supset \mathcal{H} \xrightarrow{\iota_\delta} U(\overline{n}) \subset Usp(\overline{n}, \overline{n}). \qquad (6.3.11)$$

If this condition is realized let $L(\phi)$ be a parametrization of the coset $\mathcal{G}/\mathcal{H}$ by means of coset representatives. Relying on the symplectic embedding of equation (6.3.11) we obtain a map:

$$L(\phi) \longrightarrow \mathcal{O}(\phi) = \begin{pmatrix} U_0(\phi) & U_1^*(\phi) \\ U_1(\phi) & U_0^*(\phi) \end{pmatrix} \in Usp(\overline{n}, \overline{n}) \qquad (6.3.12)$$

that associates with $L(\phi)$ a coset representative of $Usp(\overline{n}, \overline{n})/U(\overline{n})$. By construction if $\phi' \neq \phi$ *no* unitary $\overline{n} \times \overline{n}$ matrix W can exist such that:

$$\mathcal{O}(\phi') = \mathcal{O}(\phi) \begin{pmatrix} W & 0 \\ 0 & W^* \end{pmatrix}. \qquad (6.3.13)$$

On the other hand let $\xi \in \mathcal{G}$ be an element of the isometry group of $\mathcal{G}/\mathcal{H}$. Via the symplectic embedding of equation (6.3.11) we obtain a $Usp(\overline{n}, \overline{n})$ matrix

$$\mathcal{S}_\xi = \begin{pmatrix} T_\xi & V_\xi^* \\ V_\xi & T_\xi^* \end{pmatrix} \qquad (6.3.14)$$

such that

$$\mathcal{S}_\xi \mathcal{O}(\phi) = \mathcal{O}(\xi(\phi)) \begin{pmatrix} W(\xi, \phi) & 0 \\ 0 & W^*(\xi, \phi) \end{pmatrix} \qquad (6.3.15)$$

where $\xi(\phi)$ denotes the image of the point $\phi \in \mathcal{G}/\mathcal{H}$ through ξ and $W(\xi, \phi)$ is a suitable $U(\overline{n})$ compensator depending both on ξ and ϕ. Combining equations (6.3.15), (6.3.12) with equations (6.3.10) we immediately obtain:

$$U_0^\dagger(\xi(\phi)) + U_1^\dagger(\xi(\phi)) = W[U_0^\dagger(\phi)(A^T + iB^T) + U_1^\dagger(\phi)(A^T - iB^T)] \qquad (6.3.16)$$

$$U_0^\dagger(\xi(\phi)) - U_1^\dagger(\xi(\phi)) = W[U_0^\dagger(\phi)(D^T - iC^T) - U_1^\dagger(\phi)(D^T + iC^T)].$$

Setting:

$$\mathcal{N} \equiv i[U_0^\dagger + U_1^\dagger]^{-1} [U_0^\dagger - U_1^\dagger] \qquad (6.3.17)$$

and using the result of equation (6.3.17) one checks that the transformation rule (6.2.38) is verified. It is also an immediate consequence of the analogue of equations (6.3.7) satisfied by U_0 and U_1 that the matrix in equation (6.3.17) is symmetric

$$\mathcal{N}^T = \mathcal{N}. \qquad (6.3.18)$$

Equation (6.3.17) is the master formula derived in 1981 by Gaillard and Zumino [25]. It explains the structure of the gauge-field kinetic terms in all $N \geq 3$ extended supergravity theories where the scalar manifold is always a homogeneous coset manifold. (See table 6.1.) In these cases the bosonic Lagrangian (6.1.21) is completely fixed from the information provided by table 6.1. The choice of the coset scalar manifold suffices to determine the scalar kinetic term, while the choice of the symplectic embedding plus the use of the master formula (6.3.17) determines the period matrix $\mathcal{N}_{\Lambda\Sigma}$. In the case of $N = 2$ supergravity the scalar manifold is not necessarily a coset manifold. The requirement imposed by supersymmetry is that:

$$\mathcal{M}_{scalar} = \text{special Kähler manifold } \mathcal{SM} \qquad (6.3.19)$$

and for this class of manifolds there exists another formula that determines the *period matrix* $\mathcal{N}$. In the next section we shall discuss special Kähler geometry and derive such a formula. In the class of special Kähler manifolds there is a subclass that is composed of homogeneous cosets $\mathcal{G}/\mathcal{H}$. For this subclass the construction of the period matrix through the relations of special geometry and the construction through the Gaillard–Zumino master formula (6.3.17) coincide.

In the next subsection we pause to illustrate the master formula (6.3.17) with an example that is of particular relevance in all superstring related supergravities.

6.3.1 Symplectic embedding of the $\mathcal{ST}[m, n]$ homogeneous manifolds

Because of their relevance in superstring compactifications let us illustrate the general procedure with the following class of homogeneous manifolds:

$$\mathcal{ST}[m, n] \equiv \frac{SU(1, 1)}{U(1)} \otimes \frac{SO(m, n)}{SO(m) \otimes SO(n)}. \qquad (6.3.20)$$

The isometry group of the $\mathcal{ST}[m, n]$ manifolds defined in equation (6.3.20) contains a factor $(SU(1, 1))$ whose transformations act as non-perturbative S-dualities and another factor $(SO(m, n))$ whose transformations act as T-dualities, holding true at each order in string perturbation theory. The field S is obtained by combining together the *dilaton* D and the *axion* $\mathcal{A}$:

$$S = \mathcal{A} - i \exp[D] \qquad (6.3.21)$$
$$\partial^\mu \mathcal{A} \equiv \varepsilon^{\mu\nu\rho\sigma} \partial_\nu B_{\rho\sigma}$$

while t^i is the name usually given to the modulus fields of the compactified target space. Now in string and supergravity applications S will be identified with the complex coordinate on the manifold $\frac{SU(1,1)}{U(1)}$, while t^i will be the coordinates of the coset space $\frac{SO(m,n)}{SO(m)\otimes SO(n)}$. The case $\mathcal{ST}[6, n]$ is the scalar manifold in $N = 4$ supergravity, while the case $\mathcal{ST}[2, n]$ is a very interesting instance of the special Kähler manifold appearing in superstring compactifications. Although as differentiable and metric manifolds the spaces $\mathcal{ST}[m, n]$ are just direct products of two factors (corresponding to the above mentioned different physical interpretation of the coordinates S and t^i), from the point of view of the symplectic embedding and duality rotations they have to be regarded as a single entity. This is even more evident in the case $m = 2, n = $ arbitrary, where the following theorem has been proven by Ferrara and Van Proeyen [28]: $ST [2, n]$ are the only special Kähler manifolds with a direct product structure. The definition of special Kähler manifolds is given in the next section, yet the anticipation of this result should make clear that the special Kähler structure (encoding the duality rotations in the $N = 2$ case) is not a property of the individual factors but of the product as a whole. Neither factor is by itself a special manifold although the product is.

At this point comes the question of the correct symplectic embedding. Such a question has two aspects:

(i)   intrinsically inequivalent embeddings and
(ii) symplectically equivalent embeddings that become inequivalent after gauging.

The first issue in the above list is group theoretical in nature. When we say that the group G is embedded into $Sp(2\bar{n}, \mathbb{R})$ we must specify how this is done from the point of view of irreducible representations. Group theoretically the matter is settled by specifying how the fundamental representation of $Sp(2\bar{n})$ splits into irreducible representations of G:

$$2\bar{n} \xrightarrow{\ G\ } \oplus_{i=1}^{\ell} \mathbf{D}_i. \tag{6.3.22}$$

Once equation (6.3.22) is given (in supersymmetric theories such information is provided by supersymmetry) the only arbitrariness which is left is that of conjugation by arbitrary $Sp(2\bar{n}, \mathbb{R})$ matrices. Suppose we have determined an embedding ι_δ that obeys the law in equation (6.3.22), then:

$$\forall S \in Sp(2\bar{n}, \mathbb{R}) : \iota'_\delta \equiv S \circ \iota_\delta \circ S^{-1} \tag{6.3.23}$$

will obey the same law. That in equation (6.3.23) is a symplectic transformation that corresponds to an allowed duality-rotation/field-redefinition in the Abelian theory of the type in equation (6.2.5) discussed in the previous subsection. Therefore all Abelian Lagrangians related by such transformations are physically equivalent.

The matter changes in the presence of *gauging*. When we switch on the gauge coupling constant and the electric charges, symplectic transformations cease to yield physically equivalent theories. This is the second issue in the above list. The choice of a symplectic gauge becomes physically significant. The construction of supergravity theories proceeds in two steps. In the first step, one constructs the Abelian theory: at that level the only relevant constraint is that encoded in equation (6.3.22) and the choice of a symplectic gauge is immaterial. Actually one can write the entire theory in such a way that *symplectic covariance* is manifest. In the second step one *gauges* the theory. This *breaks symplectic covariance* and the choice of the correct symplectic gauge becomes a physical issue. This issue has been recently emphasized by the results in [29] where it has been shown that whether $N = 2$ supersymmetry can be spontaneously broken to $N = 1$ or not depends on the symplectic gauge.

In the applications of supergravity to the issue of BPS black holes what matters is the *Abelian ungauged theory*, so we will not further emphasize the aspects of the theory related to the gauging.

These facts being cleared we proceed to discuss the symplectic embedding of the $ST[m, n]$ manifolds.

Let η be the symmetric flat metric with signature (m, n) that defines the $SO(m, n)$ group, via the relation

$$L \in SO(m, n) \iff L^T \eta L = \eta. \tag{6.3.24}$$

Both in the $N = 4$ and in the $N = 2$ theory, the number of gauge fields in the theory is given by:

$$\text{No of vector fields} = m \oplus n \tag{6.3.25}$$

m being the number of *graviphotons* and n the number of *vector multiplets*. Hence we have to embed $SO(m, n)$ into $Sp(2m + 2n, \mathbb{R})$ and the explicit form of the decomposition in equation (6.3.22) required by supersymmetry is:

$$\mathbf{2m + 2n} \overset{SO(m,n)}{\longrightarrow} \mathbf{m + n} \oplus \mathbf{m + n} \tag{6.3.26}$$

where $\mathbf{m + n}$ denotes the fundamental representation of $SO(m, n)$. Equation (6.3.26) is easily understood in physical terms. $SO(m, n)$ must be a T-duality group, namely a symmetry holding true order by order in perturbation theory. As such it must rotate electric field strengths into electric field strengths and magnetic field strengths into magnetic field strengths. The two irreducible representations into which the fundamental representation of the symplectic group decomposes when reduced to $SO(m, n)$ correspond precisely to electric and magnetic sectors, respectively. In the *simplest gauge* the symplectic embedding satisfying equation (6.3.26) is block diagonal and takes the form:

$$\forall L \in SO(m, n) \quad \overset{\iota_\delta}{\hookrightarrow} \quad \begin{pmatrix} L & \mathbf{0} \\ \mathbf{0} & (L^T)^{-1} \end{pmatrix} \in Sp(2m + 2n, \mathbb{R}). \tag{6.3.27}$$

Consider instead the group $SU(1, 1) \sim SL(2, \mathbb{R})$. This is the factor in the isometry group of $ST[m, n]$ that is going to act by means of S-duality non-perturbative rotations. Typically it will rotate each electric field strength into its homologous magnetic one. Correspondingly supersymmetry implies that its embedding into the symplectic group must satisfy the following condition:

$$\mathbf{2m + 2n} \overset{SL(2,\mathbb{R})}{\longrightarrow} \oplus_{i=1}^{m+n} \mathbf{2} \tag{6.3.28}$$

where $\mathbf{2}$ denotes the fundamental representation of $SL(2, \mathbb{R})$. In addition it must commute with the embedding of $SO(m, n)$ in equation (6.3.27). Both conditions are fulfilled by setting:

$$\forall \begin{pmatrix} a & b \\ c & d \end{pmatrix} \in SL(2, \mathbb{R}) \quad \overset{\iota_\delta}{\hookrightarrow} \quad \begin{pmatrix} a\,\mathbb{1} & b\,\eta \\ c\,\eta & d\,\mathbb{1} \end{pmatrix} \in Sp(2m + 2n, \mathbb{R}). \tag{6.3.29}$$

Utilizing equations (6.3.9) the corresponding embeddings into the group $Usp(m + n, m + n)$ are immediately derived:

$$\forall L \in SO(m,n) \overset{\iota_\delta}{\hookrightarrow} \begin{pmatrix} \frac{1}{2}(L + \eta L\eta) & \frac{1}{2}(L - \eta L\eta) \\ \frac{1}{2}(L - \eta L\eta) & \frac{1}{2}(L + \eta L\eta) \end{pmatrix} \in Usp(m + n, m + n)$$

$$\forall \begin{pmatrix} t & v^* \\ v & t^* \end{pmatrix} \in SU(1,1) \overset{\iota_\delta}{\hookrightarrow} \begin{pmatrix} \operatorname{Re} t\, \mathbb{1} + i\operatorname{Im} t\eta & \operatorname{Re} v\, \mathbb{1} - i\operatorname{Im} v\eta \\ \operatorname{Re} v\, \mathbb{1} + i\operatorname{Im} v\eta & \operatorname{Re} t\, \mathbb{1} - i\operatorname{Im} t\eta \end{pmatrix}$$
$$\in Usp(m + n, m + n) \tag{6.3.30}$$

where the relation between the entries of the $SU(1,1)$ matrix and those of the corresponding $SL(2, \mathbb{R})$ matrix are provided by the relation in equation (6.3.10).

Equipped with these relations we can proceed to derive the explicit form of the *period matrix* $\mathcal{N}$.

The homogeneous manifold $SU(1,1)/U(1)$ can be conveniently parametrized in terms of a single complex coordinate S, whose physical interpretation will be that of *axion–dilaton*, according to equation (6.3.22). The coset parametrization appropriate for comparison with other constructions (special geometry or $N = 4$ supergravity) is given by the family of matrices:

$$M(S) \equiv \frac{1}{n(S)} \begin{pmatrix} \mathbb{1} & \frac{i-S}{i+S} \\ \frac{i+\bar{S}}{i-\bar{S}} & \mathbb{1} \end{pmatrix} \quad : \quad n(S) \equiv \sqrt{\frac{4\operatorname{Im} S}{1 + |S|^2 + 2\operatorname{Im} S}}. \tag{6.3.31}$$

To parametrize the coset $SO(m,n)/SO(m) \times SO(n)$ we can instead take the usual coset representatives (see for instance [16]):

$$L(X) \equiv \begin{pmatrix} (\mathbb{1} + XX^T)^{1/2} & X \\ X^T & (\mathbb{1} + X^T X)^{1/2} \end{pmatrix} \tag{6.3.32}$$

where the $m \times n$ real matrix X provides a set of independent coordinates. Inserting these matrices into the embedding formulae of equations (6.3.30) we obtain a matrix:

$$\iota_\delta(M(S)) \circ \iota_\delta(L(X)) = \begin{pmatrix} U_0(S, X) & U_1^*(S, X) \\ U_1(S, X) & U_0^*(S, X) \end{pmatrix} \in Usp(n + m, n + m) \tag{6.3.33}$$

that inserted into the master formula of equation (6.3.17) yields the following result:

$$\mathcal{N} = i\operatorname{Im} S\, \eta L(X)L^T(X)\eta + \operatorname{Re} S\, \eta. \tag{6.3.34}$$

Alternatively, remarking that if $L(X)$ is an $SO(m,n)$ matrix also $L(X)' = \eta L(X)\eta$ is such a matrix and represents the same equivalence class, we can rewrite (6.3.34) in the simpler form:

$$\mathcal{N} = i\operatorname{Im} S\, L(X)'L^{T\prime}(X) + \operatorname{Re} S\, \eta. \tag{6.3.35}$$

6.4 Special Kähler Geometry

As already stressed, in the case of $N = 2$ supergravity the requirements imposed by supersymmetry on the scalar manifold $\mathcal{M}_{scalar}$ of the theory is that it should be the following direct product:

$$\mathcal{M}_{scalar} = SM \otimes \mathcal{H}M$$
$$\dim_{\mathbf{C}} SM = n_v \qquad\qquad (6.4.1)$$
$$\dim_{\mathbf{R}} \mathcal{H}M = 4n_h$$

where SM, $\mathcal{H}M$ are respectively *special Kähler* and *quaternionic* and n_v, n_h are respectively the number of *vector multiplets* and *hypermultiplets* contained in the theory. The direct product structure (6.4.2) imposed by supersymmetry precisely reflects the fact that the quaternionic and special Kähler scalars belong to different supermultiplets. In the construction of BPS black holes it turns out that the hyperscalars are spectators playing no dynamical role. Hence in this set of lectures we do not discuss the hypermultiplets any further and we confine our attention to an $N = 2$ supergravity where the graviton multiplet, containing, besides the graviton $g_{\mu\nu}$, also a graviphoton A^0_μ, is coupled to $n_v = n$ *vector multiplets*. Such a theory has an action of type (6.1.21) where the number of gauge fields is $\bar{n} = 1 + n$ and the number of scalar fields is $m = 2n$. Correspondingly the indices have the following ranges

$$\Lambda, \Sigma, \Gamma, \ldots = 0, 1, \ldots, n \qquad\qquad (6.4.2)$$
$$I, I, K = 1, \ldots, 2n.$$

To make the action (6.1.21) fully explicit, we need to discuss the geometry of the vector multiplet scalars, namely special Kähler geometry. This is what we do in the next subsections. Let us begin by reviewing the notions of Kähler and Hodge–Kähler manifolds. The readers interested in the full-fledged structure of $N = 2$ supergravity can read [27] which contains its most general formulation and explains all the details of its construction.

6.4.1 Hodge–Kähler manifolds

Consider a *line bundle* $\mathcal{L} \xrightarrow{\pi} M$ over a Kähler manifold. By definition this is a holomorphic vector bundle of rank $r = 1$. For such bundles the only available Chern class is the first:

$$c_1(\mathcal{L}) = \frac{i}{2\pi} \bar{\partial}(h^{-1} \partial h) = \frac{i}{2\pi} \bar{\partial} \partial \log h \qquad\qquad (6.4.3)$$

where the 1-component real function $h(z, \bar{z})$ is some Hermitian fibre metric on $\mathcal{L}$. Let $f(z)$ be a holomorphic section of the line bundle $\mathcal{L}$: noting that under the action of the operator $\bar{\partial} \partial$ the term $\log(\bar{\xi}(\bar{z}) \xi(z))$ yields a vanishing contribution, we conclude that the formula in equation (6.4.3) for the first Chern class can be re-expressed as follows:

$$c_1(\mathcal{L}) = \frac{i}{2\pi} \bar{\partial}\,\partial \log \|\xi(z)\|^2 \tag{6.4.4}$$

where $\|\xi(z)\|^2 = h(z,\bar{z})\,\bar{\xi}(\bar{z})\,\xi(z)$ denotes the norm of the holomorphic section $\xi(z)$.

Equation (6.4.4) is the starting point for the definition of Hodge–Kähler manifolds, an essential notion in supergravity theory.

A Kähler manifold $\mathcal{M}$ is a Hodge manifold if and only if there exists a line bundle $\mathcal{L} \longrightarrow \mathcal{M}$ such that its first Chern class equals the cohomology class of the Kähler 2-form K:

$$c_1(\mathcal{L}) = [K]. \tag{6.4.5}$$

In local terms this means that there is a holomorphic section $W(z)$ such that we can write

$$K = \frac{i}{2\pi} g_{ij^*}\,dz^i \wedge d\bar{z}^{j^*} = \frac{i}{2\pi} \bar{\partial}\,\partial \log \|W(z)\|^2. \tag{6.4.6}$$

Recalling the local expression of the Kähler metric in terms of the Kähler potential $g_{ij^*} = \partial_i\,\partial_{j^*}\mathcal{K}(z,\bar{z})$, it follows from equation (6.4.6) that if the manifold $\mathcal{M}$ is a Hodge manifold, then the exponential of the Kähler potential can be interpreted as the metric $h(z,\bar{z}) = \exp(\mathcal{K}(z,\bar{z}))$ on an appropriate line bundle $\mathcal{L}$.

This structure is precisely that advocated by the Lagrangian of $N = 1$ matter coupled supergravity: the holomorphic section $W(z)$ of the line bundle $\mathcal{L}$ is what, in $N = 1$ supergravity theory, is named the superpotential and the logarithm of its norm $\log\|W(z)\|^2 = \mathcal{K}(z,\bar{z}) + \log|W(z)|^2 = G(z,\bar{z})$ is precisely the invariant function in terms of which one writes the potential and Yukawa coupling terms of the supergravity action (see [21] and for a review [16]).

6.4.2 Special Kähler manifolds: general discussion

There are in fact two kinds of special Kähler geometry: the local and the rigid one. The former describes the scalar field sector of vector multiplets in $N = 2$ supergravity while the latter describes the same sector in rigid $N = 2$ Yang–Mills theories. Since $N = 2$ includes $N = 1$ supersymmetry, local and rigid special Kähler manifolds must be compatible with the geometric structures that are respectively enforced by local and rigid $N = 1$ supersymmetry in the scalar sector. The distinction between the two cases deals with the first Chern class of the line bundle $\mathcal{L} \xrightarrow{\pi} \mathcal{M}$, whose sections are the possible superpotentials. In the local theory $c_1(\mathcal{L}) = [K]$ and this restricts $\mathcal{M}$ to be a Hodge–Kähler manifold. In the rigid theory, instead, we have $c_1(\mathcal{L}) = 0$. At the level of the Lagrangian this reflects in a different behaviour of the fermion fields. These latter are sections of $\mathcal{L}^{1/2}$ and couple to the canonical Hermitian connection defined on $\mathcal{L}$:

$$\theta \equiv h^{-1}\partial h = \frac{1}{h}\,\partial_i h\,dz^i \qquad \bar{\theta} \equiv h^{-1}\bar{\partial} h = \frac{1}{h}\,\partial_{i^*}h\,d\bar{z}^{i^*}. \tag{6.4.7}$$

In the local case where

$$[\bar{\partial}\theta] = c_1(\mathcal{L}) = [K] \tag{6.4.8}$$

the fibre metric h can be identified with the exponential of the Kähler potential and we obtain:

$$\theta = \partial \mathcal{K} = \partial_i \mathcal{K} \, dz^i \qquad \bar{\theta} = \bar{\partial}\mathcal{K} = \partial_{i^*}\mathcal{K} \, d\bar{z}^{i^*}. \tag{6.4.9}$$

In the rigid case, $\mathcal{L}$ is instead a flat bundle and its metric is unrelated to the Kähler potential. Actually one can choose a vanishing connection:

$$\theta = \bar{\theta} = 0. \tag{6.4.10}$$

The distinction between rigid and local special manifolds is the $N = 2$ generalization of this difference occurring at the $N = 1$ level.

In these lectures, since we are interested in BPS black holes and therefore in supergravity, we discuss only the case of local special geometry, leaving aside the rigid case that is relevant for $N = 2$ gauge theories. The interested reader can find further details in [26] or in [27] which contain also all the references to the original papers on special geometry.

In the $N = 2$ case, in addition to the line bundle $\mathcal{L}$ we need a flat holomorphic vector bundle $SV \longrightarrow \mathcal{M}$ whose sections can be identified with the superspace *Fermi–Fermi* components of electric and magnetic field strengths. In this way, according to the discussion of previous sections the diffeomorphisms of the scalar manifolds will be lifted to produce an action on the gauge-field strengths as well. In a supersymmetric theory where scalars and gauge fields belong to the same multiplet this is a mandatory condition. However, this symplectic bundle structure must be made compatible with the line-bundle structure already requested by $N = 1$ supersymmetry. This leads to the existence of two kinds of special geometry. Another essential distinction between the two kinds of geometry arises from the different number of vector fields in the theory. In the rigid case this number equals that of the vector multiplets so that

$$\text{No of vector fields} \equiv \bar{n} = n$$
$$\text{No of vector multiplets} \equiv n = \dim_{\mathbb{C}} \mathcal{M} \tag{6.4.11}$$
$$\text{rank}\, SV \equiv 2\bar{n} = 2n.$$

On the other hand, in the local case, in addition to the vector fields arising from the vector multiplets we have also the graviphoton coming from the graviton multiplet. Hence we conclude:

$$\text{No of vector fields} \equiv \bar{n} = n + 1$$
$$\text{No of vector multiplets} \equiv n = \dim_{\mathbb{C}} \mathcal{M} \tag{6.4.12}$$
$$\text{rank}\, SV \equiv 2\bar{n} = 2n + 2.$$

In the sequel we make extensive use of covariant derivatives with respect to the canonical connection of the line bundle $\mathcal{L}$. Let us review its normalization. As is well known there exists a correspondence between line bundles and $U(1)$-bundles. If $\exp[f_{\alpha\beta}(z)]$ is the transition function between two local trivializations of the line-bundle $\mathcal{L} \longrightarrow \mathcal{M}$, the transition function in the corresponding principal $U(1)$-bundle $\mathcal{U} \longrightarrow \mathcal{M}$ is just $\exp[\mathrm{i}\,\mathrm{Im}\,f_{\alpha\beta}(z)]$ and the Kähler potentials in two different charts are related by:

$$\mathcal{K}_\beta = \mathcal{K}_\alpha + f_{\alpha\beta} + \overline{f}_{\alpha\beta}. \tag{6.4.13}$$

At the level of connections this correspondence is formulated by setting:

$$U(1)\text{-connection} \equiv \mathcal{Q} = \mathrm{Im}\theta = -\frac{\mathrm{i}}{2}(\theta - \overline{\theta}). \tag{6.4.14}$$

If we apply the above formula to the case of the $U(1)$-bundle $\mathcal{U} \longrightarrow \mathcal{M}$ associated with the line bundle $\mathcal{L}$ whose first Chern class equals the Kähler class, we obtain

$$\mathcal{Q} = -\frac{\mathrm{i}}{2}(\partial_i \mathcal{K}\,\mathrm{d}z^i - \partial_{i*}\mathcal{K}\,\mathrm{d}\overline{z}^{i*}). \tag{6.4.15}$$

Now let $\Phi(z, \overline{z})$ be a section of $\mathcal{U}^p$. By definition its covariant derivative is

$$\nabla\Phi = (d + \mathrm{i}p\mathcal{Q})\Phi \tag{6.4.16}$$

or, in components,

$$\nabla_i \Phi = (\partial_i + \tfrac{1}{2}p\partial_i\mathcal{K})\Phi \qquad \nabla_{i*}\Phi = (\partial_{i*} - \tfrac{1}{2}p\partial_{i*}\mathcal{K})\Phi. \tag{6.4.17}$$

A covariantly holomorphic section of $\mathcal{U}$ is defined by the equation: $\nabla_{i*}\Phi = 0$. We can easily map each section $\Phi(z, \overline{z})$ of $\mathcal{U}^p$ into a section of the line bundle $\mathcal{L}$ by setting:

$$\widetilde{\Phi} = e^{-p\mathcal{K}/2}\Phi. \tag{6.4.18}$$

With this position we obtain:

$$\nabla_i \widetilde{\Phi} = (\partial_i + p\partial_i\mathcal{K})\widetilde{\Phi} \qquad \nabla_{i*}\widetilde{\Phi} = \partial_{i*}\widetilde{\Phi}. \tag{6.4.19}$$

Under the map of equation (6.4.18) covariantly holomorphic sections of $\mathcal{U}$ flow into holomorphic sections of $\mathcal{L}$ and vice versa.

6.4.3 Special Kähler manifolds: the local case

We are now ready to give the definition of local special Kähler manifolds and illustrate their properties. A first definition that does not make direct reference to the symplectic bundle is the following:

Definition 6.4.1. A Hodge–Kähler manifold is *special Kähler (of the local type)* if there exists a completely symmetric holomorphic three-index section W_{ijk} of

$(T^* \mathcal{M})^3 \otimes \mathcal{L}^2$ (and its antiholomorphic conjugate $W_{i^* j^* k^*}$) such that the following identity is satisfied by the Riemann tensor of the Levi–Civita connection:

$$\partial_{m^*} W_{ijk} = 0 \qquad \partial_m W_{i^* j^* k^*} = 0$$
$$\nabla_{[m} W_{i]jk} = 0 \qquad \nabla_{[m} W_{i^*]j^* k^*} = 0 \qquad (6.4.20)$$
$$\mathcal{R}_{i^* j \ell^* k} = g_{\ell^* j} g_{ki^*} + g_{\ell^* k} g_{ji^*} - e^{2\mathcal{K}} W_{i^* \ell^* s^*} W_{tkj} g^{s^* t}.$$

In the above equations ∇ denotes the covariant derivative with respect to both the Levi–Civita and the $U(1)$ holomorphic connection of equation (6.4.15). In the case of W_{ijk}, the $U(1)$ weight is $p = 2$.

The holomorphic sections W_{ijk} have two different physical interpretations in the case that the special manifold is utilized as scalar manifold in an $N = 1$ or $N = 2$ theory. In the first case they correspond to the Yukawa couplings of Fermi families [22]. In the second case they provide the coefficients for the anomalous magnetic moments of the gauginos, since they appear in the Pauli-terms of the $N = 2$ effective action. Out of the W_{ijk} we can construct covariantly holomorphic sections of weight 2 and -2 by setting:

$$C_{ijk} = W_{ijk} e^{\mathcal{K}} \qquad C_{i^* j^* k^*} = W_{i^* j^* k^*} e^{\mathcal{K}}. \qquad (6.4.21)$$

Next we can give the second more intrinsic definition that relies on the notion of the flat symplectic bundle. Let $\mathcal{L} \longrightarrow \mathcal{M}$ denote the complex line bundle whose first Chern class equals the Kähler form K of an n-dimensional Hodge–Kähler manifold $\mathcal{M}$. Let $SV \longrightarrow \mathcal{M}$ denote a holomorphic flat vector bundle of rank $2n + 2$ with structural group $Sp(2n + 2, \mathbb{R})$. Consider tensor bundles of the type $\mathcal{H} = SV \otimes \mathcal{L}$. A typical holomorphic section of such a bundle will be denoted by Ω and will have the following structure:

$$\Omega = \begin{pmatrix} X^{\Lambda} \\ F_{\Sigma} \end{pmatrix} \qquad \Lambda, \Sigma = 0, 1, \dots, n. \qquad (6.4.22)$$

By definition the transition functions between two local trivializations $U_i \subset \mathcal{M}$ and $U_j \subset \mathcal{M}$ of the bundle $\mathcal{H}$ have the following form:

$$\begin{pmatrix} X \\ F \end{pmatrix}_i = e^{f_{ij}} M_{ij} \begin{pmatrix} X \\ F \end{pmatrix}_j \qquad (6.4.23)$$

where f_{ij} are holomorphic maps $U_i \cap U_j \to \mathbb{C}$ while M_{ij} is a constant $Sp(2n + 2, \mathbb{R})$ matrix. For a consistent definition of the bundle the transition functions are obviously subject to the cocycle condition on a triple overlap:

$$e^{f_{ij} + f_{jk} + f_{ki}} = 1$$
$$M_{ij} M_{jk} M_{ki} = 1. \qquad (6.4.24)$$

Let $i\langle \, | \, \rangle$ be the compatible Hermitian metric on $\mathcal{H}$

$$i\langle \Omega | \overline{\Omega} \rangle \equiv -i\Omega^T \begin{pmatrix} 0 & \mathbb{1} \\ -\mathbb{1} & 0 \end{pmatrix} \overline{\Omega}. \qquad (6.4.25)$$

Definition 6.4.2. We say that a Hodge–Kähler manifold $\mathcal{M}$ is *special Kähler of the local type* if there exists a bundle $\mathcal{H}$ of the type described above such that for some section $\Omega \in \Gamma(\mathcal{H}, \mathcal{M})$ the Kähler 2-form is given by:

$$K = \frac{i}{2\pi}\partial\bar{\partial}\log(i\langle\Omega\,|\,\overline{\Omega}\rangle). \tag{6.4.26}$$

From the point of view of local properties, equation (6.4.26) implies that we have an expression for the Kähler potential in terms of the holomorphic section Ω:

$$\mathcal{K} = -\log\left(i\langle\Omega\,|\,\overline{\Omega}\rangle\right) = -\log\left[i\left(\overline{X}^\Lambda F_\Lambda - \overline{F}_\Sigma X^\Sigma\right)\right]. \tag{6.4.27}$$

The relation between the two definitions of special manifolds is obtained by introducing a non-holomorphic section of the bundle $\mathcal{H}$ according to:

$$V = \begin{pmatrix} L^\Lambda \\ M_\Sigma \end{pmatrix} \equiv e^{\mathcal{K}/2}\Omega = e^{\mathcal{K}/2}\begin{pmatrix} X^\Lambda \\ F_\Sigma \end{pmatrix} \tag{6.4.28}$$

so that equation (6.4.27) becomes:

$$1 = i\langle V\,|\,\overline{V}\rangle = i\left(\overline{L}^\Lambda M_\Lambda - \overline{M}_\Sigma L^\Sigma\right). \tag{6.4.29}$$

Since V is related to a holomorphic section by equation (6.4.28) it immediately follows that:

$$\nabla_{i^*}V = \left(\partial_{i^*} - \tfrac{1}{2}\partial_{i^*}\mathcal{K}\right)V = 0. \tag{6.4.30}$$

On the other hand, from equation (6.4.29), defining:

$$U_i = \nabla_i V = \left(\partial_i + \tfrac{1}{2}\partial_i\mathcal{K}\right)V \equiv \begin{pmatrix} f_i^\Lambda \\ h_{\Sigma|i} \end{pmatrix} \tag{6.4.31}$$

it follows that:

$$\nabla_i U_j = iC_{ijk}\,g^{k\ell^*}\,\overline{U}_{\ell^*} \tag{6.4.32}$$

where ∇_i denotes the covariant derivative containing both the Levi–Civita connection on the bundle TM and the canonical connection θ on the line bundle $\mathcal{L}$. In equation (6.4.32) the symbol C_{ijk} denotes a covariantly holomorphic ($\nabla_{\ell^*}C_{ijk} = 0$) section of the bundle $TM^3 \otimes \mathcal{L}^2$ that is totally symmetric in its indices. This tensor can be identified with the tensor of equation (6.4.21) appearing in equation (6.4.21). Alternatively, the set of differential equations:

$$\nabla_i V = U_i \tag{6.4.33}$$
$$\nabla_i U_j = iC_{ijk}g^{k\ell^*}U_{\ell^*} \tag{6.4.34}$$
$$\nabla_{i^*}U_j = g_{i^*j}V \tag{6.4.35}$$
$$\nabla_{i^*}V = 0 \tag{6.4.36}$$

with V satisfying equations (6.4.28), (6.4.29) gives yet another definition of special geometry. This is actually what one obtains from the $N = 2$ solution of superspace

Bianchi identities. In particular it is easy to find equation (6.4.21) as integrability conditions of (6.4.36). The *period matrix* is now introduced via the relations:

$$\overline{M}_\Lambda = \mathcal{N}_{\Lambda\Sigma}\overline{L}^\Sigma \qquad h_{\Sigma|i} = \mathcal{N}_{\Lambda\Sigma}f_i^\Sigma \tag{6.4.37}$$

which can be solved introducing the two $(n+1) \times (n+1)$ vectors

$$f_I^\Lambda = \begin{pmatrix} f_i^\Lambda \\ \overline{L}^\Lambda \end{pmatrix} \qquad h_{\Lambda|I} = \begin{pmatrix} h_{\Lambda|i} \\ \overline{M}_\Lambda \end{pmatrix} \tag{6.4.38}$$

and setting:

$$\mathcal{N}_{\Lambda\Sigma} = h_{\Lambda|I} \circ (f^{-1})^I{}_\Sigma. \tag{6.4.39}$$

As a consequence of its definition the matrix $\mathcal{N}$ transforms, under diffeomorphisms of the base Kähler manifold, exactly as requested by the rule in equation (6.2.38). Indeed this is the very reason why the structure of special geometry has been introduced. The existence of the symplectic bundle $\mathcal{H} \longrightarrow \mathcal{M}$ is required in order to be able to pull back the action of the diffeomorphisms on the field strengths and to construct the kinetic matrix $\mathcal{N}$.

From the previous formulae it is easy to derive a set of useful relations among which we quote the following [23]:

$$\text{Im}\,\mathcal{N}_{\Lambda\Sigma}L^\Lambda\overline{L}^\Sigma = -\tfrac{1}{2} \tag{6.4.40}$$

$$\langle V, U_i \rangle = \langle V, U_{i^*} \rangle = 0 \tag{6.4.41}$$

$$U^{\Lambda\Sigma} \equiv f_i^\Lambda f_{j^*}^\Sigma g^{ij^*} = -\tfrac{1}{2}(\text{Im}\,\mathcal{N})^{-1|\Lambda\Sigma} - \overline{L}^\Lambda L^\Sigma \tag{6.4.42}$$

$$g_{ij^*} = -i\langle U_i \mid \overline{U}_{j^*} \rangle = -2f_i^\Lambda \text{Im}\,\mathcal{N}_{\Lambda\Sigma}\overline{f}_{j^*}^\Sigma \tag{6.4.43}$$

$$C_{ijk} = \langle \nabla_i U_j \mid U_k \rangle = f_i^\Lambda \partial_j \overline{\mathcal{N}}_{\Lambda\Sigma}f_k^\Sigma = (\mathcal{N} - \overline{\mathcal{N}})_{\Lambda\Sigma}f_i^\Lambda \partial_j f_k^\Sigma. \tag{6.4.44}$$

In particular equations (6.4.44) express the Kähler metric and the anomalous magnetic moments in terms of symplectic invariants. It is clear from our discussion that nowhere we have assumed the base Kähler manifold to be a homogeneous space. So, in general, special manifolds are not homogeneous spaces. Yet there is a subclass of homogenous special manifolds. The homogeneous symmetric ones were classified by Cremmer and Van Proeyen in [24] and are displayed in table 6.2.

It goes without saying that for homogeneous special manifolds the two constructions of the period matrix, that provided by the master formula in equation (6.3.17) and that given by equation (6.4.39), must agree. In subsection 6.4.5 we verify it in the case of the manifolds $ST[2, n]$ that correspond to the second infinite family of homogeneous special manifolds displayed in table 6.2.

Anyhow, since special geometry guarantees the existence of a kinetic period matrix with the correct covariance property it is evident that to each special manifold we can associate a duality covariant bosonic Lagrangian of the type considered in equation (6.2.5). However, special geometry contains more structures than just

Table 6.2. Homogeneous symmetric special manifolds.

n	G/H	$Sp(2n+2)$	Symp. rep. of G
1	$\frac{SU(1,1)}{U(1)}$	$Sp(4)$	$\underline{4}$
n	$\frac{SU(1,n)}{SU(n)\times U(1)}$	$Sp(2n+2)$	$\mathbf{n+1}\oplus\mathbf{n+1}$
$n+1$	$\frac{SU(1,1)}{U(1)}\otimes\frac{SO(2,n)}{SO(2)\times SO(n)}$	$Sp(2n+4)$	$\underline{2}\otimes(\mathbf{n+2}\oplus\mathbf{n+2})$
six	$\frac{Sp(6,\mathbf{R})}{SU(3)\times U(1)}$	$Sp(14)$	$\underline{14}$
9	$\frac{SU(3,3)}{S(U(3)\times U(3))}$	$Sp(20)$	$\underline{20}$
15	$\frac{SO^{\ast}(12)}{SU(6)\times U(1)}$	$Sp(32)$	$\underline{32}$
27	$\frac{E_{7(-6)}}{E_6\times SO(2)}$	$Sp(56)$	$\underline{56}$

the period matrix $\mathcal{N}$ and the scalar metric $g_{ij^{\ast}}$. All the other items of the construction do have a place and play an essential role in the supergravity Lagrangian and the supersymmetry transformation rules. We shall have to manipulate them also in discussing the BPS black holes.

6.4.4 Special Kähler manifolds: the issue of special coordinates

So far no privileged coordinate system has been chosen on the base Kähler manifold $\mathcal{M}$ and no mention has been made of the holomorphic prepotential $F(X)$ that is ubiquitous in the $N = 2$ literature. The simultaneous avoidance of privileged coordinates and of the prepotential is not accidental. Indeed, when the definition of special Kähler manifolds is given in intrinsic terms, as we did in the previous subsection, the holomorphic prepotential $F(X)$ can be dispensed with. Whether a prepotential $F(X)$ exists or not depends on the choice of a symplectic gauge which is immaterial in the Abelian theory but not in the gauged one. Actually, in the local case, it appears that some physically interesting cases are precisely instances where $F(X)$ does not exist. In contrast, the prepotential $F(X)$ seems to be a necessary ingredient in the tensor calculus constructions of $N = 2$ theories that for this reason are not completely general. This happens because tensor calculus uses special coordinates from the very start. Let us then see how the notion of $F(X)$ emerges if we resort to special coordinate systems.

Note that under a Kähler transformation $\mathcal{K} \rightarrow \mathcal{K} + f(z) + \overline{f}(\overline{z})$ the holomorphic section transforms, in the local case, as $\Omega \rightarrow \Omega\,e^{-f}$, so that we have $X^{\Lambda} \rightarrow X^{\Lambda}\,e^{-f}$. This means that, at least locally, the upper half of Ω (associated with the electric field strengths) can be regarded as a set X^{Λ} of homogeneous coordinates on $\mathcal{M}$, provided that the Jacobian matrix

$$e_i^I(z) = \partial_i\left(\frac{X^I}{X^0}\right) \qquad a = 1,\ldots,n \qquad (6.4.45)$$

is invertible. In this case, for the lower part of the symplectic section Ω we obtain $F_\Lambda = F_\Lambda(X)$. Recalling equations (6.4.41), in particular:

$$0 = \langle V \mid U_i \rangle = X^\Lambda \partial_i F_\Lambda - \partial_i X^\Lambda F_\Lambda \qquad (6.4.46)$$

we obtain:

$$X^\Sigma \partial_\Sigma F_\Lambda(x) = F_\Lambda(X) \qquad (6.4.47)$$

so that we can conclude:

$$F_\Lambda(X) = \frac{\partial}{\partial X^\Lambda} F(X) \qquad (6.4.48)$$

where $F(X)$ is a homogeneous function of degree 2 of the homogeneous coordinates X^Λ. Therefore, when the determinant of the Jacobian (6.4.45) is nonvanishing, we can use the *special coordinates*:

$$t^I \equiv \frac{X^I}{X^0} \qquad (6.4.49)$$

and the whole geometric structure can be derived by a single holomorphic prepotential:

$$\mathcal{F}(t) \equiv (X^0)^{-2} F(X). \qquad (6.4.50)$$

In particular, equation (6.4.27) for the Kähler potential becomes

$$\mathcal{K}(t, \bar{t}) = -\log i \left[2 \left(\mathcal{F} - \overline{\mathcal{F}} \right) - \left(\partial_I \mathcal{F} + \partial_{I^*} \overline{\mathcal{F}} \right) \left(t^I - \bar{t}^{I^*} \right) \right] \qquad (6.4.51)$$

while equation (6.4.44) for the magnetic moments simplifies into

$$W_{IJK} = \partial_I \partial_J \partial_K \mathcal{F}(t). \qquad (6.4.52)$$

Finally we note that in the rigid case the Jacobian from a generic parametrization to special coordinates

$$e_i^I(z) = \partial_i \left(\frac{X^I}{X^0} \right) = A + B\overline{\mathcal{N}} \qquad (6.4.53)$$

cannot have zero eigenvalues, and therefore the function F always exist. In this case the matrix $\overline{\mathcal{N}}$ coincides with $\frac{\partial^2 F}{\partial X^I \partial X^J}$.

6.4.5 The special geometry of the $\mathcal{ST}[2,n]$ manifolds

When we studied the symplectic embeddings of the $\mathcal{ST}[m, n]$ manifolds, defined by equation (6.3.20), a study that led us to the general formula in equation (6.3.35), we remarked that the subclass $\mathcal{ST}[2, n]$ constitutes a family of special Kähler manifolds: actually a quite relevant one. Here we survey the special geometry of this class.

Consider a standard parametrization of the $SO(2,n)/SO(2) \times SO(n)$ manifold, for instance that in equation (6.3.32). In the $m = 2$ case we can introduce a canonical complex structure on the manifold by setting:

$$\Phi^\Lambda(X) \equiv \frac{1}{\sqrt{2}}(L_0^\Lambda + i L_1^\Lambda) \qquad (\Lambda = 0, 1, a \quad a = 2, \ldots, n+1). \quad (6.4.54)$$

The relations satisfied by the upper two rows of the coset representative (a consequence of $L(X)$ being pseudo-orthogonal with respect to the metric $\eta_{\Lambda\Sigma} = \text{diag}(+, +, -, \ldots, -)$):

$$L_0^\Lambda L_0^\Sigma \eta_{\Lambda\Sigma} = 1 \qquad L_0^\Lambda L_1^\Sigma \eta_{\Lambda\Sigma} = 0 \qquad L_1^\Lambda L_1^\Sigma \eta_{\Lambda\Sigma} = 1 \quad (6.4.55)$$

can be summarized into the complex equations:

$$\overline{\Phi}^\Lambda \Phi^\Sigma \eta_{\Lambda\Sigma} = 1 \qquad \Phi^\Sigma \eta_{\Lambda\Sigma} = 0. \quad (6.4.56)$$

Equations (6.4.56) are solved by posing:

$$\Phi^\Lambda = \frac{X^\Lambda}{\sqrt{\overline{X}^\Lambda X^\Sigma \eta_{\Lambda\Sigma}}} \quad (6.4.57)$$

where X^Λ denotes any set of complex parameters, determined up to an overall multiplicative constant and satisfying the constraint:

$$X^\Lambda X^\Sigma \eta_{\Lambda\Sigma} = 0. \quad (6.4.58)$$

In this way we have proved the identification, as differentiable manifolds, of the coset space G/H where $G = SO(2,n)$ and $H = SO(2) \times SO(n)$ with the vanishing locus of the quadric in equation (6.4.58). Taking any holomorphic solution of equation (6.4.58), for instance:

$$X^\Lambda(y) \equiv \begin{pmatrix} 1/2\,(1 + y^2) \\ i/2\,(1 - y^2) \\ y^a \end{pmatrix} \quad (6.4.59)$$

where y^a is a set of n independent complex coordinates, inserting it into equation (6.4.57) and comparing with equation (6.4.54) we obtain the relation between whatever coordinates we had previously used to write the coset representative $L(X)$ and the complex coordinates y^a. In other words we can regard the matrix L as a function of the y^a that are named the Calabi Visentini coordinates [31].

Consider in addition the *axion–dilaton* field S that parametrizes the $SU(1,1)/U(1)$ coset according to equation (6.3.31). The special geometry of the manifold $ST[2,n]$ is completely specified by writing the holomorphic symplectic section Ω as follows [32]:

$$\Omega(y, S) = \begin{pmatrix} X^\Lambda \\ F_\Lambda \end{pmatrix} = \begin{pmatrix} X^\Lambda(y) \\ S\,\eta_{\Lambda\Sigma} X^\Sigma(y) \end{pmatrix}. \quad (6.4.60)$$

Notice that with the above choice, it is not possible to describe F_Λ as derivatives of any prepotential. Yet everything else can be calculated utilizing the formulae we presented in the text. The Kähler potential is:

$$\mathcal{K} = \mathcal{K}_1(S) + \mathcal{K}_2(y) = -\log[\,i\,(\overline{S} - S)] - \log X^T \eta X. \qquad (6.4.61)$$

The Kähler metric is block diagonal:

$$g_{ij^*} = \begin{pmatrix} g_{S\overline{S}} & 0 \\ 0 & g_{a\overline{b}} \end{pmatrix} \qquad \left\{ \begin{array}{l} g_{S\overline{S}} = \partial_S \partial_{\overline{S}} \mathcal{K}_1 = \dfrac{-1}{(S - S)^2} \\ g_{a\overline{b}}(y) = \partial_a \partial_{\overline{b}} \mathcal{K}_2 \end{array} \right. \qquad (6.4.62)$$

as expected. The anomalous magnetic moment-Yukawa couplings C_{ijk} $(i = S, a)$ have a very simple expression in the chosen coordinates:

$$C_{Sab} = -\exp[\mathcal{K}]\,\delta_{ab} \qquad (6.4.63)$$

all the other components being zero.

Using the definition of the *period matrix* given in equation (6.4.39) we obtain

$$\mathcal{N}_{\Lambda\Sigma} = (S - \overline{S}) \frac{X_\Lambda \overline{X}_\Sigma + \overline{X}_\Lambda X_\Sigma}{\overline{X}^T \eta X} + \overline{S}\eta_{\Lambda\Sigma}. \qquad (6.4.64)$$

In order to see that equation (6.4.64) just coincides with equation (6.3.35) it suffices to note that as a consequence of its definition (6.4.54) and of the pseudo-orthogonality of the coset representative $L(X)$, the vector Φ^Λ satisfies the following:

$$\Phi^\Lambda \overline{\Phi}^\Sigma + \Phi^\Sigma \overline{\Phi}^\Lambda = \tfrac{1}{2} L_\Gamma^\Lambda L_\Delta^\Sigma \left(\delta^{\Gamma\Delta} + \eta^{\Gamma\Delta} \right). \qquad (6.4.65)$$

Inserting equation (6.4.65) into equation (6.4.64), formula (6.3.35) is retrieved.

This completes the proof that the choice (6.4.60) of the special geometry holomorphic section corresponds to the symplectic embedding (6.3.27) and (6.3.29) of the coset manifold $ST[2, n]$. In this symplectic gauge the symplectic transformations of the isometry group are the simplest possible ones and the entire group $SO(2, n)$ is represented by means of *classical* transformations that do not mix electric fields with magnetic fields. The disadvantage of this basis, if any, is that there is no holomorphic prepotential. To find an $F(X)$ it suffices to make a symplectic rotation to a different basis.

If we set:

$$\begin{aligned} X^1 &= \tfrac{1}{2}(1 + y^2) = -\tfrac{1}{2}(1 - \eta_{ij} t^i t^j) \\ X^2 &= i\tfrac{1}{2}(1 - y^2) = t^2 \\ X^a &= y^a = t^{2+a} \qquad a = 1, \ldots, n - 1 \\ X^{a=n} &= y^n = \tfrac{1}{2}(1 + \eta_{ij} t^i t^j) \end{aligned} \qquad (6.4.66)$$

where

$$\eta_{ij} = \mathrm{diag}\,(+, -, \ldots, -) \qquad i, j = 2, \ldots, n + 1 \qquad (6.4.67)$$

then we can show that $\exists\, \mathcal{C} \in Sp(2n+2, \mathbb{R})$ such that:

$$
\mathcal{C}\begin{pmatrix} X^\Lambda \\ S\eta_{\Lambda\Sigma}\,X^\Lambda \end{pmatrix} = \exp[\varphi(t)] \begin{pmatrix} 1 \\ S \\ t^i \\ 2\mathcal{F} - t^i \frac{\partial}{\partial t^i}\mathcal{F} - S\frac{\partial}{\partial S}\mathcal{F} \\ S\frac{\partial}{\partial S}\mathcal{F} \\ \frac{\partial}{\partial t^i}\mathcal{F} \end{pmatrix}
\tag{6.4.68}
$$

with

$$
\begin{aligned}
\mathcal{F}(S,t) &= \tfrac{1}{2}S\,\eta_{ij}t^i t^j = \tfrac{1}{2}d_{IJK}t^I t^J t^K \\
t^1 &= S \\
d_{IJK} &= \begin{cases} d_{1jk} = \eta_{ij} \\ 0 \text{ otherwise} \end{cases}
\end{aligned}
\tag{6.4.69}
$$

and

$$
W_{IJK} = d_{IJK} = \frac{\partial^3 \mathcal{F}(S,t^i)}{\partial t^I \partial t^J \partial t^K}.
\tag{6.4.70}
$$

This means that in the new basis the symplectic holomorphic section $\mathcal{C}\Omega$ can be derived from the following cubic prepotential:

$$
F(X) = \frac{1}{3!} \frac{d_{IJK}\,X^I X^J X^K}{X_0}.
\tag{6.4.71}
$$

For instance in the case $n = 1$ the matrix which does such a job is:

$$
\mathcal{C} = \begin{pmatrix}
1 & 0 & -1 & 0 & 0 & 0 \\
0 & 0 & 0 & 1 & 0 & 1 \\
0 & -1 & 0 & 0 & 0 & 0 \\
0 & 0 & 0 & \frac{1}{2} & 0 & -\frac{1}{2} \\
-\frac{1}{2} & 0 & -\frac{1}{2} & 0 & 0 & 0 \\
0 & 0 & 0 & 0 & -1 & 0
\end{pmatrix}.
\tag{6.4.72}
$$

6.5 The bosonic Lagrangian and the supersymmetry transformation rules of $N = 2$ supergravity

Using special Kähler geometry as described in the previous section we can now write both the bosonic action and the supersymmetry transformation rules of the fermions pertaining to $N = 2$ supergravity, namely the ingredients needed to write down the equations defining $N = 2$ BPS black holes.

In the complex notation appropriate to special geometry and using the definitions (6.2.9) the action can be written as follows

$$
\mathcal{L}^{SUGRA|Bose}_{ungauged} = \sqrt{-g}[R[g] + g_{ij^*}(z,\bar{z})\,\partial^\mu z^i\,\partial_\mu \bar{z}^{j^*}
$$
$$
+ i(\overline{\mathcal{N}}_{\Lambda\Sigma}\mathcal{F}^{-\Lambda}_{\mu\nu}\mathcal{F}^{-\Sigma|\mu\nu} - \mathcal{N}_{\Lambda\Sigma}\mathcal{F}^{+\Lambda}_{\mu\nu}\mathcal{F}^{+\Sigma|\mu\nu})].
\tag{6.5.1}
$$

The fermion fields of the theory are of two kinds:

(i)   the gravitino $\psi_{A\mu}$
(ii)   the gaugino λ^{iA}.

The gravitino $\psi_{A\mu}$ is an $SU(2)$ doublet ($A = 1, 2$) of spin $3/2$ spinor–vectors. Here the group $SU(2)$ is the automorphism group of the $N = 2$ supersymmetry algebra, usually called R-symmetry. Following the conventions of [16] and [27], the position of the $SU(2)$ index, whether upper or lower, is related to the chirality projection on the fields. For the gravitino doublet we have left chirality:

$$\gamma_5\, \psi_{A\mu} = \psi_{A\mu}. \tag{6.5.2}$$

The complex conjugate doublet, transforming under the conjugate representation of $SU(2)$ contains the right chirality projection of the same Majorana field:

$$\gamma_5\, \psi_\mu^A = -\psi_\mu^A. \tag{6.5.3}$$

On the other hand the gauginos λ^{iA}, are $SU(2)$ doublets of spin $1/2$ fields that carry also a contravariant world-index of the special Kähler manifold. In other words the gauginos, in addition to being sections of the spin bundle on space–time, are also sections of the holomorphic tangent bundle TSM. The chirality of the gaugino with upper $SU(2)$ doublet index is left:

$$\gamma_5\, \lambda^{iA} = \lambda^{iA}. \tag{6.5.4}$$

The right chirality projection $\lambda_A^{i^*} = -\gamma_5\, \lambda_A^{i^*}$ transforms under $SU(2)$ in the same way as the left-projection of the gravitino but it is a section of the antiholomorphic tangent bundle $\overline{T}SM$.

Correspondingly in a purely bosonic background the supersymmetry transformation rules of the fermion fields are the following ones:

$$\delta\, \Psi_{A\mu} = \mathcal{D}_\mu\, \epsilon_A + \epsilon_{AB}\, T_{\mu\nu}^-\, \gamma^\nu \epsilon^B \tag{6.5.5}$$

$$\delta\, \lambda^{iA} = \mathrm{i}\, \nabla_\mu z^i\, \gamma^\mu \epsilon^A + G_{\mu\nu}^{-i}\gamma^{\mu\nu}\epsilon_B \epsilon^{AB} \tag{6.5.6}$$

where the local supersymmetry parameter is also split into chirality projections transforming in the doublet or conjugate doublet representation of $SU(2)$:

$$\gamma_5\, \epsilon_A = \epsilon_A \qquad \gamma_5\, \epsilon^A = -\epsilon^A \tag{6.5.7}$$

and where we have defined:

$$T_{\mu\nu}^- = 2\mathrm{i}(\mathrm{Im}\,\mathcal{N})_{\Lambda\Sigma} L^\Sigma F_{\mu\nu}^{\Lambda-} \tag{6.5.8}$$

$$T_{\mu\nu}^+ = 2\mathrm{i}(\mathrm{Im}\,\mathcal{N})_{\Lambda\Sigma} \overline{L}^\Sigma F_{\mu\nu}^{\Lambda+} \tag{6.5.9}$$

$$G_{\mu\nu}^{i-} = -g^{ij^*}\overline{f}_{j^*}^{\Gamma}(\mathrm{Im}\,\mathcal{N})_{\Gamma\Lambda} F_{\mu\nu}^{\Lambda-} \tag{6.5.10}$$

$$G_{\mu\nu}^{i^*+} = -g^{i^*j} f_j^{\Gamma}(\mathrm{Im}\,\mathcal{N})_{\Gamma\Lambda} F_{\mu\nu}^{\Lambda+}. \tag{6.5.11}$$

The scalar field dependent combinations of fields strengths appearing in the fermion supersymmetry transformation rules that we listed above in equations (6.5.8), (6.5.9), (6.5.10), (6.5.11) have a profound meaning and play a key role in the physics of BPS black holes. The combination $T_{\mu\nu}^-$ is named the graviphoton field strength and its integral over a 2-sphere at infinity gives the value of *the central charge Z* of the $N = 2$ supersymmetry algebra. The combination $G_{\mu\nu}^{i-}$ is named the matter field strength. Evaluating its integral on a 2-sphere at infinity one obtains the so called *matter charges* Z^i. These are the covariant derivatives of the central charge Z with respect to the *moduli*, namely the vector multiplet scalars z^i.

This will become clear in our general discussion of the $N = 2$ BPS black holes that we are finally in a position to address, having clarified all the necessary geometric ingredients.

6.6 $N = 2$ BPS black holes: general discussion

In this section we consider the general properties of BPS saturated black holes in the context of $N = 2$ supergravity. As our analysis will reveal these properties are completely rooted in the special Kähler geometric structure of the mother theory. In particular the entropy of the black hole is related to the central charge, namely to the integral of the graviphoton field strength evaluated for very special values of the scalar fields z^i. These special values, named the *fixed scalars* z^i_{fix}, are functions solely of the electric and magnetic charges $\{q_\Sigma, p^\Lambda\}$ of the hole and are attained by the scalars $z^i(r)$ at the hole horizon $r = 0$.

To illustrate how this happens let us finally go back to the *ansatz* (6.1.1) for the metric and to the *ansatz* (6.1.2) for the vector field strengths. It is convenient to rephrase the same *ansatz* in the complex formalism well adapted to the $N = 2$ theory. To this effect we begin by constructing a 2-form which is *anti-self-dual* in the background of the metric (6.1.1) and whose integral on the 2-sphere at infinity S_∞^2 is normalized to 2π. A short calculation yields:

$$E^- = i\frac{e^{2U(r)}}{r^3}\, dt \wedge \vec{x} \cdot d\vec{x} + \tfrac{1}{2}\frac{x^a}{r^3}\, dx^b \wedge dx^c \epsilon_{abc}$$

$$2\pi = \int_{S_\infty^2} E^- \tag{6.6.1}$$

and with a little additional effort one obtains:

$$E_{\mu\nu}^- \gamma^{\mu\nu} = 2\,i\frac{e^{2U(r)}}{r^3}\, \gamma_a x^a\, \gamma_0 \tfrac{1}{2}\left[1 + \gamma_5\right] \tag{6.6.2}$$

which will prove of great help in the unfolding of the supersymmetry transformation rules.

Next, introducing the following complex combination of the magnetic charge p^Λ and of the radial function $\ell^\Sigma(r)$ defined by equation (6.1.2):

$$t^\Lambda(r) = 2\pi(p^\Lambda + i\ell^\Lambda(r)) \tag{6.6.3}$$

we can rewrite the *ansatz* (6.1.2) as:

$$F^{-|\Lambda} = \frac{t^{\Lambda}}{4\pi} E^{-}$$

$$(6.6.4)$$

and we retrieve the original formulae from:

$$\begin{aligned}
F^{\Lambda} &= 2\,\mathrm{Re}\, F^{-|\Lambda} = \frac{p^{\Lambda}}{2r^3}\epsilon_{abc}x^a\,\mathrm{d}x^b \wedge \mathrm{d}x^c \\
&\quad - \frac{\ell^{\Lambda}(r)}{r^3}\,e^{2\mathcal{U}}\,\mathrm{d}t \wedge \vec{x} \cdot \mathrm{d}\vec{x} \\
\widetilde{F}^{\Lambda} &= -2\,\mathrm{Im}\, F^{-|\Lambda} = -\frac{\ell^{\Lambda}(r)}{2r^3}\epsilon_{abc}x^a\mathrm{d}x^b \wedge \mathrm{d}x^c \\
&\quad - \frac{p^{\Lambda}}{r^3}\,e^{2\mathcal{U}}\,\mathrm{d}t \wedge \vec{x} \cdot \mathrm{d}\vec{x}.
\end{aligned}$$

$$(6.6.5)$$

Before proceeding further it is convenient to define the electric and magnetic charges of the hole as it is appropriate in any electromagnetic theory. Recalling the general form of the field equations and of the Bianchi identities as given in (6.2.15) we see that the field strengths $\mathcal{F}_{\mu\nu}$ and $\mathcal{G}_{\mu\nu}$ are both closed 2-forms, since their duals are divergenceless. Hence we can invoke Gauss theorem and claim that their integral on a closed space–like 2-sphere does not depend on the radius of the sphere. These integrals are the electric and magnetic charges of the hole that, in a quantum theory, we expect to be quantized. We set:

$$q_{\Lambda} \equiv \frac{1}{4\pi} \int_{S^2_{\infty}} G_{\Lambda|\mu\nu}\,\mathrm{d}x^{\mu} \wedge \mathrm{d}x^{\nu}$$

$$(6.6.6)$$

$$p^{\Sigma} \equiv \frac{1}{4\pi} \int_{S^2_{\infty}} F^{\Sigma}_{\mu\nu}\,\mathrm{d}x^{\mu} \wedge \mathrm{d}x^{\nu}.$$

$$(6.6.7)$$

If rather than the integral of G_{Λ} we were to calculate the integral of $\widetilde{F}^{\Lambda}$ which is not a closed form we would obtain a function of the radius:

$$4\pi\,\ell^{\Lambda}(r) = -\int_{S^2_r} \widetilde{F}^{\Lambda} = 2\,\mathrm{Im}\,t^{\Lambda}.$$

$$(6.6.8)$$

Consider now the supersymmetry transformations rules (6.5.5), (6.5.6) and write the BPS condition:

$$0 = \mathcal{D}_{\mu}\,\xi_A + \epsilon_{AB}\,T^{-}_{\mu\nu}\,\gamma^{\nu}\xi^B$$

$$(6.6.9)$$

$$0 = i\,\nabla_{\mu}\,z^i\,\gamma^{\mu}\xi^A + G^{-i}_{\mu\nu}\gamma^{\mu\nu}\xi_B\epsilon^{AB}$$

$$(6.6.10)$$

where the Killing spinor $\xi_A(r)$ satisfies equation (6.1.5) and is of the form of a single radial function times a constant spinor:

$$\begin{aligned}
\xi_A(r) &= e^{f(r)}\chi_A \qquad \chi_A = \text{constant} \\
\gamma_0\chi_A &= \pm i\epsilon_{AB}\chi^B.
\end{aligned}$$

$$(6.6.11)$$

Inserting equations (6.5.8), (6.5.10), (6.6.11) into equations (6.6.9), (6.6.10) and using the result (6.6.2), with a little work we obtain the first order differential equations:

$$\frac{dz^i}{dr} = \mp \left(\frac{e^{U(r)}}{4\pi r^2}\right) g^{ij^*}\overline{f}_{j^*}^\Lambda (\mathcal{N} - \overline{\mathcal{N}})_{\Lambda\Sigma} t^\Sigma$$

$$= \mp \left(\frac{e^{U(r)}}{4\pi r^2}\right) g^{ij^*}\nabla_{j^*}\overline{Z}(z,\overline{z},p,q) \qquad (6.6.12)$$

$$\frac{dU}{dr} = \mp \left(\frac{e^{U(r)}}{r^2}\right)(M_\Sigma p^\Sigma - L^\Lambda q_\Lambda) = \mp \left(\frac{e^{U(r)}}{r^2}\right) Z(z,\overline{z},p,q)$$

$$\qquad (6.6.13)$$

where

(i) $\mathcal{N}_{\Lambda\Sigma}(z,\overline{z})$ is the period matrix of special geometry defined by equation (6.4.37),

(ii) the vector $V = (L^\Lambda(z,\overline{z}), M_\Sigma(z,\overline{z}))$ is the covariantly holomorphic section of the symplectic bundle entering the definition of a Special Kähler manifold (see equation (6.4.28)

(iii)

$$Z(z,\overline{z},p,q) \equiv (M_\Sigma p^\Sigma - L^\Lambda q_\Lambda) \qquad (6.6.14)$$

is the local realization on the scalar manifold $\mathcal{SM}$ of the central charge of the $N = 2$ superalgebra and

(iv)

$$Z^i(z,\overline{z},p,q) \equiv g^{ij^*}\nabla_{j^*}\overline{Z}(z,\overline{z},p,q) \qquad (6.6.15)$$

are the central charges associated with the matter vectors, the so-called matter central charges.

To obtain equations (6.6.12), (6.6.13) we made use of the following two properties:

$$0 = \overline{h}_{j^*|\Lambda} t^{*\Sigma} - \overline{f}_{j^*}^\Lambda \mathcal{N}_{\Lambda\Sigma} t^{*\Sigma}$$
$$0 = M_\Sigma t^{*\Sigma} - L^\Lambda \mathcal{N}_{\Lambda\Sigma} t^{*\Sigma} \qquad (6.6.16)$$

which are a direct consequence of the definition (6.4.37) of the period matrix. The electric charges $\ell^\Lambda(r)$ defined in (6.6.8) are *modulus dependent* charges which are functions of the radial direction through the moduli z^i. On the other hand, the *modulus independent* electric charges q_Λ in equations (6.6.13), (6.6.12) are those defined by equation (6.6.6) which, together with p^Λ fulfil a Dirac quantization condition. Their definition allows them to be expressed in terms of of $t^\Lambda(r)$ as follows:

$$q_\Lambda = \frac{1}{2\pi}\text{Re}(\mathcal{N}(z(r),\overline{z}(r))t(r))_\Lambda. \qquad (6.6.17)$$

Equation (6.6.17) may be inverted in order to find the modulus dependence of $\ell_\Lambda(r)$. The independence of q_Λ on r is a consequence of one of the Maxwell's equations:

$$\partial_a \left(\sqrt{-g} \widetilde{G}^{a0|\Lambda}(r) \right) = 0 \Rightarrow \partial_r \, \mathrm{Re}(\mathcal{N}(z(r), \overline{z}(r)) t(r))^\Lambda = 0. \tag{6.6.18}$$

6.6.1   Fixed scalars and entropy

In this way we have reduced the condition that the black hole should be a BPS saturated state to a pair of first order differential equations for the metric scale factor $U(r)$ and for the scalar fields $z^i(r)$. To obtain explicit solutions one should specify the special Kähler manifold one is working with, namely the specific Lagrangian model. There are, however, some very general and interesting conclusions that can be drawn in a model-independent way. They are just consequences of the fact that these BPS conditions are *first order differential equations*. Because of that there are fixed points (see the papers [33, 35, 34]) namely values either of the metric or of the scalar fields which, once attained in the evolution parameter r (= the radial distance) persist indefinitely. The fixed point values are just the zeros of the right-hand side in either of the coupled equations (6.6.13) and (6.6.12). The fixed point for the metric equation is $r = \infty$, which corresponds to its asymptotic flatness. The fixed point for the moduli is $r = 0$. So, independently from the initial data at $r = \infty$ that determine the details of the evolution, the scalar fields flow into their fixed point values at $r = 0$, which, as we will show, turns out to be a horizon. Indeed in the vicinity of $r = 0$ also the metric takes the universal form of an $AdS_2 \times S^2$, Bertotti–Robinson metric.

Let us see this more closely.

To begin with we consider the equations determining the fixed point values for the moduli and the universal form attained by the metric at the modulus fixed point:

$$0 = -g^{ij^\star} \overline{f}^\Gamma_{j^\star} (\mathrm{Im}\mathcal{N})_{\Gamma\Lambda} t^\Lambda(0) \tag{6.6.19}$$

$$\frac{dU}{dr} \cong \mp \left(\frac{e^{U(r)}}{r^2} \right) Z(z_{fix}, \overline{z}_{fix}, p, q). \tag{6.6.20}$$

Multiplying equation (6.6.19) by f_i^Σ using the identity (6.4.42) and the definition (6.6.14) of the central charge we conclude that at the fixed point the following condition is true:

$$0 = -\frac{1}{2} \frac{t^\Lambda}{4\pi} - \frac{Z_{fix} \overline{L}^\Lambda_{fix}}{8\pi}. \tag{6.6.21}$$

In terms of the previously defined electric and magnetic charges (see equations (6.6.6), (6.6.7), (6.6.17)) equation (6.6.21) can be rewritten as:

$$p^\Lambda = \mathrm{i}(Z_{fix} \overline{L}^\Lambda_{fix} - \overline{Z}_{fix} L^\Lambda_{fix}) \tag{6.6.22}$$

$$q_\Sigma = i(Z_{fix} \overline{M}_\Sigma^{fix} - \overline{Z}_{fix} M_\Sigma^{fix}) \qquad (6.6.23)$$

$$Z_{fix} = M_\Sigma^{fix} p^\Sigma - L_{fix}^\Lambda q_\Lambda \qquad (6.6.24)$$

which can be regarded as algebraic equations determining the value of the scalar fields at the fixed point as functions of the electric and magnetic charges p^Λ, q_Σ:

$$L_{fix}^\Lambda = L^\Lambda(p, q) \longrightarrow Z_{fix} = Z(p, q) = \text{const.} \qquad (6.6.25)$$

In the vicinity of the fixed point the differential equation for the metric becomes:

$$\pm \frac{dU}{dr} = \frac{Z(p, q)}{r^2} e^{U(r)} \qquad (6.6.26)$$

which has the approximate solution:

$$\exp[U(r)] \xrightarrow{r \to 0} \text{const} + \frac{Z(p, q)}{r}. \qquad (6.6.27)$$

Hence, near $r = 0$ the metric (6.1.1) becomes of the Bertotti–Robinson type (see equation (6.1.18)) with Bertotti–Robinson mass given by:

$$m_{BR}^2 = |Z(p, q)|^2. \qquad (6.6.28)$$

In the metric (6.1.18) the surface $r = 0$ is lightlike and corresponds to a horizon since it is the locus where the Killing vector generating time translations $\frac{\partial}{\partial t}$, which is timelike at spatial infinity $r = \infty$, becomes lightlike. The horizon $r = 0$ has a finite area given by:

$$\text{area}_H = \int_{r=0} \sqrt{g_{\theta\theta} g_{\phi\phi}} \, d\theta \, d\phi = 4\pi \, m_{BR}^2. \qquad (6.6.29)$$

Hence, independently from the details of the considered model, the BPS saturated black holes in an $N = 2$ theory have a Bekenstein–Hawking entropy given by the following horizon area:

$$\frac{\text{area}_H}{4\pi} = |Z(p, q)|^2 \qquad (6.6.30)$$

the value of the central charge being determined by equations (6.6.24). Such equations can also be seen as the variational equations for the minimization of the horizon area as given by (6.6.30), if the central charge is regarded as a function of both the scalar fields and the charges:

$$\text{area}_H(z, \bar{z}) = 4\pi |Z(z, \bar{z}, p, q)|^2$$
$$\frac{\delta \text{area}_H}{\delta z} = 0 \longrightarrow z = z_{fix}. \qquad (6.6.31)$$

This viewpoint will be pursued in the next chapter where we consider the general properties of *supergravity central charges* and their extremization.

Chapter 7

The structure of supergravity central charges and the black hole entropy

7.1 Introduction to central charges

The present one is the central chapter in this series of lectures. As we explained in the introduction the relevance of extremal black holes for string theory is that in the context of a supersymmetric theory the extremality condition, that is the coincidence of the two horizons, can be reinterpreted as the statement that the black hole is a BPS saturated state where the ADM mass is equal to the modulus of the supersymmetry central charge.

Actually in N-extended supersymmetry the central charge is an antisymmetric tensor Z_{AB} that admits several skew eigenvalues (see equation (4.2.4)) so that the discussion of the BPS condition becomes more involved depending on the structure of these skew eigenvalues. In this chapter we study the general properties of the supergravity central charges and their relation with the horizon area, namely with the black hole entropy. A first instance of such a relation was already derived in the previous chapter in the context of $N = 2$ supergravity where we obtained the fundamental relations (6.6.30) and (6.6.25) expressing the horizon area as the modulus square of the central charge evaluated at the extremum value of the scalars. In the present chapter we put this $N = 2$ result in a much more general perspective and we illustrate its deep meaning. Indeed, after illustrating the general structure and properties of the supergravity central charges, we show that with any Lagrangian of type (6.1.21) we can associate an effective one-dimensional field theory that governs the radial dependence of the black hole metric and of the scalar fields. The interactions in such an effective theory are determined by a specific potential, *the geodesic potential* $V(\phi, \vec{Q})$ that depends on the scalar fields and on the electric and magnetic charges of the hole. Finite horizon area black holes are characterized by scalar fields that are regular on the horizon and reach at $r = 0$ extremum values of the geodesic potential. Actually the horizon area and, hence, the black hole entropy is given by the extremum value of the geodesic potential. When the theory is supersymmetric the geodesic potential can be reexpressed as a quadratic positive definite form in the supersymmetry central charges. Hence extremizing the potential and finding the black hole entropy means to extremize such a quadratic form in the central charges. Using the formal structure of the supergravity central charges and their invariance under the U-duality group one can derive very general properties of the horizon area. Indeed we can show that it is a topological invariant depending solely on the vector $\vec{Q}$ of *quantized electric and magnetic charges* that is also the square root of a quartic algebraic invariant of the U-duality group. This is probably the deepest and most significant aspect of black hole physics in the context of supergravity. In the next chapter we shall illustrate the concepts introduced in the present one through the explicit construction of $N = 8$ black holes. Here we keep our discussion at a very general level.

7.2   Central charges in N-extended $D = 4$ supergravity

As we have explained in table 6.1, for $N \geq 3$ the scalar manifold of $D = 4$ supergravity is always a homogeneous rank coset manifold G/H. Furthermore, as we discussed in section 6.3, the isometry group G is symplectically embedded into $Sp(2\bar{n}, \mathbb{R})$, the symbol $\bar{n}$ denoting the total number of vector fields contained in the theory that can either belong to the graviton multiplet (*graviphotons*), or to the matter multiplets, if such multiplets are allowed ($N \leq 4$). Correspondingly if $L(\phi)$ is the coset representative of G/H in some representation of G, we denote by

$$\mathbb{L}_{Sp}(\phi) = \begin{pmatrix} A & B \\ C & D \end{pmatrix} \tag{7.2.1}$$

the same coset representative embedded in the fundamental $2\bar{n}$ real representation of $Sp(2\bar{n}, \mathbb{R})$. Its blocks $A(\phi), B(\phi), C(\phi), D(\phi)$ can be constructed in terms of the original coset representative $L(\phi)$. Performing the isomorphism transformation (6.3.9) we can rather use a coset representative that is embedded in the fundamental $2\bar{n}$ representation of the group $Usp(\bar{n}, \bar{n})$. We name such a matrix $\mathbb{L}_{Usp}(\phi)$ and for convenience in later manipulations we parametrize it as follows:

$$\mathbb{L}_{Usp} = \frac{1}{\sqrt{2}} \begin{pmatrix} f + ih & \bar{f} + i\bar{h} \\ f - ih & \bar{f} - i\bar{h} \end{pmatrix} = \mathcal{C}\mathbb{L}_{Sp}\mathcal{C}^{-1}. \tag{7.2.2}$$

Explicitly the relation between the $\bar{n} \times \bar{n}$ complex matrices f and h defined by equation (7.2.2) and the real blocks of the symplectic coset representative is written below:

$$f = \frac{1}{\sqrt{2}}(A - iB)$$

$$h = \frac{1}{\sqrt{2}}(C - iD). \tag{7.2.3}$$

Explicitly the condition $\mathbb{L}_{Usp} \in Usp(\bar{n}, \bar{n})$ reads as follows:

$$\begin{cases} i(f^{\dagger}h - h^{\dagger}f) &= \mathbb{1} \\ (f^{t}h - h^{t}f) &= 0. \end{cases} \tag{7.2.4}$$

The $\bar{n} \times \bar{n}$ sub-blocks of $\mathbf{L}_{Usp}$ are submatrices f, h which can be decomposed with respect to the isotropy subgroup $H_{aut} \times H_{matter} \subset G$.

Indeed in all supergravity theories $N \geq 3$, where the scalar manifold is a homogeneous coset G/H, the isotropy subgroup H has the following direct product structure:

$$H = H_{aut} \otimes H_{matter}$$
$$H_{aut} = SU(N) \otimes U(1) \tag{7.2.5}$$

where H_{aut} is the automorphism group of the supersymmetry N-extended algebra, usually called R-symmetry, while H_{matter} is a group related to the structure of the matter multiplets one couples to supergravity. Actually in $D = 4$ there are just two cases: $H_{matter} = SU(n)$ in $N = 3$ and $H_{matter} = SO(n)$ in $N = 4$, as the reader can see by looking at table 6.1. The decomposition we mentioned is the following one:

$$f = (f^\Lambda_{AB}, f^\Lambda_I)$$
$$h = (h_{\Lambda AB}, h_{\Lambda I}) \tag{7.2.6}$$

where AB is a pair of indices in the antisymmetric representation of $H_{aut} = SU(N) \times U(1)$ and I is an index in the fundamental representation of H_{matter}. Upper $SU(N)$ indices label objects in the complex conjugate representation of $SU(N)$, as we have already emphasized while discussing the $N = 2$ case (see for example formula (6.5.2) and following ones):

$$(f^\Lambda_{AB})^* = f^{\Lambda AB} \quad \text{and similarly for the other cases.} \tag{7.2.7}$$

Note that we can consider $(f^\Lambda_{AB}, h_{\Lambda AB})$ and $(f^\Lambda_I, h_{\Lambda I})$ as symplectic sections of an $Sp(2\bar{n}, \mathbb{R})$ bundle over the G/H base manifold. We will prove in the following that this bundle is actually flat. The real embedding given by $\mathbf{L}_{Sp}$ is appropriate for duality transformations of $\mathcal{F}^\pm$ and their duals $\mathcal{G}^\pm$, according to equations (6.2.19), while the complex embedding in the matrix $\mathbf{L}_{Usp}$ is appropriate in writing down the supersymmetry transformation rules of the fermions and of the vectors. The kinetic matrix $\mathcal{N}$, given by the Gaillard–Zumino master formula (6.3.17), can be rewritten as follows:

$$\mathcal{N} = hf^{-1} \qquad \mathcal{N} = \mathcal{N}^t. \tag{7.2.8}$$

Using (7.2.4) and (7.2.8) we find

$$(f^t)^{-1} = i(\mathcal{N} - \overline{\mathcal{N}})\overline{f} \tag{7.2.9}$$

that, more explicitly, reads

$$f_{AB\Lambda} \equiv (f^{-1})_{AB\Lambda} = i(\mathcal{N} - \overline{\mathcal{N}})_{\Lambda\Sigma}\overline{f}^\Sigma_{AB} \tag{7.2.10}$$
$$f_{I\Lambda} \equiv (f^{-1})_{I\Lambda} = i(\mathcal{N} - \overline{\mathcal{N}})_{\Lambda\Sigma}\overline{f}^\Sigma_I. \tag{7.2.11}$$

Similarly to the $N = 2$ case, the graviphoton field strengths and the matter field strengths, whose integrals yield the central and matter charges, are defined by the symplectic invariant, scalar field dressed, combinations of $F^\Lambda_{\mu\nu}$ fields appearing in the fermion supersymmetry transformation rules. In other words there exists a generalization of formulae (6.5.5), (6.5.6) and (6.5.8), (6.5.9), (6.5.10), (6.5.11) that we presently describe. In extended supergravity there are three kinds of fermion:

(i)   The gravitino $\psi_{A\mu}$ which carries an R-symmetry index A running in the fundamental of $SU(N)$, the chirality assignments being as described in equations (6.5.2), (6.5.3).

(ii) The dilatino fields χ_{ABC}, that are spin $1/2$ members of the graviton multiplet (for $N \geq 3$) and transform in the three times antisymmetric representation of $SU(N)$. Their chirality assignment is:

$$\gamma_5 \, \chi_{ABC} = \chi_{ABC}$$

$$\gamma_5 \, \chi^{ABC} = -\chi^{ABC}. \tag{7.2.12}$$

(iii) The gauginos λ^I_A that are spin $1/2$ members of the vector multiplets and transform in the fundamental representation of the R-symmetry $SU(N)$. For $N = 3, 4$ they are also in the fundamental representation of the H_{matter} group, respectively $SU(n)$ or $SO(n)$. In the $N = 2$ case, where the scalar multiplet is not necessarily a coset manifold G/H, the index I is replaced by the world index i^*, since the gaugino transforms as a section of the antiholomorphic tangent bundle (see equations (6.5.4))

The supersymmetry transformation laws of these fermion fields on bosonic backgrounds are

$$\delta\psi_{A\mu} = \nabla_\mu \epsilon_A \; - \tfrac{1}{4} T^{(-)}_{AB|\rho\sigma} \gamma^{\rho\sigma} \gamma_\mu \epsilon^B = 0 \tag{7.2.13}$$

$$\delta\chi_{ABC} = 4i P_{ABCD|i} \partial_\mu \Phi^i \gamma^\mu \epsilon^D - 3T^{(-)}_{[AB|\rho\sigma} \gamma^{\rho\sigma} \epsilon_{C]} = 0 \tag{7.2.14}$$

for the gravitino and dilatino, respectively, and

$$\delta\lambda^I_A = a P^I_{AB,i} \partial_a \phi^i \gamma^a \epsilon^B + b \, T^{-I}_{ab} \gamma^{ab} \epsilon_A \tag{7.2.15}$$

for the gaugino, a, b being some numerical coefficients, depending on the model and unessential for the purpose of the present discussion. Furthermore the dressed graviphoton and matter field strengths are defined by the following equations, completely analogous to equations (6.5.8) and subsequent ones:

$$T^-_{AB} = i(\overline{f}^{-1})_{AB\Lambda} F^{-\Lambda} = f^\Lambda_{AB}(\mathcal{N} - \overline{\mathcal{N}})_{\Lambda\Sigma} F^{-\Sigma} = h_{\Lambda AB} F^{-\Lambda} - f^\Lambda_{AB} \mathcal{G}^-_\Lambda$$

$$T^-_I = i(\overline{f}^{-1})_{I\Lambda} F^{-\Lambda} = f^\Lambda_I(\mathcal{N} - \overline{\mathcal{N}})_{\Lambda\Sigma} F^{-\Sigma} = h_{\Lambda I} F^{-\Lambda} - f^\Lambda_I \mathcal{G}^-_\Lambda \tag{7.2.16}$$

$$\overline{T}^{+AB} = (T^-_{AB})^*$$

$$\overline{T}^{+I} = (T^-_I)^*.$$

Obviously, for $N > 4$, $f_{\Lambda I} = T_I = 0$. Note also that the right-hand side of equations (7.2.17) makes explicit the symplectic invariance of the dressed field strengths.

Finally, in order to complete our description of the supersymmetry transformation rules, let us explain the meaning of the H_{aut} antisymmetric tensors $P_{ABCD|i}$, $P_{AB\,I|i}$ (the index i is a world index labelling G/H coordinates). As will be clear in the following these correspond to a decomposition of the G/H coset manifold vielbein into irreducible representations of the isotropy subgroup H. What is general in this decomposition is that, for all N-cases, we always obtain the required representations needed to make sense of the supersymmetry transformation rules (7.2.13)–(7.2.15). This happens because the choice of G/H is completely determined by supersymmetry.

In full analogy to what we did in the $N = 2$ case we can now construct *field dependent central and matter charges*, by integrating the dressed field strengths $T_{AB} = T_{AB}^+ + T_{AB}^-$ and (for $N = 3, 4$) $T_I = T_I^+ + T_I^-$ on a 2-sphere of radius r. As before these objects depend on r since the corresponding field strength is not conserved. Note that as already discussed in the $N = 2$ case we always assume spherical symmetry and hence purely radial dependence of the scalar and vector fields. For this purpose we note that

$$T_{AB}^+ = h_{\Lambda AB}F^{+\Lambda} - f_{AB}^\Lambda \mathcal{G}_\Lambda^+ = 0 \tag{7.2.17}$$

$$T_I^+ = h_{\Lambda I}F^{+\Lambda} - f_I^\Lambda \mathcal{G}_\Lambda^+ = 0 \tag{7.2.18}$$

as a consequence of equations (7.2.8) and (6.2.22). Therefore we have:

$$Z_{AB} = \int_{S^2} T_{AB} = \int_{S^2}(T_{AB}^+ + T_{AB}^-) = \int_{S^2} T_{AB}^- = h_{\Lambda AB}(r)p^\Lambda - f_{AB}^\Lambda(r)q_\Lambda \tag{7.2.19}$$

$$Z_I = \int_{S^2} T_I = \int_{S^2}(T_I^+ + T_I^-) = \int_{S^2} T_I^- = h_{\Lambda I}(r)p^\Lambda - f_I^\Lambda(r)q_\Lambda \quad (N \le 4) \tag{7.2.20}$$

where:

$$q_\Lambda = \int_{S^2} \mathcal{G}_\Lambda \qquad p^\Lambda = \int_{S^2} F^\Lambda \tag{7.2.21}$$

are the conserved quantized charges defined as in equations (6.6.6) and (6.6.7). We see that the central and matter charges are given by symplectic invariants and that the presence of dyons in $D = 4$ is related to the symplectic embedding.

7.2.1 Differential relations obeyed by the dressed charges

Let $\Gamma = \mathbf{L}_{Usp}^{-1}\,d\mathbf{L}_{Usp}$ be the $Usp(\bar{n}, \bar{n})$ Lie algebra left invariant 1-form satisfying the Maurer Cartan equation:

$$d\Gamma + \Gamma \wedge \Gamma = 0. \tag{7.2.22}$$

Note that the above equation (7.2.22) actually implies the flatness of the symplectic bundle over G/H we have already mentioned.

In terms of the matrices (f, h) the 1-form Γ reads:

$$
\Gamma \equiv \mathbb{L}_{Usp}^{-1} \, d\mathbb{L}_{Usp} = \begin{pmatrix} i(f^\dagger \, dh - h^\dagger \, df) & i(f^\dagger \, d\overline{h} - h^\dagger \, d\overline{f}) \\ -i(f^t \, dh - h^t \, df) & -i(f^t \, d\overline{h} - h^t \, d\overline{f}) \end{pmatrix}
$$

$$
\equiv \begin{pmatrix} \Omega^{(H)} & \overline{\mathcal{P}} \\ \mathcal{P} & \overline{\Omega}^{(H)} \end{pmatrix} \tag{7.2.23}
$$

where the $\overline{n} \times \overline{n}$ sub-blocks $\Omega^{(H)}$ and $\mathcal{P}$ embed the H connection and the G/H vielbein, respectively. This identification follows from the Cartan decomposition of the $Usp(\overline{n}, \overline{n})$ Lie algebra. Explicitly, if we define the $H_{aut} \times H_{matter}$-covariant derivative of a vector $V = (V_{AB}, V_I)$ as:

$$
\nabla V = dV - V\omega \qquad \omega = \begin{pmatrix} \omega^{AB}_{CD} & 0 \\ 0 & \omega^I_J \end{pmatrix} \tag{7.2.24}
$$

from (7.2.2) and (7.2.23), we obtain the $(\overline{n} \times \overline{n})$ matrix equation:

$$
\nabla(\omega)(f + ih) = (\overline{f} + i\overline{h})\mathcal{P}
$$

$$
\nabla(\omega)(f - ih) = (\overline{f} - i\overline{h})\mathcal{P} \tag{7.2.25}
$$

together with their complex conjugates. Using further the definition (7.2.7) we have:

$$
\nabla(\omega) f^\Lambda_{AB} = \overline{f}^\Lambda_I \, P^I_{AB} + \tfrac{1}{2}\overline{f}^{\Lambda CD} P_{ABCD}
$$

$$
\nabla(\omega) f^\Lambda_I = \tfrac{1}{2}\overline{f}^{\Lambda AB} P_{ABI} + \overline{f}^{\Lambda J} P_{JI} \tag{7.2.26}
$$

where we have decomposed the embedded vielbein $\mathcal{P}$ as follows:

$$
\mathcal{P} = \begin{pmatrix} P_{ABCD} & P_{ABJ} \\ P_{ICD} & P_{IJ} \end{pmatrix} \tag{7.2.27}
$$

the sub-blocks being related to the vielbein of G/H, $P = L^{-1}\nabla^{(H)}L$, written in terms of the indices of $H_{aut} \times H_{matter}$. Note that, since f belongs to the unitary matrix $\mathbb{L}_{Usp}$, we have: $(f^\Lambda_{AB}, f^\Lambda_I)^* = (\overline{f}^{\Lambda AB}, \overline{f}^{\Lambda I})$. Obviously, the same differential relations that we wrote for f hold true for the dual matrix h as well.

Using the definition of the charges (7.2.19), (7.2.20) we then obtain the following differential relations among charges:

$$
\nabla(\omega) Z_{AB} = \overline{Z}_I P^I_{AB} + \tfrac{1}{2}\overline{Z}^{CD} P_{ABCD}
$$

$$
\nabla(\omega) Z_I = \tfrac{1}{2}\overline{Z}^{AB} P_{ABI} + \overline{Z}_J P^J_I . \tag{7.2.28}
$$

Depending on the coset manifold, some of the blocks of (7.2.27) can be actually zero. For example in $N = 3$ the vielbein of $G/H = SU(3, n)/SU(3) \times SU(n) \times U(1)$ [36] is P_{IAB} (AB antisymmetric), $I = 1, \ldots, n$; $A, B = 1, 2, 3$ and it turns out that $P_{ABCD} = P_{IJ} = 0$.

In $N = 4$, $G/H = SU(1, 1)/U(1) \otimes O(6, n)/O(6) \times O(n)$ [37], and we have $P_{ABCD} = \epsilon_{ABCD} P$, $P_{IJ} = \overline{P} \delta_{IJ}$, where P is the Kählerian vielbein of $\frac{SU(1,1)}{U(1)}$, ($A, \ldots, D$ $SU(4)$ indices and I, J $O(n)$ indices) and P_{IAB} is the vielbein of $O(6, n)/O(6) \times O(n)$.

For $N > 4$ (no matter indices) we have that $\mathcal{P}$ coincides with the vielbein P_{ABCD} of the relevant G/H.

7.2.2 Sum rules satisfied by central charges

Besides the differential relations (7.2.29), the charges also satisfy a sum rule that we presently describe.

The sum rule has the following form:

$$\frac{1}{2} Z_{AB} \overline{Z}^{AB} + Z_I \overline{Z}^I = -\frac{1}{2} Q^t \mathcal{M}(\mathcal{N}) Q \qquad (7.2.29)$$

where $\mathcal{M}(\mathcal{N})$ and Q are:

$$\mathcal{M} = \begin{pmatrix} \mathbb{1} & -\mathrm{Re}\mathcal{N} \\ 0 & \mathbb{1} \end{pmatrix} \begin{pmatrix} \mathrm{Im}\mathcal{N} & 0 \\ 0 & \mathrm{Im}\mathcal{N}^{-1} \end{pmatrix} \begin{pmatrix} \mathbb{1} & 0 \\ -\mathrm{Re}\mathcal{N} & \mathbb{1} \end{pmatrix} \qquad (7.2.30)$$

$$Q = \begin{pmatrix} p^\Lambda \\ q_\Lambda \end{pmatrix}. \qquad (7.2.31)$$

In order to obtain this result we just need to observe that from the fundamental identities (7.2.4) and from the definition of the kinetic matrix given in (7.2.8) it follows:

$$ff^\dagger = -\mathrm{i}\left(\mathcal{N} - \overline{\mathcal{N}}\right)^{-1} \qquad (7.2.32)$$

$$hh^\dagger = -\mathrm{i}(\overline{\mathcal{N}}^{-1} - \mathcal{N}^{-1})^{-1} \equiv -\mathrm{i}\mathcal{N}(\mathcal{N} - \overline{\mathcal{N}})^{-1}\overline{\mathcal{N}} \qquad (7.2.33)$$

$$hf^\dagger = \mathcal{N} ff^\dagger \qquad (7.2.34)$$

$$fh^\dagger = ff^\dagger \overline{\mathcal{N}}. \qquad (7.2.35)$$

Note that if we go back to the $N = 2$ case treated in the previous chapter, the sum rule on the central and matter charges (7.2.29) takes the form:

$$|Z|^2 + |Z_i|^2 \equiv |Z|^2 + Z_i g^{i\overline{j}} \overline{Z}_{\overline{j}} = -\frac{1}{2} Q^t \mathcal{M} Q \qquad (7.2.36)$$

with the same right-hand side as in the previous discussion.

7.3 The geodesic potential

At this point we can use the information we have so far collected on the central and matter charges in a different way that makes no direct reference to supersymmetry and enlightens the significance of the scalar field functional:

$$V(\phi, Q) = -\tfrac{1}{2} Q^t \, \mathcal{M}(\mathcal{N}) \, Q \tag{7.3.1}$$

introduced in the sum rule (7.2.29), (7.2.36).

To this effect we consider the general action (6.1.21) and the *ansatz* (6.1.1) for the metric, (6.1.2) for the vector fields and (6.1.3) for the scalars. From such an action one can derive the field equations varying with respect to the metric, the vector fields and the scalars. Inserting the above *ansatz*, such equations reduce to a system of second order ordinary differential equations in the variable r for the unknown functions $U(r)$ and $\phi^I(r)$. We can think of such equations as the Euler–Lagrange equations derived from an effective action which is quite straightforward to write down. Such an effective action has the following explicit form:

$$S_{eff} \equiv \int \mathcal{L}_{eff}(\tau) \, d\tau \qquad \tau = -\frac{1}{r}$$

$$\mathcal{L}_{eff}(\tau) = \left(\frac{dU}{d\tau}\right)^2 + g_{IJ} \frac{d\phi^I}{d\tau} \frac{d\phi^J}{d\tau} + e^{2U} \, V(\phi, Q) \tag{7.3.2}$$

where the potential $V(\phi)$ appearing in (7.3.2) is the same object as defined in equation (7.3.1). Actually the field equations from the original action are equivalent to the variational equations obtained from the effective action (7.3.2) provided we add to them also the following constraint:

$$\left(\frac{dU}{d\tau}\right)^2 g_{IJ} \frac{d\phi^I}{d\tau} \frac{d\phi^J}{d\tau} - e^{2U} \, V(\phi, Q) \cong 0. \tag{7.3.3}$$

Let us first consider a simple case where we assume that the scalar fields are constants from *horizon* $r = 0$ to *infinity* $r = \infty$:

$$\phi^I = \text{constant} = \phi^I_\infty. \tag{7.3.4}$$

Extremal black holes satisfying such an additional simplifying condition (7.3.4) are named *double extreme black holes*.

It follows from (7.3.3), upon use of (7.3.4), that:

$$\left(\frac{dU}{d\tau}\right)^2 = e^{2U} \, V(\phi, Q). \tag{7.3.5}$$

At the horizon $\tau \to -\infty$, the metric (6.1.1) must approach the Bertotti–Robinson metric (6.1.18), as we already discussed in complete generality, so that we have a boundary condition for the differential equation (7.3.5):

$$e^{2U(\tau)} \xrightarrow{\tau \to \infty} \frac{1}{m_{BR}^2} \frac{1}{\tau^2} = \frac{4\pi}{\text{area}_H} \frac{1}{\tau^2} \tag{7.3.6}$$

where m_{BR} is the Bertotti–Robinson mass and area_H is the area of the horizon, related to the first by equation (6.6.29). Using this information, for double extreme black holes we have:

$$V(\phi_H, Q) = \frac{\text{area}_H}{4\pi} \tag{7.3.7}$$

where ϕ_H denotes the value of the scalar fields at the horizon, which in this case is the same as their value at infinity. Actually the result (7.3.7) is more general and it is true also for generic extremal black holes where we relax condition (7.3.4) but we still assume that the kinetic term of the scalars ϕ^I (that in the original Lagrangian (6.1.21) and not that in the effective action) should be finite at the horizon:

$$\lim_{\tau \to \infty} g_{IJ} \frac{d\phi^I}{d\tau} \frac{d\phi^J}{d\tau} e^{2U} \tau^4 < \infty. \tag{7.3.8}$$

Indeed, using again the boundary condition (7.3.6) on U we find:

$$g_{IJ} \frac{d\phi^I}{d\tau} \frac{d\phi^J}{d\tau} \frac{4\pi}{\text{area}_H} \tau^2 \overset{\tau \to \infty}{\longrightarrow} X^2 = \text{finite quantity}. \tag{7.3.9}$$

However, we must have $X^2 = 0$ if the moduli ϕ are assumed to be finite at the horizon. Indeed, if $X^2 \neq 0$, near the horizon we can write:

$$\tau \frac{d\phi^I}{d\tau} \overset{\tau \to \infty}{\longrightarrow} \text{const} \quad \to \quad \phi^I \sim \text{const} \times \log \tau. \tag{7.3.10}$$

Hence also for generic extremal black holes the same conclusion as reached before in equation (7.3.7) holds true.

Hence the area of the horizon is expressed in terms of the finite value reached by the scalars at the horizon. But what is such a value? We can easily show that ϕ_H^I is determined by the following extremization of the potential (7.3.1):

$$\left. \frac{\partial V}{\partial \phi^I} \right|_H = 0. \tag{7.3.11}$$

Indeed considering the variational equation for the scalar fields derived from the effective action (7.3.2) we have:

$$\frac{D^2}{d\tau^2} \phi^I = \frac{1}{2} e^{2U} \frac{\partial V}{\partial \phi^I} g^{IJ}. \tag{7.3.12}$$

Near the horizon the contribution from the quadratic terms proportional to the Levi–Civita connection $\Gamma^I_{JK} d_\tau \phi^J d_\tau \phi^K$ vanishes so that equation (7.3.12) reduces to:

$$\frac{d^2}{d\tau^2} \phi^I \simeq \frac{1}{2} \frac{\partial V}{\partial \phi^I} g^{IJ} \frac{4\pi}{\text{area}_H} \frac{1}{\tau^2} \tag{7.3.13}$$

whose solution is:

$$\phi^I \sim \frac{2\pi}{\text{area}_H} \frac{\partial V}{\partial \phi^I} g^{IJ} \log \tau + \phi_H^I. \tag{7.3.14}$$

Invoking once again the finiteness of the scalar fields ϕ at the horizon we conclude that their the extremum condition (7.3.11) must be true in order to be consistent with equation (7.3.14).

In this way we have reached, for the general case of an extremal black hole with a finite horizon area, the same conclusion we had reached in section 6.6.1 for such a black hole in the context of $N = 2$ supergravity. Namely the scalars flow at the horizon at a fixed point that is determined as the extremum of a potential. In the $N = 2$ case we obtained the equations (6.6.24) determining the horizon values by starting from the BPS conditions (6.6.12) and (6.6.13) and imposing that we are at the fixed point of the first order differential equations. We already anticipated that the same result could be obtained by extremizing a potential, whose value at the horizon is the horizon area. We can now verify explicitly such a statement. To this effect we should stress that in the supersymmetric case the *geodesic potential* appearing in the effective action (7.3.2) and defined in equation (7.3.1) is the right-hand side of the *central and matter charge sum rule* of equation (7.2.29) or (7.2.36) for the $N = 2$ case. Hence extremizing the geodesic potential, in supersymmetric cases, is the same thing as extremizing the left-hand side of the sum rule. For the $N = 2$ case we find:

$$\begin{aligned}
0 &= \frac{\mathrm{d}}{\mathrm{d}z^i}(Z\,\overline{Z} + g^{ij^*}\, Z_i\, \overline{Z}_{j^*}) \\
&= \nabla_i Z\,\overline{Z} + \nabla_i Z_j\, \overline{Z}_{j^*}\, g^{jk^*} + Z_j\, \nabla_i\, \overline{Z}_{k^*}\, g^{jk^*} \\
&= \nabla_i Z\,\overline{Z} + i\, C_{ijm}\, \overline{Z}_{\ell^*}\, \overline{Z}_{k^*}\, g^{\ell^* m}\, g^{jk^*} + Z_j\, Z\, \delta_j^i
\end{aligned} \tag{7.3.15}$$

where we have used the special geometry identities (6.4.36) and the already quoted identification (see (6.6.14)and following ones):

$$Z_i = \nabla_i Z. \tag{7.3.16}$$

Indeed the special geometry identities holding true on the symplectic sections and their derivatives are automatically extended to the central and matter charges that are linear combinations of such sections with constant coefficients (the quantized charges). From equation (7.3.15) it immediately follows that the extremum condition translates into equation (6.6.19) since it implies:

$$Z_i = \nabla_i Z = 0 \quad \longrightarrow \quad \partial_i |Z|^2 = 0. \tag{7.3.17}$$

Hence the horizon area is indeed a minimum and, as we claimed in section 6.6.1, the fixed values of the scalars are obtained by extremizing such an area.

7.4   General properties of central charges at the extremum

The geodesic potential (7.3.1), whose extrema correspond to the finite values reached by the scalar fields at the horizon of an extremal black hole, can be rewritten as the left-hand side of equation (7.2.29) in the case of a supersymmetric theory. This statement must be interpreted in the following way. We are free to consider an arbitrary theory of gravity coupled to scalar and vector fields described by an action of type (6.1.21). If supersymmetry is not advocated no special relation exists on the number of scalars relative to the number of vectors and any symplectically embedded scalar manifold is allowed. For any such theory there is a *period matrix* $\mathcal{N}$ and correspondingly we can construct the potential (7.3.1). We are also free to look for extremal black solutions of such a theory and, according to the discussion presented in the previous section, the value attained by the scalars at the horizon is determined by equation (7.3.11), provided we look for *finite horizon area* solutions. Yet it is only in the supersymmetric case that there exists a concept of central charges of the supersymmetry algebra and in that case the geodesic potential can be rewritten as a sum of squares of such charges:

$$V^{SUSY}(\phi, \vec{Q}) = \tfrac{1}{2} Z_{AB} \overline{Z}^{AB} + Z_I \overline{Z}^I. \tag{7.4.1}$$

We can reconsider extremization of the geodesic potential from the point of view of its supersymmetric reinterpretation (7.4.1). Writing the extremum condition as the vanishing of the exterior differential, we immediately obtain

$$
\begin{aligned}
0 &= \mathrm{d}V^{SUSY}(\phi, \vec{Q}) \\
&= \tfrac{1}{2} \nabla Z_{AB} \, \overline{Z}^{AB} + \tfrac{1}{2} Z_{AB} \, \nabla \overline{Z}^{AB} + \nabla Z_I \, \overline{Z}^I + Z_I \, \nabla \overline{Z}^I
\end{aligned} \tag{7.4.2}
$$

and inserting the differential relations (7.2.29) into (7.4.2) we obtain the following condition:

$$
\begin{aligned}
0 = {}& [\tfrac{1}{2}(\overline{Z}^I \, P_{I|AB} + \tfrac{1}{2}\overline{Z}^{CD} \, P_{ABCD}) \overline{Z}^{AB} + CC] \\
&+ [\tfrac{1}{2}(\overline{Z}^{AB} \, P_{AB|I} + \overline{Z}^J \, P_{J|I}) \overline{Z}^J + CC].
\end{aligned} \tag{7.4.3}
$$

Since the scalar vielbein provides a frame of independent 1-forms on the scalar manifold, equation (7.4.3) can be true only if the coefficients of each vielbein component vanish independently. This implies that, at the extremum, the following conditions on the central and matter charges have to be true:

$$Z_I^{fix} = 0 \tag{7.4.4}$$

$$Z_{[AB}^{fix} Z_{CD]}^{fix} = 0. \tag{7.4.5}$$

Equations (7.4.4) and (7.4.5) are the higher N generalization of the extremum condition (7.3.17) we found in the case of $N = 2$ supergravity. As we see for all values of N the matter charges Z_I vanish at the horizon. In the $N = 2$ case, where the matter charges can be interpreted as the moduli covariant derivatives of the unique central charge, condition (7.4.4) suffices to determine the extremum. In the higher N case, where the central charge is an antisymmetric $N \times N$ matrix, the extremum is characterized by the additional algebraic condition (7.4.5). Its meaning becomes clear if we make a local $SU(N)$ R-symmetry transformation that reduces the central charge tensor to its *normal frame* where it is skew diagonal (see equation (4.2.4)). Focusing on the $N = 2p =$ even case, in the normal frame the only non-vanishing entries are

$$Z_1^{fix} \equiv Z_{12}^{fix}, \ Z_2^{fix} \equiv Z_{34}^{fix}, \ \ldots, \ Z_{N/2}^{fix} \equiv Z_{N-1\,N}^{fix} \qquad (7.4.6)$$

and equation (7.4.5) implies that either all vanish, or at most one of them, say $Z_1^{fix} = Z_{12}^{fix}$, is non-zero while all the others vanish. Hence the constraint (7.4.5) yields two possibilities:

$$\text{solutions of the constraint} = \begin{cases} \text{1st solution:} \quad Z_i^{fix} = 0 \qquad (i = 1, \ldots, p) \\ \text{2nd solution:} \quad \begin{cases} Z_1^{fix} \neq 0 \\ Z_i^{fix} \qquad (i = 2, \ldots, p). \end{cases} \end{cases}$$
$$(7.4.7)$$

On the other hand, recalling (7.3.7) we have:

$$\frac{\text{area}_H}{4\pi} = \sum_{i=1}^{p} \left| Z_i^{fix} \right|^2 \qquad (7.4.8)$$

so that if the horizon area (and hence the entropy) of the black hole has to be finite then we can exclude the first possibility in equation (7.4.7) and we find:

$$\frac{\text{area}_H}{4\pi} = \left| Z_1^{fix} \right|^2 > 0. \qquad (7.4.9)$$

Equation (7.4.9) is the higher N generalization of equation (6.6.30) holding true in the $N = 2$ case. It leads to the following very important conclusion:

Statement 7.4.1. BPS saturated black holes are classified by the skew-eigenvalue structure of their central charge Z_{AB} evaluated at the horizon. The only BPS black holes that have a non vanishing entropy area$_H/4\pi > 0$ are those admitting a single non-zero skew-eigenvalue Z_1^{fix} while all the other Z_i^{fix} vanish.

7.5   The horizon area as a topological U duality invariant

As we just saw the horizon area can be found by extremizing the geodesic potential and then replacing the fixed values of the scalars in equation (7.3.7). Since the only parameters appearing in the geodesic potential (7.3.1) are the quantized charges $\{p^\Lambda, q_\Sigma\}$, it follows that the fixed values of the scalars will depend only on such quantized charges and so will the horizon area:

$$\frac{\text{area}_H}{4\pi} = S(p, q). \qquad (7.5.1)$$

On the other hand we recall that, by construction, the geodesic potential is a symplectic invariant and hence an invariant under U-duality transformations. Consider for instance the $N > 4$ supersymmetric case where the geodesic potential is expressed as in equation (7.4.1) with no matter charges Z_I. Introduce the charge vector (7.2.31) that transforms in the appropriate *real symplectic representation* of the U-duality group which defines the symplectic embedding of the coset manifold U/H. Then recall the definition (7.2.19) of the central charges and equation (7.2.2) which relates the complex Usp-realization $\mathbb{L}_{Usp}(\phi)$ of the coset representative to its real symplectic version $\mathbb{L}_{Sp}(\phi)$. By straightforward algebra we verify that the geodesics potential is nothing else but the following bi-quadratic form in the charge vector and in the coset representative:

$$V(\phi, \vec{Q}) \equiv \tfrac{1}{2}\overline{Z}^{AB}(\phi)\, Z_{AB}(\phi)$$

$$= \tfrac{1}{2}\vec{Q}^T [\mathbb{L}_{Sp}^{-1}(\phi)]^T \mathbb{L}_{Sp}^{-1}(\phi)\, \vec{Q}. \tag{7.5.2}$$

The invariance under the U-transformations is verified in the following way. By definition of coset representative, if $A \in U \subset Sp(2\bar{n}, \mathbb{R})$ we have:

$$\mathbb{L}_{Sp}(A\phi) = A\,\mathbb{L}_{Sp}(\phi)\, W_A(\phi) \tag{7.5.3}$$

$$W_A(\phi) \in H \subset U(\bar{n}) \subset Sp(2\bar{n}, \mathbb{R}) \implies [W_A(\phi)]^T\, W_A(\phi) = \mathbb{1} \tag{7.5.4}$$

where $A\phi$ denotes the non-linear action of the U group element A on the scalar fields ϕ and $W_A(\phi)$ is the H-compensator that in the symplectic representation is an orthogonal matrix (indeed we have $U(\bar{n}) = SO(2\bar{n}) \cap Sp(2\bar{n}, \mathbb{R})$). Relying on equation (7.5.4) and on the representation (7.5.2) of the geodesic potential we find:

$$V(A\phi, A\vec{Q}) = \vec{Q}^T\, A^T [\mathbb{L}_{Sp}^{-1}(A\phi)]^T \mathbb{L}_{Sp}^{-1}(A\phi) A\, \vec{Q}$$

$$= \vec{Q}^T\, A^T\, [A^{-1}]^T [\mathbb{L}_{Sp}^{-1}(\phi)]^T$$

$$[W_A^{-1}(\phi)]^T\, W_A^{-1}(\phi) \mathbb{L}_{Sp}^{-1}(\phi) A^{-1}\, A\, \vec{Q}$$

$$= V(\phi, \vec{Q}) \tag{7.5.5}$$

which explicitly proves the U-invariance of of the geodesic potential. It follows that also the horizon area $S(p, q)$ obtained by substitution of the fixed scalar values in $V(\phi, Q)$ will be a U-invariant. Not only that. Equation (7.5.2) shows that the invariant $I(p, q)$ must be homogenous of order two in the charge vector $\vec{Q}$. At this point we can just rely on group theory. The geodesic potential is bi-quadratic in the charge vector and in the coset representative. Hence $S(p, q)$ cannot be just a quadratic invariant since it is obtained by substitution of $\phi^{fix} = \phi(p, q)$ into an expression that is already quadratic in the charge vector. The other possibility is that $S(p, q)$ be the square root of a quartic invariant. Hence we can guess the following formula:

$$\frac{\text{area}_H}{4\pi} \equiv S(p, q) = \sqrt{Q^\Lambda Q^\Sigma Q^\Delta Q^\Gamma d_{\Lambda\Sigma\Delta\Gamma}} \tag{7.5.6}$$

where $d_{\Lambda\Sigma\Delta\Gamma}$ is a *four-index symmetric invariant tensor* of the U-duality group in the real representation that defines the symplectic embedding of the manifold U/H.

For instance in the case of $N = 8$ supergravity where $U = E_{7(7)}$ a four-index symmetric invariant tensor does indeed exist in the fundamental **56** representation. The horizon area of an $N = 8$ BPS black hole is correctly given by equation (7.5.6). We explicitly verify this statement in the next chapter by solving the appropriate differential equations and computing the horizon area. Here we pursue a more abstract discussion. Having established that the square of the horizon area is a quartic U-invariant we can try to construct it as an appropriate linear combination of H-invariants. We start from the central charges $Z_{AB}(p, q, \phi)$ that are linear in the charge vector $\vec{Q}$ and that transform covariantly under the compensating subgroup H. We consider the possible H-invariants $I_i \, (Z(p, q, \phi))$ that are quartic in the central charges Z_{AB} and we write the *ansatz*:

$$S^2(p, q) = I = \sum_i \alpha_i \, I_i(Z(p, q, \phi)) \tag{7.5.7}$$

where α_i are numerical coefficients. For arbitrary α_i values I is a function of the scalar fields ϕ through the scalar field dependence of the central charges. Yet for special choices of the coefficients α_i the H-invariant I can be independent from the ϕ^I. We argue that such a linear combination of H-invariants is the quartic U-invariant we look for. To find the appropriate α_i it suffices to impose:

$$\frac{\partial}{\partial \phi^I} I = 0. \tag{7.5.8}$$

This is a very much indirect way of constructing U-invariants that has a distinctive advantage. It puts into evidence the relation between the black hole horizon area and the eigenvalues of the central charge. As an example we consider the $N = 8$ case.

7.5.1 Construction of the quartic $E_{7(7)}$ invariant in the $N = 8$ theory

In the $N = 8$ case we have $U = E_{7(7)}$, $H = SU(8)$ and the central charge Z_{AB} is in the complex **28** representation of $SU(8)$. There are three quartic $SU(8)$-invariants that we can write:

$$I_1 \equiv \left(Z_{AB} \, \overline{Z}^{BA} \right)^2$$

$$I_2 \equiv Z_{AB} \, \overline{Z}^{BC} \, Z_{CD} \, \overline{Z}^{DA}$$

$$I_3 \equiv \frac{1}{2} \left(\frac{1}{2!4!} \epsilon^{ABCDEFGH} \, Z_{AB} \, Z_{CD} \, Z_{EF} \, Z_{GH} + \text{CC} \right). \tag{7.5.9}$$

We can easily determine the coefficients $\alpha_1, \alpha_2, \alpha_3$ by going to the normal frame. Naming the four skew-eigenvalues as in equation (7.4.6) we obtain:

$$I_1 = 4 \left(\sum_{i=1}^{4} |Z_i|^2 \right)^2$$

$$I_2 = 2 \sum_{i<j} |Z_i|^2 |Z_j|^2$$

$$I_3 = \frac{1}{2} \left(\prod_{i=1}^{4} Z_i + \prod_{i=1}^{4} \overline{Z}_i \right). \tag{7.5.10}$$

On the other hand, naming:

$$P_1 = P_{1234} \qquad P_2 = P_{1256} \qquad P_3 = P_{1278} \tag{7.5.11}$$

the differential relations (7.2.29) reduce in the normal frame to:

$$\nabla Z_1 = P_1 \overline{Z}_2 + P_2 \overline{Z}_3 + P_3 \overline{Z}_4$$
$$\nabla Z_2 = P_1 \overline{Z}_1 + P_3 \overline{Z}_3 + P_2 \overline{Z}_4$$
$$\tag{7.5.12}$$
$$\nabla Z_3 = P_2 \overline{Z}_1 + P_3 \overline{Z}_2 + P_1 \overline{Z}_4$$
$$\nabla Z_4 = P_3 \overline{Z}_1 + P_2 \overline{Z}_2 + P_1 \overline{Z}_3.$$

Inserting equations (7.5.13) and (7.5.10) in the condition

$$d \left(\sum_{i=1}^{3} \alpha_i I_i \right) = 0 \tag{7.5.13}$$

and cancelling separately the terms proportional to P_i ($i = 1, 2, 3$) we obtain $\alpha_1 = 1/4$, $\alpha_2 = -1$, $\alpha_3 = 8$.

Hence we can finally write:

$$\left(\frac{\text{area}_H}{4\pi} \right)^2 = \frac{1}{4} I_1 - I_2 + 8 I_3$$

$$= \sum_{i=1}^{4} \left(|Z_i|^2 \right)^2 - 2 \sum_{i<j} |Z_i|^2 |Z_j|^2 + 4 \left(\prod_{i=1}^{4} Z_i + \prod_{i=1}^{4} \overline{Z}_i \right)$$

$$= (\rho_1 + \rho_2 + \rho_3 + \rho_4)(\rho_1 + \rho_2 - \rho_3 - \rho_4)$$
$$\times (\rho_1 - \rho_2 + \rho_3 - \rho_4)(\rho_1 - \rho_2 - \rho_3 + \rho_4)$$
$$+ 8 \rho_1 \rho_2 \rho_3 \rho_4 (\cos \theta - 1). \tag{7.5.14}$$

The last equality in equation (7.5.14) has been obtained by writing the four skew-eigenvalues Z_i in the following way:

$$Z_i = \rho_i \exp[i\theta] \qquad \rho_i \in \mathbb{R} \tag{7.5.15}$$

where θ is a common phase. The central charge Z_{AB} can always be reduced to such a form because after the matrix has been skew-diagonalized we still have enough $SU(8)$ transformations to gauge away the three relative phases of the eigenvalues. An alternative and more intrinsic way of seeing that the truly independent parameters are **5**, namely the four moduli ρ_i plus the overall phase θ, is to observe that the stability subgroup of Z_{AB} in the normal form is $(SU(2))^4$. Then it suffices to count the dimensionality of the coset

$$\dim_{\mathbf{R}} \frac{SU(8)}{(SU(2))^4} = 51 \tag{7.5.16}$$

to conclude that 51 of the 56 real components of Z_{AB} can be gauged away leaving a residual five.

Statement 7.5.1. The most general $N = 8$ BPS black solution depends on five intrinsic parameters and its horizon area is given by equation (7.5.14) in terms of such parameters.

From equations (7.5.14), (7.5.15) we see that there are three possible cases:

$$\begin{array}{lll}
\rho_1 > \rho_2 > \rho_3 > \rho_4 \geq 0 & \text{area}_H > 0 & \frac{1}{8} \text{ BPS} \\
\rho_1 = \rho_2 > \rho_3 = \rho_4 \geq 0 & \text{area}_H = 0 & \frac{1}{4} \text{ BPS} \\
\rho_1 = \rho_2 = \rho_3 = \rho_4 \geq 0 & \text{area}_H = 0 & \frac{1}{2} \text{ BPS.}
\end{array} \tag{7.5.17}$$

The last column in table (7.5.17) is a prediction about the fraction of preserved supersymmetries that we shall verify in the next chapter.

Let us finally clarify the meaning of equation (7.5.14) in comparison with our previous conclusion in statement 7.4.1 that at the horizon there is at most one non-vanishing central charge eigenvalue and that the horizon area is essentially the square of that eigenvalue (see equation (7.4.9)). Superficially equations (7.5.14) and (7.4.9) might seem to be contradictory, but this is not so.

Statement 7.5.2. The numerical value of the horizon area is a *topological* quantity depending only on the vector of quantized charges $\vec{Q}$ that is also a group theoretical invariant of the U-duality group. This numerical value was named $S(p, q)$ since it is interpreted as the black hole entropy in black hole thermodynamics and its quantized charge dependence is given by equation (7.5.6).

By means of the construction described in the present section we have tried to reexpress the topological invariant $S(p, q)$ in terms of the central charges skew-eigenvalues Z_i. These latter depend on the scalar fields ϕ^I and therefore are not constant in space–time. In a black hole solution they have a radial dependence: $Z_i = Z_i(r)$ (see equation (7.2.19)). However the combination of central charges we have constructed is such that its numerical value does not depend on the scalar fields. Hence equation (7.5.14) is true at any point of space–time and expresses the value of the black entropy in terms of the central charge skew-eigenvalue evaluated at that point. Thus it happens that if we go to radial infinity $r = \infty$, by integrating the graviphoton field strength on a 2-sphere of very large radius we obtain, in principle, **5** different parameters, namely:

$$\rho_1^\infty \neq \rho_2^\infty \neq \rho_3^\infty \neq \rho_4^\infty \qquad \theta^\infty. \tag{7.5.18}$$

On the other hand if we evaluate the central charges at the horizon, for a finite area black hole we find only one parameter:

$$\rho_1^H \neq 0 \qquad \rho_2^H = \rho_3^H = \rho_4^H = 0 \qquad \theta^H = 0. \qquad (7.5.19)$$

Yet whether we use the parameters in equation (7.5.18) or those in equation (7.5.19) the topological invariant corresponding to the horizon area is always expressed by equation (7.5.14), the numerical value being that provided by equation (7.5.6). In particular we obtain the relation:

$$\begin{aligned}
\left(\rho_1^H\right)^2 &= (\rho_1^\infty + \rho_2^\infty + \rho_3^\infty + \rho_4^\infty)(\rho_1^\infty + \rho_2^\infty - \rho_3^\infty - \rho_4^\infty) \\
&\quad \times (\rho_1^\infty - \rho_2^\infty + \rho_3^\infty - \rho_4^\infty)(\rho_1^\infty - \rho_2^\infty - \rho_3^\infty + \rho_4^\infty) \\
&\quad + 8\rho_1^\infty \rho_2^\infty \rho_3^\infty \rho_4^\infty (\cos\theta^\infty - 1)
\end{aligned} \qquad (7.5.20)$$

that applies to finite area $N = 8$ black holes. The **5** parameters parametrizing the central charge on a large 2-sphere depend, besides the quantized charges $\vec{Q}$, also on the *black hole moduli*, namely on the values:

$$\mu^I \equiv \phi^I(\infty) \qquad (7.5.21)$$

attained by the scalar fields at radial infinity. These moduli are completely arbitrary. Yet the combination (7.5.14) of central charges is such that it actually depends only on the charges $\vec{Q}$ and not on the moduli. On the other hand, at the horizon the scalar fields flow to their fixed value that depends only on the charges and is insensitive to the boundary conditions at radial infinity.

The relevance of this discussion is that it puts into evidence that the five *moduli* characterizing the most general finite area $N = 8$ black hole can be parametrized by the skew eigenvalues of the central charge at radial infinity.

In the next chapter we address the problem of the explicit construction of $N = 8$ BPS black holes.

Chapter 8

$N = 8$ supergravity: BPS black holes with 1/2, 1/4 and 1/8 of supersymmetry

8.1 Introduction to $N = 8$ BPS black holes

In this chapter we consider BPS extremal black holes in the context of $N = 8$ supergravity.

$N = 8$ supergravity is the four-dimensional effective Lagrangian of both type IIA and type IIB superstrings compactified on a torus T^6. Alternatively it can be viewed as the 4D effective Lagrangian of 11-dimensional M-theory compactified on a torus T^7. For this reason its U-duality group $E_{7(7)}(\mathbb{Z})$, which is defined as the discrete part of the isometry group of its scalar manifold:

$$\mathcal{M}_{scalar}^{(N=8)} = \frac{E_{7(7)}}{SU(8)} \tag{8.1.1}$$

unifies all superstring dualities relating the various consistent superstring models. The *non-perturbative BPS states* one needs to adjoin to the string states in order to complete linear representations of the U-duality group are, generically, *BPS black holes*.

These latter can be viewed as intersections of several p-brane solutions of the higher dimensional theory *wrapped* on the homology cycles of the T^6 (or T^7) torus. Depending on how many p-branes intersect, the residual supersymmetry can be:

(i)   1/2 of the original supersymmetry
(ii) 1/4 of the original supersymmetry
(iii) 1/8 of the original supersymmetry.

As we have already emphasized in section 6.1 and further discussed in chapter 7 the distinction between these three kinds of BPS solution can be considered directly in a four-dimensional setup and it is related to the structure of the central charge eigenvalues and to the behaviour of the scalar fields at the horizon. BPS black holes with a finite horizon area are those for which the scalar fields are regular at the horizon and reach a fixed value there. These can only be the 1/8-type black holes, whose structure is that of $N = 2$ black holes embedded into the $N = 8$ theory. For 1/2 and 1/4 black holes the scalar fields always diverge at the horizon and the entropy is zero.

The nice point, in this respect, is that we can make a complete classification of all BPS black holes belonging to the three possible types. Indeed the distinction of the solutions into these three classes can be addressed *a priori* and, as we are going to see, corresponds to a classification into different orbits of the possible **56**-dimensional vectors $Q = \{p^\Lambda, q_\Sigma\}$ of magnetic–electric charges of the hole. Indeed $N = 8$ supergravity contains **28** gauge fields A_μ^Λ and correspondingly the hole can carry **28** magnetic p^Λ and **28** electric q_Σ charges. Through the symplectic embedding of the scalar manifold (8.1.1) it follows that the field strengths $F_{\mu\nu}^\Lambda$ plus their duals $G_{\Sigma|\mu\nu}$ transform in the fundamental **56** representation of $E_{7(7)}$ and the same is true of their integrals, namely the charges.

The Killing spinor equation (6.1.6) that imposes preservation of either 1/2, or 1/4, or 1/8 of the original supersymmetries enforces two consequences different in the three cases:

(i)   a different decomposition of the scalar field manifold into two sectors:

  • a sector of *dynamical scalar fields* that evolve in the radial parameter r
  • a sector of *spectator scalar fields* that do not evolve in r and are constant in the BPS solution

(ii) a different orbit structure for the charge vector Q.

Then, up to U-duality transformations, for each case one can write a fully general *generating solution* that contains the minimal necessary number of excited dynamical fields and the minimal necessary number of non-vanishing charges. All other solutions of the same supersymmetry type can be obtained from the generating one by the action of $E_{7(7)}$-rotations.

Such an analysis is clearly group theoretical and requires the use of appropriate techniques. The basic one is provided by the solvable Lie algebra representation of the scalar manifold.

8.2   Solvable Lie algebra description of the supergravity scalar manifold

As we have already explained in chapter 6 it has been known for many years [38] that the scalar field manifold of both pure and matter coupled $N > 2$ extended supergravities in $D = 10 - r$ ($r = 6, 5, 4, 3, 2, 1$) is a non-compact homogenous symmetric manifold $U_{(D,N)}/H_{(D,N)}$, where $U_{(D,N)}$ (depending on the space–time dimensions and on the number of supersymmetries) is a non-compact Lie group and $H_{(D,N)} \subset U_{(D,N)}$ is a maximal compact subgroup. For instance in the physical $D = 4$ case the situation was summarized in table 6.1. In the case of maximally extended supergravities in $D = 10 - r$ dimensions the scalar manifold has a universal structure:

$$\frac{U_D}{H_D} = \frac{E_{r+1(r+1)}}{H_{r+1}} \tag{8.2.1}$$

where the Lie algebra of the U_D-group $E_{r+1(r+1)}$ is the maximally non-compact real section of the exceptional E_{r+1} series of the simple complex Lie algebras and H_{r+1} is its maximally compact subalgebra [39]. This series of homogeneous spaces is summarized in table 8.1.

Table 8.1. U-duality groups and maximal compact subgroups of maximally extended supergravities.

$D = 9$	$E_{2(2)} \equiv SL(2, \mathbf{R}) \otimes O(1, 1)$	$H = O(2)$	$\dim_{\mathbf{R}}(U/H) = 3$
$D = 8$	$E_{3(3)} \equiv SL(3, \mathbf{R}) \otimes Sl(2, \mathbf{R})$	$H = O(2) \otimes O(3)$	$\dim_{\mathbf{R}}(U/H) = 7$
$D = 7$	$E_{4(4)} \equiv SL(5, \mathbf{R})$	$H = O(5)$	$\dim_{\mathbf{R}}(U/H) = 14$
$D = 6$	$E_{5(5)} \equiv O(5, 5)$	$H = O(5) \otimes O(5)$	$\dim_{\mathbf{R}}(U/H) = 25$
$D = 5$	$E_{6(6)}$	$H = Usp(8)$	$\dim_{\mathbf{R}}(U/H) = 42$
$D = 4$	$E_{7(7)}$	$H = SU(8)$	$\dim_{\mathbf{R}}(U/H) = 70$
$D = 3$	$E_{8(8)}$	$H = O(16)$	$\dim_{\mathbf{R}}(U/H) = 128$

All the homogeneous coset manifolds appearing in the various extended supergravity models share the very important property of being non-compact Riemannian manifolds $\mathcal{M}$ admitting a solvable Lie algebra description, i.e. they can be expressed as Lie group manifolds generated by a solvable Lie algebra $Solv$:

$$\mathcal{M} = \exp(Solv). \tag{8.2.2}$$

This property is of great relevance for the physical interpretation of supergravity since it allows an intrinsic algebraic characterization of the scalar fields of these theories which, without such a characterization, would simply be undistinguishable coordinates of a manifold. Furthermore the choice of the solvable parametrization provides essential technical advantages with respect to the choice of other parametrizations and plays a crucial role in the solution of BPS black hole equations. For this reason we devote the present section to explaining the basic features of such a solvable Lie algebra description of supergravity scalars.

8.2.1 Solvable Lie algebras: the machinery

Let us start by giving few preliminary definitions. A *solvable* Lie algebra $\mathbb{G}_s$ is a Lie algebra whose nth order (for some $n \geq 1$) derivative algebra vanishes:

$$\mathcal{D}^{(n)}\mathbb{G}_s = 0$$

$$\mathcal{D}\mathbb{G}_s = [\mathbb{G}_s, \mathbb{G}_s] \qquad \mathcal{D}^{(k+1)}\mathbb{G}_s = [\mathcal{D}^{(k)}\mathbb{G}_s, \mathcal{D}^{(k)}\mathbb{G}_s].$$

A *metric* Lie algebra $(\mathbb{G}, h)$ is a Lie algebra endowed with a Euclidean metric h. An important theorem states that if a Riemannian manifold $(\mathcal{M}, g)$ admits a transitive group of isometries $\mathcal{G}_s$ generated by a solvable Lie algebra $\mathbb{G}_s$ of the same dimension as $\mathcal{M}$, then:

$$\mathcal{M} \sim \mathcal{G}_s = \exp(\mathbb{G}_s)$$
$$g_{|e \in \mathcal{M}} = h$$

where h is a Euclidean metric defined on $\mathbb{G}_s$. Therefore there is a one to one correspondence between Riemannian manifolds fulfilling the hypothesis stated above and solvable metric Lie algebras $(\mathbb{G}_s, h)$.

Consider now a homogeneous coset manifold $\mathcal{M} = \mathcal{G}/\mathcal{H}$, $\mathcal{G}$ being a non-compact real form of a semisimple Lie group and $\mathcal{H}$ its maximal compact subgroup. If $\mathbb{G}$ is the Lie algebra generating $\mathcal{G}$, the so called Iwasawa decomposition ensures the existence of a solvable Lie subalgebra $\mathbb{G}_s \subset \mathbb{G}$, acting transitively on $\mathcal{M}$, such that [40]:

$$\mathbb{G} = \mathbb{H} + \mathbb{G}_s \qquad \dim \mathbb{G}_s = \dim \mathcal{M} \qquad (8.2.3)$$

$\mathbb{H}$ being the maximal compact subalgebra of $\mathbb{G}$ generating $\mathcal{H}$. Note that the sum (8.2.3) is not required to be an orthogonal decomposition, namely the elements of $\mathbb{G}_s$ are not requested to be orthogonal to the elements of $\mathbb{H}$ but simply linearly independent from them.

By virtue of the previously stated theorem, $\mathcal{M}$ may be expressed as a solvable group manifold generated by $\mathbb{G}_s$. The algebra $\mathbb{G}_s$ is constructed as follows [40]. Consider the Cartan orthogonal decomposition

$$\mathbb{G} = \mathbb{H} \oplus \mathbb{K}. \qquad (8.2.4)$$

Let us denote by $\mathcal{H}_K$ the maximal Abelian subspace of $\mathbb{K}$ and by $\mathcal{H}$ the Cartan subalgebra of $\mathbb{G}$. It can be proven [40] that $\mathcal{H}_K = \mathcal{H} \cap \mathbb{K}$, that is it consists of all non-compact elements of $\mathcal{H}$. Furthermore let h_{α_i} denote the elements of $\mathcal{H}_K$, $\{\alpha_i\}$ being a subset of the positive roots of $\mathbb{G}$ and Φ^+ the set of positive roots β not orthogonal to all the α_i (i.e. the corresponding 'shift" operators E_β do not commute with $\mathcal{H}_K$). It can be demonstrated that the solvable algebra $\mathbb{G}_s$ defined by the Iwasawa decomposition may be expressed in the following way:

$$\mathbb{G}_s = \mathcal{H}_K \oplus \left\{ \sum_{\alpha \in \Phi^+} E_\alpha \cap \mathbb{G} \right\} \qquad (8.2.5)$$

where the intersection with $\mathbb{G}$ means that $\mathbb{G}_s$ is generated by those suitable complex combinations of the 'shift' operators which belong to the real form of the isometry algebra $\mathbb{G}$.

The *rank* of a homogeneous coset manifold is defined as the maximum number of commuting semisimple elements of the non-compact subspace $\mathbb{K}$. Therefore it coincides with the dimension of $\mathcal{H}_K$, i.e. the number of non-compact Cartan generators of $\mathbb{G}$. A coset manifold is *maximally non-compact* if $\mathcal{H} = \mathcal{H}_K \subset \mathbb{G}_s$. The relevance of maximally non-compact coset manifolds relies on the fact that they are spanned by the scalar fields in the maximally extended supergravity theories. In the solvable representation of a manifold the local coordinates of the manifold are the parameters of the generating Lie algebra, therefore adopting this parametrization of scalar manifolds in supergravity implies the definition of a one to one correspondence between the scalar fields and the generators of *Solv* [41, 42].

Special Kähler manifolds and quaternionic manifolds admitting such a description were classified in the 70s by Alekseevskii *et al* [43].

8.2.2   Solvable Lie algebras: the simplest example

The simplest example of solvable Lie algebra parametrization is the case of the two-dimensional manifold $\mathcal{M} = SL(2, \mathbb{R})/SO(2)$ which may be described as the exponential of the following solvable Lie algebra:

$$
\begin{aligned}
SL(2, \mathbb{R})/SO(2) &= \exp(Solv) \\
Solv &= \{\sigma_3, \sigma_+\} \\
[\sigma_3, \sigma_+] &= 2\sigma_+
\end{aligned}
\tag{8.2.6}
$$

$$
\sigma_3 = \begin{pmatrix} 1 & 0 \\ 0 & -1 \end{pmatrix} \qquad \sigma_+ = \begin{pmatrix} 0 & 1 \\ 0 & 0 \end{pmatrix}.
$$

From (8.2.2) we can see a general feature of *Solv*, i.e. it may always be expressed as the direct sum of semisimple (the non-compact Cartan generators of the isometry group) and nilpotent generators, which in a suitable basis are represented respectively by diagonal and upper triangular matrices. This property, as we shall see, is one of the advantages of the solvable Lie algebra description since it allows us to express the coset representative of a homogeneous manifold as a solvable group element which is the product of a diagonal matrix and the exponential of a nilpotent matrix, which is a polynomial in the parameters. The simple solvable algebra represented in (8.2.2) is called *key algebra* and will be denoted by F.

8.2.3   Another example: the solvable Lie algebra of the STU-model

As a second example we illustrate the solvable Lie algebra parametrization of a coset manifold that will play a crucial role in the discussion of 1/8 preserving BPS black holes. The coset manifold we refer to is an instance of special Kähler homogeneous manifold, namely

$$
\begin{aligned}
ST[2, 2] &= \frac{SU(1, 1)}{U(1)} \otimes \frac{SO(2, 2)}{SO(2) \times SO(2)} \\
&\sim \frac{SU(1, 1)}{U(1)} \otimes \frac{SU(1, 1)}{U(1)} \otimes \frac{SU(1, 1)}{U(1)} \\
&\sim \left(\frac{SL(2, \mathbb{R})}{SO(2)} \right)^3.
\end{aligned}
\tag{8.2.7}
$$

This is the scalar manifold of an $N = 2$ supergravity theory containing three vector multiplets. The corresponding three complex scalar fields are usually named S, T and U and for this reason such a model is named the STU-model. Hence the scalar manifold of the STU-model is a special Kähler manifold generated by a solvable Lie algebra which is the sum of three commuting key algebras:

$$\mathcal{M}_{STU} = \left(\frac{SL(2, \mathbb{R})}{SO(2)}\right)^3 = \exp\left(Solv_{STU}\right)$$
$$Solv_{STU} = F_1 \oplus F_2 \oplus F_3$$
$$F_i = \{h_i, g_i\} \qquad [h_i, g_i] = 2g_i$$
$$\left[F_i, F_j\right] = 0 \qquad\qquad\qquad (8.2.8)$$

the parameters of the Cartan generators h_i are the dilatons of the theory, while the parameters of the nilpotent generators g_i are the axions. The three $SO(2)$ isotropy groups of the manifold are generated by the three compact generators $\widetilde{g}_i = g_i - g_i^\dagger$.

8.2.4 A third example: solvable Lie algebra description of the manifold $\mathcal{M}_{T^6/Z_3}$

If we compactify the type IIA superstring on the orbifold of a six-torus modulo some discrete group we obtain a low energy effective supergravity that corresponds to a truncation of the original $N = 8$ supergravity, generally with lower supersymmetry. For instance if the six extra dimensions are compactified on the orbifold $T^6/\mathbb{Z}_3$, the resulting theory is $N = 2$ supergravity coupled to **9** vector multiplets. The 18 scalars belonging to these vector multiplets are the coordinates of a special Kähler manifold that we name $\mathcal{M}_{T^6/Z_3}$ since it can be geometrically interpreted as the modulus space of Kähler class deformations of the orbifold. This space is one of the exceptional special Kähler homogeneous spaces of the Cremmer–Van Proeyen classification: indeed it is $SU(3, 3)/SU(3) \times U(3)$ (see table 6.2). It plays an important role in deriving the *generating solution* of BPS black holes of the $1/8$ type. For this reason we anticipate here its solvable Lie algebra description as an illustration of the general theory by means of a third example. We find that this 18-dimensional special Kähler manifold is generated by a solvable algebra whose structure is slightly more involved than that considered in the previous examples; indeed it contains the solvable Lie algebra of the STU-model plus some additional nilpotent generators. Explicitly we have:

$$\mathcal{M}_{T^6/Z_3} = \frac{SU(3, 3)}{SU(3) \times U(3)} = \exp\left(Solv\right)$$
$$Solv = Solv_{STU} \oplus \mathbf{X} \oplus \mathbf{Y} \oplus \mathbf{Z}. \qquad (8.2.9)$$

The four-dimensional subspaces $\mathbf{X}, \mathbf{Y}, \mathbf{Z}$ consist of nilpotent generators, while the only semisimple generators are the three Cartan generators contained in $Solv_{STU}$ which define the rank of the manifold. The algebraic structure of $Solv$ together with the details of the construction of the $SU(3, 3)$ generators in the representation **20** is what we explain next. This corresponds to the explicit construction of the symplectic embedding of the coset manifold implied by special geometry. Such a construction will be particularly relevant to the construction of the 1/8 supersymmetry preserving black holes. Indeed the **20**-dimensional representation of $SU(3, 3)$ is that which accommodates the **10** $\oplus$ **10** electric plus magnetic field strengths of $N = 2$ supergravity coupled to **9** vector multiplets. In a first step we shall argue that the $N = 8$ black holes of type 1/8 are actually solutions of such an $N = 2$ theory embedded into the $N = 8$ one. In a second step we shall even argue that the *generating solution* is actually a solution of a further truncation of this $N = 2$ theory to an STU-model.

Anticipating these physical motivations, in describing the solvable Lie algebra structure of the $\mathcal{M}_{T^6/Z_3}$ manifold we pay particular attention to how it incorporates the STU solvable algebra. Indeed applying the decomposition (8.2.5) to the manifold $\mathcal{M}_{T^6/Z_3}$ one obtains:

$$SU(3, 3) = [SU(3)_1 \oplus SU(3)_2 \oplus U(1)] \oplus Solv$$
$$Solv = F_1 \oplus F_2 \oplus F_3 \oplus \mathbf{X} \oplus \mathbf{Y} \oplus \mathbf{Z}$$
$$F_i = \{h_i , g_i\} \quad i = 1, 2, 3$$
$$\mathbf{X} = \mathbf{X}^+ \oplus \mathbf{X}^- , \ \mathbf{Y} = \mathbf{Y}^+ \oplus \mathbf{Y}^- , \ \mathbf{Z} = \mathbf{Z}^+ \oplus \mathbf{Z}^-$$
$$\left[h_i , g_i\right] = 2g_i \quad i = 1, 2, 3$$
$$\left[F_i , F_j\right] = 0 \quad i \neq j$$
$$\left[h_3 , \mathbf{Y}^\pm\right] = \pm \mathbf{Y}^\pm \qquad \left[h_3 , \mathbf{X}^\pm\right] = \pm \mathbf{X}^\pm$$
$$\left[h_2 , \mathbf{Z}^\pm\right] = \pm \mathbf{Z}^\pm \qquad \left[h_2 , \mathbf{X}^\pm\right] = \mathbf{X}^\pm \qquad (8.2.10)$$
$$\left[h_1 , \mathbf{Z}^\pm\right] = \mathbf{Z}^\pm \qquad \left[h_1 , \mathbf{Y}^\pm\right] = \mathbf{Y}^\pm$$
$$\left[g_1 , \mathbf{X}\right] = \left[g_1 , \mathbf{Y}\right] = \left[g_1 , \mathbf{Z}\right] = 0$$
$$\left[g_2 , \mathbf{X}\right] = \left[g_2 , \mathbf{Y}\right] = \left[g_2 , \mathbf{Z}^+\right] = 0 \qquad \left[g_2 , \mathbf{Z}^-\right] = \mathbf{Z}^+$$
$$\left[g_3 , \mathbf{Y}^+\right] = \left[g_3 , \mathbf{X}^+\right] = \left[g_3 , \mathbf{Z}\right] = 0$$
$$\left[g_3 , \mathbf{Y}^-\right] = \mathbf{Y}^+ \qquad \left[g_3 , \mathbf{X}^-\right] = \mathbf{X}^+$$
$$[F_1 , \mathbf{X}] = [F_2 , \mathbf{Y}] = [F_3 , \mathbf{Z}] = 0$$
$$\left[\mathbf{X}^- , \mathbf{Z}^-\right] = \mathbf{Y}^-$$

where the solvable subalgebra $Solv_{STU} = F_1 \oplus F_2 \oplus F_3$ is the solvable algebra generating $\mathcal{M}_{STU}$. Denoting by $\alpha_i, i = 1, 2, \ldots, 5$ the simple roots of $SU(3, 3)$, using the *canonical* basis for the $SU(3, 3)$ algebra, the generators in (8.2.11) have the following form:

$$h_1 = H_{\alpha_1} \qquad g_1 = iE_{\alpha_1}$$
$$h_2 = H_{\alpha_3} \qquad g_2 = iE_{\alpha_3}$$
$$h_3 = H_{\alpha_5} \qquad g_3 = iE_{\alpha_5}$$
$$\mathbf{X}^+ = \begin{pmatrix} \mathbf{X}_1^+ = i(E_{-\alpha_4} + E_{\alpha_3+\alpha_4+\alpha_5}) \\ \mathbf{X}_2^+ = E_{\alpha_3+\alpha_4+\alpha_5} - E_{-\alpha_4} \end{pmatrix}$$

$$\mathbf{X}^- = \begin{pmatrix} \mathbf{X}_1^- = i(E_{\alpha_3+\alpha_4} + E_{-(\alpha_4+\alpha_5)}) \\ \mathbf{X}_2^- = E_{\alpha_3+\alpha_4} - E_{-(\alpha_4+\alpha_5)} \end{pmatrix} \qquad (8.2.11)$$

$$\mathbf{Y}^+ = \begin{pmatrix} \mathbf{Y}_1^+ = i(E_{\alpha_1+\alpha_2+\alpha_3+\alpha_4+\alpha_5} + E_{-(\alpha_2+\alpha_3+\alpha_4)}) \\ \mathbf{Y}_2^+ = E_{\alpha_1+\alpha_2+\alpha_3+\alpha_4+\alpha_5} - E_{-(\alpha_2+\alpha_3+\alpha_4)} \end{pmatrix}$$

$$\mathbf{Y}^- = \begin{pmatrix} \mathbf{Y}_1^- = i(E_{\alpha_1+\alpha_2+\alpha_3+\alpha_4} + E_{-(\alpha_2+\alpha_3+\alpha_4+\alpha_5)}) \\ \mathbf{Y}_2^- = E_{\alpha_1+\alpha_2+\alpha_3+\alpha_4} - E_{-(\alpha_2+\alpha_3+\alpha_4+\alpha_5)} \end{pmatrix}$$

$$\mathbf{Z}^+ = \begin{pmatrix} \mathbf{Z}_1^+ = i(E_{\alpha_1+\alpha_2+\alpha_3} + E_{-\alpha_2}) \\ \mathbf{Z}_2^+ = E_{\alpha_1+\alpha_2+\alpha_3} - E_{-\alpha_2} \end{pmatrix}$$

$$\mathbf{Z}^- = \begin{pmatrix} \mathbf{Z}_1^- = i(E_{\alpha_1+\alpha_2} + E_{-(\alpha_2+\alpha_3)}) \\ \mathbf{Z}_2^- = E_{\alpha_1+\alpha_2} - E_{-(\alpha_2+\alpha_3)} \end{pmatrix}.$$

As far as the embedding of the isotropy group $SO(2)^3$ of $\mathcal{M}_{STU}$ inside the $\mathcal{M}_{T^6/Z_3}$ isotropy group $SU(3)_1 \times SU(3)_2 \times U(1)$ is concerned, the three generators of the former ($\{\tilde{g}_1, \tilde{g}_2, \tilde{g}_3\}$) are related to the Cartan generators of the latter in the following way:

$$\tilde{g}_1 = \frac{1}{2}\left(\lambda + \frac{1}{2}\left(H_{c_1} - H_{d_1} + H_{c_1+c_2} - H_{d_1+d_2}\right)\right)$$

$$\tilde{g}_2 = \frac{1}{2}\left(\lambda + \frac{1}{2}\left(H_{c_1} - H_{d_1} - 2(H_{c_1+c_2} - H_{d_1+d_2})\right)\right) \qquad (8.2.12)$$

$$\tilde{g}_3 = \frac{1}{2}\left(\lambda + \frac{1}{2}\left(-2(H_{c_1} - H_{d_1}) + (H_{c_1+c_2} - H_{d_1+d_2})\right)\right)$$

where $\{c_i\}$, $\{d_i\}$, $i = 1, 2$ are the simple roots of $SU(3)_1$ and $SU(3)_2$ respectively, while λ is the generator of $U(1)$.

In order to perform the truncation to an STU-model, one needs to know also which of the $10 + 10$ field strengths should be set to zero in order to be left with the $4 + 4$ of the STU-model. This information is provided by the decomposition of the **20** of $SU(3, 3)$ with respect to the isometry group of the STU-model, namely $[SL(2, \mathbb{R})]^3$:

$$\mathbf{20} \overset{SL(2,\mathbb{R})^3}{\longrightarrow} (\mathbf{2, 2, 2}) \oplus 2 \times [(\mathbf{2, 1, 1}) \oplus (\mathbf{1, 2, 1}) \oplus (\mathbf{1, 1, 2})]. \qquad (8.2.13)$$

Then, skew diagonalizing the five Cartan generators of $SU(3)_1 \times SU(3)_2 \times U(1)$ on the **20** we obtain the ten positive weights of the representation as five component vectors $\vec{v}^{\Lambda'}$ ($\Lambda' = 0, \ldots, 9$):

$$\{C(n)\} = \left\{\frac{H_{c_1}}{2}, \frac{H_{c_1+c_2}}{2}, \frac{H_{d_1}}{2}, \frac{H_{d_1+d_2}}{2}, \lambda\right\}$$

$$C(n) \cdot |v_x^{\Lambda'}\rangle = v_{(n)}^{\Lambda'}|v_y^{\Lambda'}\rangle$$

$$C(n) \cdot |v_y^{\Lambda'}\rangle = -v_{(n)}^{\Lambda'}|v_x^{\Lambda'}\rangle. \qquad (8.2.14)$$

Using the relation (8.2.13) we compute the value of the weights $v^{\Lambda'}$ on the three generators $\tilde{g}_i$ and find out which are the four positive weights $\vec{v}^{\Lambda}$ ($\Lambda = 0, \ldots, 3$)

of the $(\mathbf{2}, \mathbf{2}, \mathbf{2})$ in (8.2.13). The complete list of weights $\vec{v}^{\Lambda'}$ and their eigenvectors $|v_{x,y}^{\Lambda'}\rangle$ are worked out below.

Evaluated on the Cartan subalgebra $\mathcal{H}$ of $SU(3)_1 \oplus SU(3)_2 \oplus U(1)$ the weights $\vec{v}^{\Lambda'}$ read as follows:

$$\vec{v}^{\Lambda'} = v^{\Lambda'} \left(\frac{H_{c_1}}{2}, \frac{H_{c_1+c_2}}{2}, \frac{H_{d_1}}{2}, \frac{H_{d_1+d_2}}{2}, \lambda \right)$$

$$v^0 = \{0, 0, 0, 0, \tfrac{3}{2}\}$$
$$v^1 = \{\tfrac{1}{2}, \tfrac{1}{2}, -\tfrac{1}{2}, -\tfrac{1}{2}, -\tfrac{1}{2}\}$$
$$v^2 = \{0, \tfrac{1}{2}, 0, -\tfrac{1}{2}, \tfrac{1}{2}\}$$
$$v^3 = \{\tfrac{1}{2}, 0, -\tfrac{1}{2}, 0, \tfrac{1}{2}\}$$
$$v^4 = \{\tfrac{1}{2}, 0, 0, -\tfrac{1}{2}, \tfrac{1}{2}\}$$
$$v^5 = \{0, \tfrac{1}{2}, -\tfrac{1}{2}, 0, \tfrac{1}{2}\} \tag{8.2.15}$$
$$v^6 = \{\tfrac{1}{2}, 0, \tfrac{1}{2}, \tfrac{1}{2}, \tfrac{1}{2}\}$$
$$v^7 = \{\tfrac{1}{2}, \tfrac{1}{2}, \tfrac{1}{2}, 0, -\tfrac{1}{2}\}$$
$$v^8 = \{0, \tfrac{1}{2}, \tfrac{1}{2}, \tfrac{1}{2}, \tfrac{1}{2}\}$$
$$v^9 = \{\tfrac{1}{2}, \tfrac{1}{2}, 0, \tfrac{1}{2}, -\tfrac{1}{2}\}.$$

The weights (8.2.16) have been ordered in such a way that the first four define the $(\mathbf{2}, \mathbf{2}, \mathbf{2})$ of the subgroup $SL(2, \mathbb{R})^3 \subset SU(3, 3)$ and in the physical interpretation of this algebraic construction the weight $\vec{v}^0$ will be associated with the graviphoton since its restriction to the Cartan generators $H_{c_1}, H_{c_1+c_2}, H_{d_1}, H_{d_1+d_2}$ of the matter isotropy group $H_{matter} = SU(3)_1 \oplus SU(3)_2$ is trivial.

As promised, while constructing the $\mathcal{M}_{T^6/\mathbb{Z}_3}$ we have completely defined the way it incorporates the STU-model. Indeed we obtained an algebraic recipe to perform the truncation to the STU-model: setting to zero all the scalars parametrizing the 12 generators $\mathbf{X} \oplus \mathbf{Y} \oplus \mathbf{Z}$ in (8.2.9) and the six vector fields corresponding to the weights $v^{\Lambda'}$, $\Lambda' = 4, \ldots, 9$. Restricting the action of the $[SL(2, \mathbb{R})]^3$ generators $(h_i, g_i, \tilde{g}_i)$ inside $SU(3, 3)$ to the eight eigenvectors $|v_{x,y}^{\Lambda}\rangle$ $(\Lambda = 0, \ldots, 3)$ the embedding of $[SL(2, \mathbb{R})]^3$ in $Sp(8)$ is automatically obtained[1].

After performing the restriction to the $Sp(8)_D$ representation of $[SL(2, \mathbb{R})]^3$ we have just described, the orthonormal basis $|v_{x,y}^{\Lambda}\rangle$ $(\Lambda = 0, 1, 2, 3)$ is:

$$|v_x^1\rangle = \{0, 0, 0, 0, \tfrac{1}{2}, -\tfrac{1}{2}, \tfrac{1}{2}, -\tfrac{1}{2}\}$$
$$|v_x^2\rangle = \{0, 0, 0, 0, \tfrac{1}{2}, \tfrac{1}{2}, \tfrac{1}{2}, \tfrac{1}{2}\}$$

---

[1] In the $Sp(8)$ representation of the U-duality group $[SL(2, \mathbb{R})]^3$ we shall use, the non-compact Cartan generators h_i are diagonal. Such a representation will be denoted by $Sp(8)_D$, where the subscript D stands for 'Dynkin'. This notation was introduced in [44] to distinguish the representation $Sp(8)_D$ from $Sp(8)_Y$ (Y standing for 'Young') where in contrast the Cartan generators of the compact isotropy group (in our case $\tilde{g}_i$) are diagonal. The two representations are related by an orthogonal transformation.

$$|v_x^3\rangle = \{-\tfrac{1}{2}, \tfrac{1}{2}, \tfrac{1}{2}, -\tfrac{1}{2}, 0, 0, 0, 0\}$$

$$|v_x^4\rangle = \{\tfrac{1}{2}, \tfrac{1}{2}, -\tfrac{1}{2}, -\tfrac{1}{2}, 0, 0, 0, 0\}$$

$$|v_y^1\rangle = \{\tfrac{1}{2}, -\tfrac{1}{2}, \tfrac{1}{2}, -\tfrac{1}{2}, 0, 0, 0, 0\} \tag{8.2.16}$$

$$|v_y^2\rangle = \{\tfrac{1}{2}, \tfrac{1}{2}, \tfrac{1}{2}, \tfrac{1}{2}, 0, 0, 0, 0\}$$

$$|v_y^3\rangle = \{0, 0, 0, 0, -\tfrac{1}{2}, \tfrac{1}{2}, \tfrac{1}{2}, -\tfrac{1}{2}\}$$

$$|v_y^4\rangle = \{0, 0, 0, 0, \tfrac{1}{2}, \tfrac{1}{2}, -\tfrac{1}{2}, -\tfrac{1}{2}\}.$$

The $Sp(8)_D$ representation of the generators of $Solv_{STU}$ reads as:

$$h_1 = \frac{1}{2} \begin{pmatrix} 1 & 0 & 0 & 0 & 0 & 0 & 0 & 0 \\ 0 & -1 & 0 & 0 & 0 & 0 & 0 & 0 \\ 0 & 0 & 1 & 0 & 0 & 0 & 0 & 0 \\ 0 & 0 & 0 & -1 & 0 & 0 & 0 & 0 \\ 0 & 0 & 0 & 0 & -1 & 0 & 0 & 0 \\ 0 & 0 & 0 & 0 & 0 & 1 & 0 & 0 \\ 0 & 0 & 0 & 0 & 0 & 0 & -1 & 0 \\ 0 & 0 & 0 & 0 & 0 & 0 & 0 & 1 \end{pmatrix}$$

$$g_1 = \frac{1}{2} \begin{pmatrix} 0 & 0 & 0 & 0 & 0 & 0 & 1 & 0 \\ 0 & 0 & 0 & 0 & 0 & 0 & 0 & 0 \\ 0 & 0 & 0 & 0 & 1 & 0 & 0 & 0 \\ 0 & 0 & 0 & 0 & 0 & 0 & 0 & 0 \\ 0 & 0 & 0 & 0 & 0 & 0 & 0 & 0 \\ 0 & 0 & 0 & -1 & 0 & 0 & 0 & 0 \\ 0 & 0 & 0 & 0 & 0 & 0 & 0 & 0 \\ 0 & -1 & 0 & 0 & 0 & 0 & 0 & 0 \end{pmatrix}$$

$$h_2 = \frac{1}{2} \begin{pmatrix} -1 & 0 & 0 & 0 & 0 & 0 & 0 & 0 \\ 0 & -1 & 0 & 0 & 0 & 0 & 0 & 0 \\ 0 & 0 & 1 & 0 & 0 & 0 & 0 & 0 \\ 0 & 0 & 0 & 1 & 0 & 0 & 0 & 0 \\ 0 & 0 & 0 & 0 & 1 & 0 & 0 & 0 \\ 0 & 0 & 0 & 0 & 0 & 1 & 0 & 0 \\ 0 & 0 & 0 & 0 & 0 & 0 & -1 & 0 \\ 0 & 0 & 0 & 0 & 0 & 0 & 0 & -1 \end{pmatrix}$$

$$g_2 = \frac{1}{2} \begin{pmatrix} 0 & 0 & 0 & 0 & 0 & 0 & 0 & 0 \\ 0 & 0 & 0 & 0 & 0 & 0 & 0 & 0 \\ 0 & 0 & 0 & 0 & 0 & 0 & 0 & -1 \\ 0 & 0 & 0 & 0 & 0 & 0 & -1 & 0 \\ 0 & 1 & 0 & 0 & 0 & 0 & 0 & 0 \\ 1 & 0 & 0 & 0 & 0 & 0 & 0 & 0 \\ 0 & 0 & 0 & 0 & 0 & 0 & 0 & 0 \\ 0 & 0 & 0 & 0 & 0 & 0 & 0 & 0 \end{pmatrix}$$

$$h_3 = \frac{1}{2} \begin{pmatrix} 1 & 0 & 0 & 0 & 0 & 0 & 0 & 0 \\ 0 & -1 & 0 & 0 & 0 & 0 & 0 & 0 \\ 0 & 0 & -1 & 0 & 0 & 0 & 0 & 0 \\ 0 & 0 & 0 & 1 & 0 & 0 & 0 & 0 \\ 0 & 0 & 0 & 0 & -1 & 0 & 0 & 0 \\ 0 & 0 & 0 & 0 & 0 & 1 & 0 & 0 \\ 0 & 0 & 0 & 0 & 0 & 0 & 1 & 0 \\ 0 & 0 & 0 & 0 & 0 & 0 & 0 & -1 \end{pmatrix}$$

$$g_3 = \frac{1}{2} \begin{pmatrix} 0 & 0 & 0 & 0 & 0 & 0 & 0 & -1 \\ 0 & 0 & 0 & 0 & 0 & 0 & 0 & 0 \\ 0 & 0 & 0 & 0 & 0 & 0 & 0 & 0 \\ 0 & 0 & 0 & 0 & -1 & 0 & 0 & 0 \\ 0 & 0 & 0 & 0 & 0 & 0 & 0 & 0 \\ 0 & 0 & 1 & 0 & 0 & 0 & 0 & 0 \\ 0 & 1 & 0 & 0 & 0 & 0 & 0 & 0 \\ 0 & 0 & 0 & 0 & 0 & 0 & 0 & 0 \end{pmatrix}.$$

8.2.5 Solvable Lie algebras: R–R and NS–NS scalars in maximally extended supergravities

At the beginning of this section we stated that every non-compact homogeneous space $\mathcal{G}/\mathcal{H}$ is actually a solvable group manifold and that its generating solvable Lie algebra $Solv\,(\mathcal{G}/\mathcal{H})$ can be constructed utilizing roots and Dynkin diagram techniques. As we stressed this is important in string theory since it offers the possibility of introducing an intrinsic algebraic characterization of the supergravity scalars. In particular this yields a group-theoretical definition of Ramond–Ramond and Neveu–Schwarz scalars. It goes as follows. The same supergravity Lagrangian admits different interpretations as low energy theory of different superstrings related by duality transformations or of M-theory. The identification of the Ramond and Neveu Schwarz sectors is different in the different interpretations. Algebraically this corresponds to inequivalent decompositions of the solvable Lie algebra $Solv\,(\mathcal{G}/\mathcal{H})$ with respect to different subalgebras. Each string theory admits a T-duality and an S-duality group whose product $S \otimes T$ constitutes a subgroup of the U-duality group, namely of the isometry group $U \equiv \mathcal{G}$ of the homogeneous scalar manifold $\mathcal{G}/\mathcal{H}$. Physically S is a non-perturbative symmetry acting on the *dilaton* while T is a perturbative symmetry acting on the 'radii' of the compactification. There exist also two compact subgroups $\mathcal{H}_S \subset S$ and $\mathcal{H}_T \subset T$ whose product $\mathcal{H}_S \otimes \mathcal{H}_T \subset H$ is contained in the maximal compact subgroup $\mathcal{H} \subset U$ such that we can write:

$$Solv\,(\mathcal{U}/\mathcal{H}) = Solv\,(S/\mathcal{H}_S) \oplus Solv\,(T/\mathcal{H}_T) \oplus \mathcal{W} \qquad (8.2.17)$$

the three addends being all subalgebras of $Solv\,(\mathcal{U}/\mathcal{H})$. The first two addends constitute the Neveu–Schwarz sector while the last subalgebra $\mathcal{W}$ which is not only solvable but also *nilpotent* constitutes the Ramond sector relative to the chosen superstring interpretation.

An example of this way of reasoning is provided by maximal supergravities in $D = 10 - r$ dimensions. For such Lagrangians the scalar sector is given by $\mathcal{M}_{scalar} = E_{r+1(r+1)}/\mathcal{H}_{r+1}$ where the group $E_{r+1(r+1)}$ is obtained exponentiating the maximally non-compact real form of the exceptional rank $r + 1$ Lie algebra E_{r+1} and $\mathcal{H}_{r+1}$ is the corresponding maximal compact subgroup (see table 8.1). If we interpret supergravity as the low energy theory of a Type IIA superstring compactified on a torus T^r, then the appropriate S-duality group is $O(1, 1)$ and the appropriate T-duality group is $SO(r, r)$. Correspondingly we obtain the decomposition:

$$Solv\left(E_{r+1(r+1)}/\mathcal{H}_{r+1}\right) = O(1, 1) \oplus Solv\left(\frac{SO(r, r)}{SO(r) \times SO(r)}\right) \oplus \mathcal{W}_{n_{r+1}}$$

$$(8.2.18)$$

where the Ramond subalgebra $\mathcal{W}_{n_{r+1}} \equiv spin[r, r]$ is nothing else but the chiral spinor representation of $SO(r, r)$. In the four-dimensional case $r = 6$, equation (8.2.18) takes the exceptional form:

$$Solv\left(\frac{E_{7(7)}}{SU(8)}\right) = Solv\left(\frac{SL(2, R)}{O(2)}\right) \oplus Solv\left(\frac{SO(6, 6)}{SO(6) \times SO(6)}\right) \oplus \mathcal{W}_{32}.$$

$$(8.2.19)$$

The 38 Neveu–Schwarz scalars are given by the first two addends in (8.2.19), while the 32 Ramond scalars in the algebra $\mathcal{W}_{32}$ transform in the spinor representation of $SO(6, 6)$ as in all the other cases.

Alternatively we can interpret maximal supergravity in $D = 10 - r$ as the compactification on a torus T^r of a type IIB superstring. In this case the ST-duality group is different. We just have:

$$S \otimes T = O(1, 1) \otimes GL(r). \qquad (8.2.20)$$

Correspondingly we write the solvable Lie algebra decomposition:

$$Solv\left(E_{r+1(r+1)}/\mathcal{H}_{r+1}\right) = O(1, 1) \oplus Solv\left(\frac{GL(r)}{SO(r)}\right) \oplus \widetilde{\mathcal{W}}_{n_{r+1}} \qquad (8.2.21)$$

where $\widetilde{\mathcal{W}}_{r+1}$ is the new algebra of Ramond scalars with respect to the type IIB interpretation. Actually, as is well known, type IIB theory already admits an $SL(2, R)$ U-duality symmetry in ten dimensions that mixes Ramond and Neveu–Schwarz states. The proper S-duality group $O(1, 1)$ is just a maximal subgroup of such $SL(2, R)$. Correspondingly equation (8.2.21) can be restated as:

$$Solv_{r+1} \equiv Solv\left(E_{r+1(r+1)}/\mathcal{H}_{r+1}\right)$$
$$= Solv\left(SL(2, R)/O(2)\right) \oplus Solv\left(\frac{GL(r)}{SO(r)}\right) \oplus \overline{\mathcal{W}}_{n_{r+1}}.$$

$$(8.2.22)$$

Finally a third decomposition of the same solvable Lie algebra can be written if the same supergravity Lagrangian is interpreted as compactification on a torus T^7 of M-theory. For the details of this and other decompositions of the scalar sector that keep track of the sequential compactifications on multiple tori we refer the reader to the original papers [41, 42].

8.3   Summary of $N = 8$ supergravity

Having reviewed the solvable Lie algebra description of the supergravity scalar manifolds we next proceed to summarize the structure of $N = 8$ supergravity. The action is of the general form (6.1.21) with g_{IJ} being the invariant metric of $E_{7(7)}/SU(8)$ and the period matrix $\mathcal{N}$ being determined from the Gaillard–Zumino master formula (6.3.17) via the appropriate symplectic embedding of $E_{7(7)}$ into $Sp(56, \mathbb{R})$ (see table 6.1). Hence, according to the general formalism discussed in chapter 7 and to equation (7.2.2) we introduce the coset representative $\mathbb{L}$ of $\frac{E_{7(7)}}{SU(8)}$ in the **56** representation of $E_{7(7)}$:

$$\mathbb{L} = \frac{1}{\sqrt{2}} \left(\begin{array}{c|c} f + ih & \overline{f} + i\overline{h} \\ f - ih & \overline{f} - i\overline{h} \end{array} \right) \tag{8.3.1}$$

where the submatrices (h, f) are 28×28 matrices labelled by antisymmetric pairs Λ, Σ, A, B ($\Lambda, \Sigma = 1, \ldots, 8$, $A, B = 1, \ldots, 8$), the first pair transforming under $E_{7(7)}$ and the second one under $SU(8)$:

$$(h, f) = \left(h_{\Lambda\Sigma|AB}, f^{\Lambda\Sigma}_{AB} \right). \tag{8.3.2}$$

As expected from the general formalism we have $\mathbb{L} \in Usp(28, 28)$. The vielbein P_{ABCD} and the $SU(8)$ connection $\Omega_A{}^B$ of $\frac{E_{7(7)}}{SU(8)}$ are computed from the left invariant 1-form $\mathbb{L}^{-1} d\mathbb{L}$:

$$\mathbb{L}^{-1} d\mathbb{L} = \left(\begin{array}{c|c} \delta^{[A}_{[C} \Omega^{B]}_{D]} & \overline{P}^{ABCD} \\ \hline P_{ABCD} & \delta^{[C}_{[A} \overline{\Omega}^{D]}_{B]} \end{array} \right) \tag{8.3.3}$$

where $P_{ABCD} \equiv P_{ABCD,i} \, d\Phi^i$ ($i = 1, \ldots, 70$) is completely antisymmetric and satisfies the reality condition

$$P_{ABCD} = \tfrac{1}{24} \epsilon_{ABCDEFGH} \overline{P}^{EFGH}. \tag{8.3.4}$$

The bosonic Lagrangian of $N = 8$ supergravity is [39]

$$\mathcal{L} = \int \sqrt{-g} \, d^4 x \left(2R + \mathrm{Im} \, \mathcal{N}_{\Lambda\Sigma|\Gamma\Delta} F^{\Lambda\Sigma}_{\mu\nu} F^{\Gamma\Delta|\mu\nu} \right.$$
$$\left. + \frac{1}{6} P_{ABCD,i} \overline{P}^{ABCD}_j \partial_\mu \Phi^i \partial^\mu \Phi^j + \frac{1}{2} \mathrm{Re} \mathcal{N}_{\Lambda\Sigma|\Gamma\Delta} \frac{\epsilon^{\mu\nu\rho\sigma}}{\sqrt{-g}} F^{\Lambda\Sigma}_{\mu\nu} F^{\Gamma\Delta}_{\rho\sigma} \right) \tag{8.3.5}$$

where the curvature 2-form is defined as

$$R^{ab} = d\omega^{ab} - \omega^a{}_c \wedge \omega^{cb} \qquad (8.3.6)$$

and the kinetic matrix $\mathcal{N}_{\Lambda\Sigma|\Gamma\Delta}$ is given by the usual general formula:

$$\mathcal{N} = hf^{-1} \quad \rightarrow \quad \mathcal{N}_{\Lambda\Sigma|\Gamma\Delta} = h_{\Lambda\Sigma|AB} f^{-1}{}^{AB}_{\Gamma\Delta}. \qquad (8.3.7)$$

The same matrix relates the (anti-) self-dual electric and magnetic 2-form field strengths, namely, setting

$$F^{\pm\,\Lambda\Sigma} = \tfrac{1}{2}(F \pm \mathrm{i} \star F)^{\Lambda\Sigma} \qquad (8.3.8)$$

according to the general formulae (6.2.22) one has

$$\begin{aligned}
G^-_{\Lambda\Sigma} &= \overline{\mathcal{N}}_{\Lambda\Sigma|\Gamma\Delta} F^{-\,\Gamma\Delta} \\
G^+_{\Lambda\Sigma} &= \mathcal{N}_{\Lambda\Sigma|\Gamma\Delta} F^{+\,\Gamma\Delta}
\end{aligned} \qquad (8.3.9)$$

where the 'dual' field strengths $G^\pm_{\Lambda\Sigma}$, according to the general formalism (see equation (6.2.12)), are defined as $G^\pm_{\Lambda\Sigma} = \frac{\mathrm{i}}{2}\frac{\delta\mathcal{L}}{\delta F^{\pm\,\Lambda\Sigma}}$. Note that the 56-dimensional (anti-) self-dual vector $\left(F^{\pm\,\Lambda\Sigma}, G^\pm_{\Lambda\Sigma}\right)$ transforms covariantly under $U \in Sp\,(56, \mathbb{R})$

$$\begin{aligned}
U\begin{pmatrix} F \\ G \end{pmatrix} &= \begin{pmatrix} F' \\ G' \end{pmatrix} \qquad\qquad U = \begin{pmatrix} A & B \\ C & D \end{pmatrix} \\
A^t C - C^t A &= 0 \\[4pt]
B^t D - D^t B &= 0 \\
A^t D - C^t B &= 1.
\end{aligned} \qquad (8.3.10)$$

The matrix transforming the coset representative $\mathbf{L}$ from the $Usp\,(28, 28)$ basis, equation (8.3.1), to the real $Sp\,(56, \mathbb{R})$ basis is the Cayley matrix:

$$\mathbf{L}_{Usp} = \mathcal{C}\mathbf{L}_{Sp}\mathcal{C}^{-1} \qquad \mathcal{C} = \begin{pmatrix} \mathbb{1} & \mathrm{i}\mathbb{1} \\ \mathbb{1} & -\mathrm{i}\mathbb{1} \end{pmatrix} \qquad (8.3.11)$$

implying equations (7.2.4). Having established our definitions and notations, let us now write down the Killing spinor equations obtained by equating to zero the SUSY transformation laws of the gravitino $\psi_{A\mu}$ and dilatino χ_{ABC} fields of $N = 8$ supergravity in a purely bosonic background:

$$\delta\chi_{ABC} = 4\mathrm{i}\,P_{ABCD|i}\,\partial_\mu\Phi^i \gamma^\mu \epsilon^D - 3T^{(-)}_{[AB|\rho\sigma}\gamma^{\rho\sigma}\epsilon_{C]} = 0 \qquad (8.3.12)$$

$$\delta\psi_{A\mu} = \nabla_\mu\epsilon_A - \tfrac{1}{4}T^{(-)}_{AB|\rho\sigma}\gamma^{\rho\sigma}\gamma_\mu\epsilon^B = 0 \qquad (8.3.13)$$

where ∇_μ denotes the derivative covariant both with respect to Lorentz and $SU\,(8)$ local transformations

$$\nabla_\mu\epsilon_A = \partial_\mu\epsilon_A - \tfrac{1}{4}\gamma_{ab}\,\omega^{ab}\epsilon_A - \Omega_A{}^B\epsilon_B \qquad (8.3.14)$$

and where $T_{AB}^{(-)}$ is the 'dressed graviphoton' 2-form, defined according to the general formulae (7.2.17)

$$T_{AB}^{(-)} = (h_{\Lambda\Sigma AB}(\Phi)F^{-\Lambda\Sigma} - f_{AB}^{\Lambda\Sigma}(\Phi)G_{\Lambda\Sigma}^-). \tag{8.3.15}$$

From equations (8.3.7), (8.3.9) we have the following identities that are the particular $N = 8$ instance of equation (7.2.17):

$$T_{AB}^+ = 0 \rightarrow T_{AB}^- = T_{AB} \qquad \overline{T}_{AB}^- = 0 \rightarrow \overline{T}_{AB}^+ = \overline{T}_{AB}.$$

Following the general procedure indicated by equation (7.2.19) we can define the central charge:

$$Z_{AB} = \int_{S^2} T_{AB} = h_{\Lambda\Sigma|AB}p^{\Lambda\Sigma} - f_{AB}^{\Lambda\Sigma}q_{\Lambda\Sigma} \tag{8.3.16}$$

which in our case is an antisymmetric tensor transforming in the **28** irreducible representation of $SU(8)$. In equation (8.3.16) the integral of the 2-form T_{AB} is evaluated on a large 2-sphere at infinity and the quantized charges $(p_{\Lambda\Sigma}, q^{\Lambda\Sigma})$ are defined, following the general equations (7.2.21) by

$$p^{\Lambda\Sigma} = \int_{S^2} F^{\Lambda\Sigma}$$

$$q_{\Lambda\Sigma} = \int_{S^2} \mathcal{N}_{\Lambda\Sigma|\Gamma\Delta} \star F^{\Gamma\Delta}. \tag{8.3.17}$$

8.4 The Killing spinor equation and its covariance group

In order to translate equation (8.3.12) and (8.3.13) into first order differential equations on the bosonic fields of supergravity we consider a configuration where all the fermionic fields are zero and a SUSY parameter that satisfies the following conditions:

$$\begin{aligned} \chi^\mu \gamma_\mu \epsilon_A &= i\mathbb{C}_{AB}\,\epsilon^B & A, B &= 1, \dots, n_{max} \\ \epsilon_A &= 0 & A &> n_{max}. \end{aligned} \tag{8.4.1}$$

Here χ^μ is a timelike Killing vector for the space–time metric (in the following we just write $\chi^\mu \gamma_\mu = \gamma^0$) and ϵ_A, ϵ^A denote the two chiral projections of a single Majorana spinor: $\gamma_5\,\epsilon_A = \epsilon_A$, $\gamma_5\,\epsilon^A = -\epsilon^A$. This is just the particularization to the $N = 8$ case of equations (6.1.5) defining the Killing spinor ξ_A and inserting an ϵ_A with the properties (8.4.1) into equations (8.3.12), (8.3.13) we obtain the $N = 8$ instance of the general equation (6.1.6). We name such an equation the *Killing spinor equation* and the investigation of its group-theoretical structure is the main task we face in order to derive the three possible types of BPS black hole, those preserving 1/2 or 1/4 or 1/8 of the original supersymmetry. To appreciate the distinction among the three types of $N = 8$ black hole solution we need to recall the results of [47] where a classification was given of the **56**-vectors of quantized electric and magnetic charges $\vec{Q}$ characterizing such solutions. The basic argument is provided by the reduction of the central charge skew-symmetric tensor $\mathbb{Z}_{AB}$ to normal form. The reduction can always be obtained by means of local $SU(8)$ transformations, but the structure of the skew eigenvalues depends on the orbit type of the **56**-dimensional charge vector which can be described by means of its stabilizer subgroup $G_{stab}(\vec{Q}) \subset E_{7(7)}$:

$$g \in G_{stab}(\vec{Q}) \subset E_{7(7)} \quad \Longleftrightarrow \quad g\,\vec{Q} = \vec{Q}. \tag{8.4.2}$$

There are three possibilities:

SUSY	Central charge	Stabilizer $\equiv G_{stab}$	Normalizer $\equiv G_{norm}$
1/2	$Z_1 = Z_2 = Z_3 = Z_4$	$E_{6(6)}$	$O(1, 1)$
1/4	$Z_1 = Z_2 \neq Z_3 = Z_4$	$SO(5, 5)$	$SL(2, \mathbb{R}) \times O(1, 1)$
1/8	$Z_1 \neq Z_2 \neq Z_3 \neq Z_4$	$SO(4, 4)$	$SL(2, \mathbb{R})^3$

$$\tag{8.4.3}$$

where the normalizer $G_{norm}(\vec{Q})$ is defined as the subgroup of $E_{7(7)}$ that commutes with the stabilizer:

$$[G_{norm}, G_{stab}] = 0. \tag{8.4.4}$$

The main result of [44] is that the most general 1/8 black hole solution of $N = 8$ supergravity is related to the normalizer group $SL(2, \mathbb{R})^3$. In the subsequent paper [45] the 1/2 and 1/4 cases were completely worked out. Finally in [46] the explicit form of the generating solutions was discussed for the 1/8 case. In these lectures we review all these results in detailed form.

In all three cases the Killing spinor equation has two features which we want presently to stress:

(i)   It requires an efficient parametrization of the scalar field sector.
(ii) It breaks the original $SU(8)$ automorphism group of the supersymmetry algebra to the subgroup $Usp(2\,n_{max}) \times SU(8 - 2\,n_{max}) \times U(1)$.

The first feature is the reason why the use of the rank 7 solvable Lie algebra $Solv_7$ associated with $E_{7(7)}/SU(8)$ is of great help in this problem. The second feature is the reason why the solvable Lie algebra $Solv_7$ has to be decomposed in a way appropriate to the decomposition of the isotropy group $SU(8)$ with respect to the subgroup $Usp(2\,n_{max}) \times SU(8 - 2\,n_{max}) \times U(1)$.

This decomposition of the solvable Lie algebra is a close relative of the decomposition of $N = 8$ supergravity into multiplets of the lower supersymmetry $N' = 2n_{max}$. This is easily understood by recalling that close to the horizon of the black hole one doubles the supersymmetries holding in the bulk of the solution. Hence the near horizon supersymmetry is now precisely $N' = 2n_{max}$ and the black solution can be interpreted as a soliton that interpolates between *ungauged* $N = 8$ supergravity at infinity and some form of the N' supergravity at the horizon.

8.4.1 The 1/2 SUSY case

Here we have $n_{max} = 8$ and correspondingly the covariance subgroup of the Killing spinor equation is $Usp(8) \subset SU(8)$. Indeed condition (8.4.1) can be rewritten as follows:

$$\gamma^0 \epsilon_A = i\mathbb{C}_{AB} \epsilon^B \qquad A, B = 1, \dots, 8 \qquad (8.4.5)$$

where $\mathbb{C}_{AB} = -\mathbb{C}_{BA}$ denotes an 8×8 antisymmetric matrix satisfying $\mathbb{C}^2 = -\mathbb{1}$. The group $Usp(8)$ is the subgroup of unimodular, unitary 8×8 matrices that are also symplectic, namely that preserve the matrix $\mathbb{C}$. Relying on equation (8.4.3) we see that in the present case $G_{stab} = E_{6(6)}$ and $G_{norm} = O(1, 1)$. Furthermore we have the following decomposition of the **70** irreducible representation of $SU(8)$ into irreducible representations of $Usp(8)$:

$$\mathbf{70} \xrightarrow{Usp(8)} \mathbf{42} \oplus \mathbf{1} \oplus \mathbf{27}. \qquad (8.4.6)$$

We are accordingly led to decompose the solvable Lie algebra as

$$Solv_7 = Solv_6 \oplus O(1, 1) \oplus \mathbb{D}_6 \qquad (8.4.7)$$
$$70 = 42 + 1 + 27 \qquad (8.4.8)$$

where, following the notation established in [44, 41]:

$$Solv_7 \equiv Solv\left(\frac{E_{7(7)}}{SU(8)}\right)$$

$$Solv_6 \equiv Solv\left(\frac{E_{6(6)}}{Usp(8)}\right)$$

$$\begin{array}{ll} \dim Solv_7 = 70 & \text{rank } Solv_7 = 7 \\ \dim Solv_6 = 42 & \text{rank } Solv_6 = 6. \end{array} \qquad (8.4.9)$$

In equation (8.4.7) $Solv_6$ is the solvable Lie algebra that describes the scalar sector of $D = 5$, $N = 8$ supergravity, while the 27-dimensional Abelian ideal $\mathbb{D}_6$ corresponds to those $D = 4$ scalars that originate from the 27-vectors of supergravity 1-dimension above [42]. Furthermore, we can also decompose the **56** charge representation of $E_{7(7)}$ with respect to $O(1, 1) \times E_{6(6)}$ obtaining

$$56 \xrightarrow{Usp(8)} (\mathbf{1}, \mathbf{27}) \oplus (\mathbf{1}, \mathbf{27}) \oplus (\mathbf{2}, \mathbf{1}). \qquad (8.4.10)$$

In order to single out the content of the first order Killing spinor equations we need to decompose them into irreducible $Usp(8)$ representations. The gravitino equation (8.3.13) is an **8** of $SU(8)$ that remains irreducible under $Usp(8)$ reduction. On the other hand the dilatino equation (8.3.12) is a **56** of $SU(8)$ that reduces as follows:

$$56 \xrightarrow{Usp(8)} \mathbf{48} \oplus \mathbf{8}. \qquad (8.4.11)$$

Hence altogether we have that three Killing spinor equations in the representations **8, 8′, 48** constraining the scalar fields parametrizing the three subalgebras **42, 1** and **27**. Working out the consequences of these constraints and deciding which scalars are set to constants, which are instead evolving and how many charges are different from zero is what we will do in a later section 8.5. As it will be explicitly seen there the content of the Killing spinor equations, after $Usp(8)$ decomposition, is such as to set to a constant 69 scalar fields parametrizing $Solv_6 \oplus \mathbb{D}_6$, thus confirming the SLA analysis discussed in the above: indeed in this case $G_{norm} = O(1, 1)$ and $H_{norm} = \mathbf{1}$, so that there is just one surviving field parametrizing $G_{norm} = O(1, 1)$. Moreover, the same Killing spinor equations tell us that the 54 belonging to the two $(\mathbf{1}, \mathbf{27})$ representations of equation (8.4.10) are actually zero, leaving only two non-vanishing charges transforming as a doublet of $O(1, 1)$.

8.4.1.1   *The 1/4 SUSY case*

Here we have $n_{max} = 4$ and correspondingly the covariance subgroup of the Killing spinor equation is $Usp(4) \times SU(4) \times U(1) \subset SU(8)$. Indeed condition (8.4.1) can be rewritten as follows:

$$\gamma^0 \epsilon_a = i\mathbb{C}_{ab} \epsilon^b \qquad a, b = 1, \ldots, 4$$
$$\epsilon_X = 0 \qquad X = 5, \ldots, 8 \qquad (8.4.12)$$

where $\mathbb{C}_{ab} = -\mathbb{C}_{ba}$ denotes a 4×4 antisymmetric matrix satisfying $\mathbb{C}^2 = -\mathbb{1}$. The group $Usp(4)$ is the subgroup of unimodular, unitary 4×4 matrices that are also symplectic, namely that preserve the matrix $\mathbb{C}$.

We are accordingly led to decompose the solvable Lie algebra in the way we describe below. Recalling the decomposition with respect to the $S \otimes T$ duality subgroups given in equation (8.2.17) and choosing the type IIA interpretation of $N = 8$ supergravity we can start from equation (8.2.19), which we can summarize as follows:

$$Solv_7 = Solv_S \oplus Solv_T \oplus W_{32} \qquad (8.4.13)$$
$$70 = 2 + 36 + 32 \qquad (8.4.14)$$

where we have adopted the shorthand notation

$$Solv_S \equiv Solv \left(\frac{SL(2, R)}{U(1)} \right)$$

$$Solv_T \equiv Solv \left(\frac{SO(6, 6)}{SO(6) \times SO(6)} \right)$$

$$\dim Solv_S = 2 \qquad \text{rank } Solv_S = 1$$
$$\dim Solv_T = 36 \qquad \text{rank } Solv_T = 6.$$

(8.4.15)

As discussed in section 8.2.5 and more extensively explained in [41] and [42], the solvable Lie algebras $Solv_S$ and $Solv_T$ describe the dilaton–axion sector and the six torus moduli, respectively, in the interpretation of $N = 8$ supergravity as the compactification of type IIA theory on a 6-torus T^6 [42]. The rank zero Abelian subalgebra $\mathcal{W}_{32}$ is instead composed by the 32 Ramond–Ramond scalars.

Introducing the decomposition (8.4.13), (8.4.14) we have succeeded in singling out a holonomy subgroup $SU(4) \times SU(4) \times U(1) \subset SU(8)$. Indeed we have $SO(6) \equiv SU(4)$. This is a step forward but it is not yet the end of the story since we actually need a subgroup $Usp(4) \times SU(4) \times U(1)$ corresponding to the invariance group of the Killing spinor equation (8.3.13), (8.3.12) with parameter (8.4.13). This means that we must further decompose the solvable Lie algebra $Solv_T$. This latter is the manifold of the scalar fields associated with vector multiplets in an $N = 4$ decomposition of the $N = 8$ theory. Indeed the decomposition (8.4.13) with respect to the ST-duality subalgebra is the appropriate decomposition of the scalar sector according to $N = 4$ multiplets.

The further SLA decomposition we need is

$$Solv_T = Solv_{T5} \oplus Solv_{T1}$$
$$Solv_{T5} \equiv Solv \left(\frac{SO(5, 6)}{SO(5) \times SO(6)} \right)$$
$$Solv_{T1} \equiv Solv \left(\frac{SO(1, 6)}{SO(6)} \right)$$

(8.4.16)

where we rely on the isomorphism $Usp(4) \equiv SO(5)$ and we have taken into account that the **70** irreducible representation of $SU(8)$ decomposes with respect to $Usp(4) \times SU(4) \times U(1)$ as follows

$$\mathbf{70} \xrightarrow{Usp(4) \times SU(4) \times U(1)} \left(\mathbf{1, 1, 1 + \bar{1}} \right) \oplus (\mathbf{5, 6, 1}) \oplus (\mathbf{1, 6, 1}) \oplus (\mathbf{4, 4, 1}) \oplus (\mathbf{4, 4, 1}).$$

(8.4.17)

Hence, altogether we can write:

$$Solv_7 = Solv_S \oplus Solv_{T5} \oplus Solv_{T1} \oplus \mathcal{W}_{32}$$
$$70 = 2 + 30 + 6 + 32.$$

(8.4.18)

Just as in the previous case we should now single out the content of the first order Killing spinor equations by decomposing them into irreducible $Usp(4) \times SU(4) \times U(1)$ representations. The dilatino equations $\delta \chi_{ABC} = 0$, and the gravitino equation $\delta \psi_A = 0$, $A, B, C = 1, \ldots, 8$ ($SU(8)$ indices), decompose as follows

$$\mathbf{56} \xrightarrow{Usp(4) \times SU(4)} (\mathbf{4}, \mathbf{1}) \oplus 2(\mathbf{1}, \mathbf{4},) \oplus (\mathbf{5}, \mathbf{4}) \oplus (\mathbf{4}, \mathbf{6}) \tag{8.4.19}$$

$$\mathbf{8} \xrightarrow{Usp(4) \times SU(4)} (\mathbf{4}, \mathbf{1}) \oplus (\mathbf{1}, \mathbf{4}). \tag{8.4.20}$$

As we shall see explicitly in section 8.6, the content of the reduced Killing spinor equations is such that only two scalar fields are essentially dynamical, all the others being set to constant up to U-duality transformations. Moreover 52 charges are set to zero leaving four charges transforming in the $(2, 2)$ representation of $Sl(2, \mathbb{R}) \times O(1, 1)$. Note that in the present case on the basis of the SLA analysis given above, one would expect three scalar fields parametrizing $G_{norm}/H_{norm} = \frac{Sl(2, \mathbb{R})}{U(1)} \times O(1, 1)$; however, the relevant Killing spinor equation gives an extra reality constraint on the $\frac{Sl(2, \mathbb{R})}{U(1)}$ field thus reducing the number of non-trivial scalar fields to two.

8.4.1.2   The 1/8 SUSY case

Here we have $n_{max} = 2$ and $Solv_7$ must be decomposed according to the decomposition of the isotropy subgroup: $SU(8) \longrightarrow SU(2) \times U(6)$. We showed in [44] that the corresponding decomposition of the solvable Lie algebra is the following one:

$$Solv_7 = Solv_3 \oplus Solv_4 \tag{8.4.21}$$

$$Solv_3 \equiv Solv \left(SO^\star(12)/U(6) \right)$$
$$\text{rank } Solv_3 = 3$$
$$\dim Solv_3 = 30$$

$$Solv_4 \equiv Solv \left(E_{6(4)}/SU(2) \times SU(6) \right)$$
$$\text{rank } Solv_4 = 4 \tag{8.4.22}$$
$$\dim Solv_4 = 40.$$

The rank three Lie algebra $Solv_3$ defined above describes the 30-dimensional scalar sector of $N = 6$ supergravity, while the rank four solvable Lie algebra $Solv_4$ contains the remaining 40 scalars belonging to $N = 6$ spin 3/2 multiplets. It should be noted that, individually, both manifolds $\exp[Solv_3]$ and $\exp[Solv_4]$ have also an $N = 2$ interpretation since we have:

$$\exp[Solv_3] = \text{homogeneous special Kähler}$$

$$\exp[Solv_4] = \text{homogeneous quaternionic} \qquad (8.4.23)$$

so that the first manifold can describe the interaction of 15 vector multiplets, while the second can describe the interaction of ten hypermultiplets. Indeed if we decompose the $N = 8$ graviton multiplet in $N = 2$ representations we find:

$$N = 8 \,\textbf{spin 2} \xrightarrow{N=2} \textbf{spin 2} + 6 \times \textbf{spin 3/2} + 15 \times \textbf{vect. mult.}$$
$$+ 10 \times \textbf{hypermult.}$$

$$(8.4.24)$$

Introducing the decomposition (8.4.21) we found in [44] that the 40 scalars belonging to $Solv_4$ are constants independent of the radial variable r. Only the 30 scalars in the Kähler algebra $Solv_3$ can be radially dependent. In fact their radial dependence is governed by a first order differential equation that can be extracted from a suitable component of the Killing spinor equation. The result in this case is that 64 of the scalar fields are actually constant while six are dynamical. Moreover 48 charges are annihilated leaving six non-zero charges transforming in the representation $(2, 2, 2)$ of the normalizer $G_{norm} = [Sl(2, \mathbb{R})]^3$. More precisely we obtained the following result. Up to U-duality transformations the most general $N = 8$ black hole is actually an $N = 2$ black hole corresponding to a very specific choice of the special Kähler manifold, namely $\exp[Solv_3]$ as in equations (8.4.24) and (8.4.23). Furthermore up to the duality rotations of $SO^\star(12)$ this general solution is actually determined by the so called STU-model studied in [48] and based on the solvable subalgebra:

$$Solv_{STU} \equiv Solv\left(\frac{SL(2, \mathbb{R})^3}{U(1)^3}\right) \subset Solv_3. \qquad (8.4.25)$$

In other words the only truly independent degrees of freedom of the black hole solution are given by three complex scalar fields, S, T, U. This is the result we already anticipated in section 8.2.3 where we used the solvable Lie algebra $Solv_{STU}$ as a preferred example in one illustration of the general concept. The real parts of the scalar fields S, T and U correspond to the three Cartan generators of $Solv_3$ and have the physical interpretation of radii of the torus compactification from $D = 10$ to $D = 4$. The imaginary parts of these complex fields are generalized theta angles.

A more detailed argument leading to the conclusion that the only relevant scalar fields in the 1/8 solution are those of the STU-model goes as follows. Let

$$\vec{Q} \equiv \begin{pmatrix} p^{\vec{\Lambda}} \\ q_{\vec{\Sigma}} \end{pmatrix} \qquad (8.4.26)$$

be the vector of electric and magnetic charges that transforms in the **56**-dimensional real representation of the U-duality group $E_{7(7)}$. Through the Cayley matrix we can convert it to the **Usp(56)** basis namely to:

$$\begin{pmatrix} t^{\vec{\Lambda}_1} = p^{\vec{\Lambda}_1} + i q_{\vec{\Lambda}_1} \\ \bar{t}_{\vec{\Lambda}_1} = p^{\vec{\Lambda}_1} - i q_{\vec{\Lambda}_1} \end{pmatrix}. \tag{8.4.27}$$

Acting on $\vec{Q}$ by means of suitable $E_{7(7)}$ transformations, we can reduce it to a *normal* form:

$$\vec{Q} \to \vec{Q}^N \equiv \begin{pmatrix} t^0_{(1,1,1)} \\ t^1_{(1,1,15)} \\ t^2_{(1,1,15)} \\ t^3_{(1,1,15)} \\ 0 \\ \cdots \\ 0 \\ \bar{t}^0_{(1,1,1)} \\ \bar{t}^1_{(1,1,15)} \\ \bar{t}^2_{(1,1,15)} \\ \bar{t}^3_{(1,1,15)} \\ 0 \\ \cdots \\ 0 \end{pmatrix} \tag{8.4.28}$$

where there are only four complex (alternatively eight real) independent charges and these charges are located in the representation $(\mathbf{1}, \mathbf{1}, \mathbf{15})$ of the subgroup:

$$U(1) \times SU(2) \times SU(6) \subset SU(8) \subset E_{7(7)}. \tag{8.4.29}$$

Indeed the general decomposition of the fundamental 56-representation of $E_{7(7)}$ with respect to the subgroup (8.4.29) is:

$$\mathbf{56}_{real} = (\mathbf{1}, \mathbf{1}, \mathbf{1})_{comp.} \oplus (\mathbf{1}, \mathbf{2}, \mathbf{6})_{comp.} \oplus (\mathbf{1}, \mathbf{1}, \mathbf{15})_{comp.}. \tag{8.4.30}$$

Consequently also the central charge $\vec{Z} \equiv \left(Z^{AB}, Z_{CD} \right)$, which depends on $\vec{Q}$ through the coset representative in a symplectic-invariant way, through a suitable $SU(8)$ transformation will be brought to its *normal* form where it has only $4 + 4$ non-vanishing components. Since the decomposition of the complex **28**-representation of $SU(8)$ under the subgroup (8.4.29) is the same as the decomposition of the real **56** of $E_{7(7)}$, namely equation (8.4.30), we can also write:

$$\vec{Z} \rightarrow \vec{Z}^N \equiv \begin{pmatrix} z^0_{(1,1,1)} \\ z^1_{(1,1,15)} \\ z^2_{(1,1,15)} \\ z^3_{(1,1,15)} \\ 0 \\ \cdots \\ 0 \\ \bar{z}^0_{(1,1,1)} \\ \bar{z}^1_{(1,1,15)} \\ \bar{z}^2_{(1,1,15)} \\ \bar{z}^3_{(1,1,15)} \\ 0 \\ \cdots \\ 0 \end{pmatrix} . \tag{8.4.31}$$

It is an easy consequence of elementary group theory, as shown in [49], [50], that $\vec{Q}^N$ is invariant with respect to the action of an $O(4, 4)$ subgroup of $E_{7(7)}$ and its *normalizer* is an $SL(2, \mathbb{R})^3 \subset E_{7(7)}$ commuting with it. Indeed it turns out that the eight real parameters in $\vec{Q}^N$ are singlets with respect to $O(4, 4)$ and in a $(2, 2, 2)$ irreducible representation of $SL(2, \mathbb{R})^3$ as shown in the following decomposition of the **56** with respect to $O(4, 4) \otimes SL(2, \mathbb{R})^3$:

$$\mathbf{56} \rightarrow (\mathbf{8_v}, \mathbf{2}, \mathbf{1}, \mathbf{1}) \oplus (\mathbf{8_s}, \mathbf{1}, \mathbf{2}, \mathbf{1}) \oplus (\mathbf{8_{s'}}, \mathbf{1}, \mathbf{1}, \mathbf{2}) \oplus (\mathbf{1}, \mathbf{2}, \mathbf{2}, \mathbf{2}). \tag{8.4.32}$$

The corresponding subgroup of $SU(8)$ leaving $\vec{Z}^N$ invariant is therefore $SU(2)^4 \cong SO(4) \times SO(4)$ which is the maximal compact subgroup of $O(4, 4)$. The reader should compare this argument with the discussion in section 7.5.1 where we have shown that $SU(2)^4$ is the stabilizer of the central charge written in normal form and where we have argued that the most general form of the 1/8 solution should contain **5** parameters, that is:

$$5 = 56 - \dim \frac{SU(8)}{SU(2)^4}. \tag{8.4.33}$$

Indeed we can write the generic **56**-dimensional charge vector $\vec{Q}$ in terms of five normal frame parameters plus 51 'angles' which parametrize the 51 dimensional compact space $\frac{SU(8)}{SU(2)^4}$, where $SU(2)^4$ is the maximal compact subgroup of the stability group $O(4, 4)$ [51]. This same counting is achieved by arguing at the level of the normalizer $SL(2, \mathbb{R})^3$. This latter contains a $U(1)^3 \subset SU(8)$ that can be used to gauge away three of the four phases of the four complex charges $t^\Lambda = p^\Lambda + iq_\Lambda$. Independently from the path used to reach it the conclusion is

Statement 8.4.1. The generating solution of $1/8$ supersymmetry preserving $N = 8$ BPS black holes depends on five essential parameters, namely four complex numbers with the same phase.

Consider now the scalar *geodesic potential* defined in equation (7.2.29), which identically can be rewritten as follows:

$$V(\phi) \equiv \tfrac{1}{2}\overline{Z}^{AB}(\phi)\, Z_{AB}(\phi)$$

$$= \tfrac{1}{2}\vec{Q}^T [\mathbf{L}^{-1}(\phi)]^T\, \mathbf{L}^{-1}(\phi)\, \vec{Q}. \tag{8.4.34}$$

As explained in section 7.3 the minimization of the potential (8.4.34) determines the fixed values of the scalar fields at the horizon of the black hole. Because of its invariance properties the scalar potential $V(\phi)$ depends on $\vec{Z}$ and therefore on $\vec{Q}$ only through their normal forms. Since the fixed scalars at the horizon of the black hole are obtained minimizing $V(\phi)$, it can be inferred that the most general solution of this kind will depend (modulo duality transformations) only on those scalar fields associated with the *normalizer* of the normal form $\vec{Q}^N$. Indeed the dependence of $V(\phi)$ on a scalar field is achieved by acting on $\vec{Q}$ in the expression of $V(\phi)$ by means of the transformations in $Solv_7$ associated with that field. Since at any point of the scalar manifold $V(\phi)$ can be made to depend only on $\vec{Q}^N$, its minimum will be defined only by those scalars that correspond to transformations acting on the non-vanishing components of the normal form (*normalizer* of $\vec{Q}^N$). Indeed all the other isometries were used to rotate $\vec{Q}$ to the normal form $\vec{Q}^N$. Among those scalars which are not determined by the fixed point conditions there are the *flat direction fields* namely those on which the scalar potential does not depend at all:

$$\text{flat direction field } q_f \quad \leftrightarrow \quad \frac{\partial}{\partial q_f}\, V(\phi) = 0. \tag{8.4.35}$$

Some of these fields parametrize $Solv\,(O(4,4)/O(4)\times O(4))$ since they are associated with isometries leaving $\vec{Q}^N$ invariant, and the remaining ones are obtained from the latter by means of duality transformations. In order to identify the scalars which are *flat* directions of $V(\phi)$, let us consider the way in which $Solv\,(O(4,4))$ is embedded into $Solv_7$. To this effect we start by reviewing the algebraic structure of the solvable Lie algebras $Solv_3$ and $Solv_4$ defined by equations (8.4.23). Since $Solv_3$ and $Solv_4$ respectively define a special Kähler and a quaternionic manifold, it is useful to describe them in Alekseevski's formalism [43].

$Solv_3$:

$$Solv_3 = F_1 \oplus F_2 \oplus F_3 \oplus \mathbf{X} \oplus \mathbf{Y} \oplus \mathbf{Z}$$

$$F_i = \{h_i\,,\, g_i\} \qquad i = 1, 2, 3$$

$$\mathbf{X} = \mathbf{X}^+ \oplus \mathbf{X}^- = \mathbf{X}_{NS} \oplus \mathbf{X}_{RR}$$

$$\mathbf{Y} = \mathbf{Y}^+ \oplus \mathbf{Y}^- = \mathbf{Y}_{NS} \oplus \mathbf{Y}_{RR}$$

$$\mathbf{Z} = \mathbf{Z}^+ \oplus \mathbf{Z}^- = \mathbf{Z}_{NS} \oplus \mathbf{Z}_{RR}$$

$$Solv\,(SU(3,3)_1) = F_1 \oplus F_2 \oplus F_3 \oplus \mathbf{X}_{NS} \oplus \mathbf{Y}_{NS} \oplus \mathbf{Z}_{NS}$$

$$Solv\left(SL(2,\mathbb{R})^3\right) = F_1 \oplus F_2 \oplus F_3$$

$$\mathcal{W}_{12} = \mathbf{X}_{RR} \oplus \mathbf{Y}_{RR} \oplus \mathbf{Z}_{RR}$$

$$\dim(F_i) = 2 \qquad \dim(\mathbf{X}_{NS/RR}) = \dim(\mathbf{X}^{\pm}) = 4$$

$$\dim(\mathbf{Y}_{NS/RR}) = \dim(\mathbf{Y}^{\pm}) = \dim(\mathbf{Z}_{NS/RR}) = \dim(\mathbf{Z}^{\pm}) = 4$$

$$\left[\mathrm{h}_i\,,\,\mathrm{g}_i\right] = \mathrm{g}_i \qquad i = 1,2,3$$

$$\left[F_i\,,\,F_j\right] = 0 \qquad i \neq j$$

$$\left[\mathrm{h}_3\,,\,\mathbf{Y}^{\pm}\right] = \pm\tfrac{1}{2}\mathbf{Y}^{\pm}$$

$$\left[\mathrm{h}_3\,,\,\mathbf{X}^{\pm}\right] = \pm\tfrac{1}{2}\mathbf{X}^{\pm}$$

$$\left[\mathrm{h}_2\,,\,\mathbf{Z}^{\pm}\right] = \pm\tfrac{1}{2}\mathbf{Z}^{\pm}$$

$$\left[\mathrm{g}_3\,,\,\mathbf{Y}^{+}\right] = \left[\mathrm{g}_2\,,\,\mathbf{Z}^{+}\right] = \left[\mathrm{g}_3\,,\,\mathbf{X}^{+}\right] = 0$$

$$\left[\mathrm{g}_3\,,\,\mathbf{Y}^{-}\right] = \mathbf{Y}^{+} \qquad \left[\mathrm{g}_2\,,\,\mathbf{Z}^{-}\right] = \mathbf{Z}^{+} \qquad \left[\mathrm{g}_3\,,\,\mathbf{X}^{-}\right] = \mathbf{X}^{+}$$

$$[F_1\,,\,\mathbf{X}] = [F_2\,,\,\mathbf{Y}] = [F_3\,,\,\mathbf{Z}] = 0$$

$$\left[\mathbf{X}^{-}\,,\,\mathbf{Z}^{-}\right] = \mathbf{Y}^{-} \tag{8.4.36}$$

$$Solv_4:$$

$$Solv_4 = F_0 \oplus F_1' \oplus F_2' \oplus F_2' \oplus \mathbf{X}_{NS}' \oplus \mathbf{Y}_{NS}' \oplus \mathbf{Z}_{NS}' \oplus \mathcal{W}_{20}$$

$$Solv\,(SL(2,\mathbb{R})) \oplus Solv\,(SU(3,3)_2) = [F_0] \oplus [F_1' \oplus F_2' \oplus F_2' \oplus \mathbf{X}_{NS}' \oplus]$$

$$F_0 = \{\mathrm{h}_0, \mathrm{g}_0\} \qquad \left[\mathrm{h}_0, \mathrm{g}_0\right] = \mathrm{g}_0$$

$$F_i' = \{\mathrm{h}_i', \mathrm{g}_i'\} \qquad i = 1,2,3$$

$$[F_0, Solv\,(SU(3,3)_2)] = 0 \qquad [\mathrm{h}_0, \mathcal{W}_{20}] = \tfrac{1}{2}\mathcal{W}_{20}$$

$$\left[\mathrm{g}_0, \mathcal{W}_{20}\right] = \left[\mathrm{g}_0, Solv\,(SU(3,3)_2)\right] = 0$$

$$[Solv\,(SL(2,\mathbb{R})) \oplus Solv\,(SU(3,3)_2)\,,\,\mathcal{W}_{20}] = \mathcal{W}_{20}. \tag{8.4.37}$$

The operators h_i, $i = 1,2,3$, are the Cartan generators of $SO^{\star}(12)$ and g_i the corresponding axions which together with h_i complete the solvable algebra $Solv_{STU} \equiv Solv\left(SL(2,\mathbb{R})^3\right)$. Referring to the description of $Solv_4$ given in equations (8.4.37):

$$Solv\,(O(4,4)) \subset Solv_4$$

$$Solv\,(O(4,4)) = F_0 \oplus F_1' \oplus F_2' \oplus F_3' \oplus \mathcal{W}_8 \tag{8.4.38}$$

where the R–R part $\mathcal{W}_8$ of $Solv\,(O(4,4))$ is the quaternionic image of $F_0 \oplus F_1' \oplus F_2' \oplus F_3'$ in $\mathcal{W}_{20}$. Therefore $Solv\,(O(4,4))$ is parametrized by the four *hypermultiplets* containing the Cartan fields of $Solv\left(E_{6(4)}\right)$. One finds that the other flat directions are all the remaining parameters of $Solv_4$, that is all the hyperscalars.

Alternatively we can observe that since the hypermultiplet scalars are flat directions of the potential, then we can use the solvable Lie algebra $Solv_4$ to set them to zero at the horizon. Since we know from the Killing spinor equations that these 40 scalars are constants it follows that we can safely set them to zero and forget about their existence (modulo U-duality transformations). Hence the non-zero scalars required for a general solution have to be looked for among the vector multiplet scalars that is in the solvable Lie algebra $Solv_3$. In other words the most general $N = 8$ black hole (up to U-duality rotations) is given by the most general $N = 2$ black hole based on the 15-dimensional special Kähler manifold:

$$\mathcal{SK}_{15} \equiv \exp[Solv_3] = \frac{SO^\star(12)}{U(1) \times SU(6)}. \tag{8.4.39}$$

Having determined the little group of the normal form enables us to decide which among the above 30 scalars have to be kept alive in order to generate the most general BPS black hole solution (modulo U-duality).

We argue as follows. The *normalizer* of the normal form is contained in the largest subgroup of $E_{7(7)}$ commuting with $O(4, 4)$. Indeed, a necessary condition for a group G^N to be the *normalizer* of $\vec{Q}^N$ is to commute with the *little group* $G^L = O(4, 4)$ of $\vec{Q}^N$:

$$\vec{Q}'^N = G^N \cdot \vec{Q}^N \quad \vec{Q}^N = G^L \cdot \vec{Q}^N$$

$$\vec{Q}'^N = G^L \cdot \vec{Q}'^N \Rightarrow [G^N, G^L] = 0. \tag{8.4.40}$$

As previously mentioned, it was proven that $G^N = SL(2, \mathbb{R})^3 \subset SO^\star(12)$ whose solvable algebra is defined by the last of equations (8.4.36). Moreover G^N coincides with the largest subgroup of $Solv_7$ commuting with G^L.

The duality transformations associated with the $SL(2, \mathbb{R})^3$ isometries act only on the eight non-vanishing components of $\vec{Q}^N$ and therefore belong to $\mathbf{Sp(8)}$.

In conclusion the most general $N = 8$ black hole solution is described by the six scalars parametrizing $Solv_{STU} \equiv Solv\left(SL(2, \mathbb{R})^3\right)$, which are the only ones involved in the fixed point conditions at the horizon.

Another way of seeing this is to notice that all the other 64 scalars are either the 16 parameters of $Solv\,(O(4, 4))$ which are flat directions of $V\,(\phi)$, or coefficients of the $48 = 56 - 8$ transformations needed to rotate $\vec{Q}$ into $\vec{Q}^N$, that is to set 48 components of $\vec{Q}$ to zero as shown in equation (8.4.28).

Therefore in section 8.7 we shall reduce our attention to the Cartan vector multiplet sector, namely to the six vectors corresponding to the solvable Lie algebra $Solv\,(SL(2, \mathbb{R}))$.

8.5 Detailed study of the 1/2 case

As established in section 8.4, the $N = 1/2$ SUSY preserving black hole solution of $N = 8$ supergravity has four equal skew eigenvalues in the normal frame for the central charges. The stabilizer of the normal form is $E_{6(6)}$ and the normalizer of this latter in $E_{7(7)}$ is $O(1, 1)$:

$$E_{7(7)} \supset E_{6(6)} \times O(1, 1). \tag{8.5.1}$$

According to our previous discussion, the relevant subgroup of the $SU(8)$ holonomy group is $Usp(8)$, since the BPS Killing spinor conditions involve supersymmetry parameters ϵ_A, ϵ^A satisfying equation (8.4.5). Relying on this information, we can write the solvable Lie algebra decomposition (8.4.7), (8.4.8) of the σ-model scalar coset $\frac{E_{7(7)}}{SU(8)}$.

As discussed in the introduction, it is natural to guess that modulo U-duality transformations the complete solution is given in terms of a single scalar field parametrizing $O(1, 1)$.

Indeed, we can now demonstrate that according to the previous discussion there is just one scalar field, parametrizing the normalizer $O(1, 1)$, which appears in the final Lagrangian, since the Killing spinor equations imply that 69 out of the 70 scalar fields are actually constants. In order to achieve this result, we have to decompose the $SU(8)$ tensors appearing in the equations (8.3.12), (8.3.13) with respect to $Usp(8)$ irreducible representations. According to the decompositions

$$\mathbf{70} \stackrel{Usp(8)}{=} \mathbf{42} \oplus \mathbf{27} \oplus \mathbf{1}$$

$$\mathbf{28} \stackrel{Usp(8)}{=} \mathbf{27} \oplus \mathbf{1} \tag{8.5.2}$$

we have

$$P_{ABCD} = \overset{\circ}{P}_{ABCD} + \tfrac{3}{2}C_{[AB} \overset{\circ}{P}_{CD]} + \tfrac{1}{16}C_{[AB}C_{CD]}P$$

$$T_{AB} = \overset{\circ}{T}_{AB} + \tfrac{1}{8}C_{AB}T \tag{8.5.3}$$

where the notation $\overset{\circ}{t}_{A_1...,A_n}$ means that the antisymmetric tensor is $Usp(8)$ irreducible, namely has vanishing C-traces: $C^{A_1 A_2} \overset{\circ}{t}_{A_1 A_2...,A_n} = 0$.

Starting from equation (8.3.12) and using equation (8.4.5) we easily find:

$$4P_{,a}\gamma^a\gamma^0 - 6T_{ab}\gamma^{ab} = 0 \tag{8.5.4}$$

where we have twice contracted the free $Usp(8)$ indices with the $Usp(8)$ metric C_{AB}. Next, using the decomposition (8.5.3), equation (8.3.12) reduces to

$$-4\left(\overset{\circ}{P}_{ABCD,a} + \tfrac{3}{2} \overset{\circ}{P}_{[CD,a} C_{AB]}\right) C^{DL}\gamma^a\gamma^0 - 3 \overset{\circ}{T}_{[AB} \delta^L_{C]}\gamma^{ab} = 0. \tag{8.5.5}$$

Now we may alternatively contract equation (8.5.5) with C^{AB} or δ^L_C obtaining two relations on $\overset{\circ}{P}_{AB}$ and $\overset{\circ}{T}_{AB}$ which imply that they are separately zero:

$$\overset{\circ}{P}_{AB} = \overset{\circ}{T}_{AB} = 0 \tag{8.5.6}$$

which also imply, taking into account (8.5.5)

$$\overset{\circ}{P}_{ABCD} = 0. \tag{8.5.7}$$

Thus we have reached the conclusion

$$\overset{\circ}{P}_{ABCD|i} \, \partial_\mu \Phi^i \gamma^\mu \epsilon^D = 0$$

$$\overset{\circ}{P}_{AB|i} \, \partial_\mu \Phi^i \gamma^\mu \epsilon^B = 0 \tag{8.5.8}$$

$$\overset{\circ}{T}_{AB} = 0 \tag{8.5.9}$$

implying that 69 out the 70 scalar fields are actually constant, while the only surviving central charge is that associated with the singlet 2-form T. Since T_{AB} is a complex combination of the electric and magnetic field strengths (8.3.15), it is clear that equation (8.5.9) implies the vanishing of 54 of the quantized charges $p^{\Lambda\Sigma}, q_{\Lambda\Sigma}$, the surviving two charges transforming as a doublet of $O(1, 1)$ according to equation (8.4.10). The only non-trivial evolution equation relates P and T as follows:

$$\left(\widehat{P} \partial_\mu \Phi \gamma^\mu - \tfrac{3}{2} i T^{(-)}_{\rho\sigma} \gamma^{\rho\sigma} \gamma^0 \right) \epsilon_A = 0 \tag{8.5.10}$$

where we have set $P = \widehat{P} \, d\Phi$ and Φ is the unique non-trivial scalar field parametrizing $O(1, 1)$.

In order to make this equation explicit we perform the usual static *ansätze*. For the metric we set the *ansatz* (6.1.1). The scalar fields are assumed to be radially dependent as generally stated in equation (6.1.3) and for the vector field strengths we assume the *ansatz* of equation (6.6.4) which adapted to the $E_{7(7)}$ notation reads as follows:

$$F^{-\Lambda\Sigma} = \frac{1}{4\pi} t^{\Lambda\Sigma}(r) E^{(-)} \tag{8.5.11}$$

$$t^{\Lambda\Sigma}(r) = 2\pi (g + i\ell(r))^{\Lambda\Sigma}. \tag{8.5.12}$$

The anti-self-dual form $E^{(-)}$ was defined in (6.6.1). Using (8.3.8), (8.3.9), (8.3.15), (6.6.1), (8.5.11) we have

$$T^-_{ab} = i \, t^{\Lambda\Sigma}(r) E^-_{ab} C^{AB} \, \text{Im} \, \mathcal{N}_{\Lambda\Sigma,\Gamma\Delta} f^{\Gamma\Delta}_{AB} . \tag{8.5.13}$$

A simple gamma matrix manipulation gives further

$$\gamma_{ab} E^{\mp}_{ab} = 2i \frac{e^{2U}}{r^3} x^i \gamma^0 \gamma^i \left(\frac{\pm 1 + \gamma_5}{2} \right) \tag{8.5.14}$$

and we arrive at the final equation

$$\frac{d\Phi}{dr} = -\frac{\sqrt{3}}{4} \ell(r)^{\Lambda\Sigma} \, \text{Im} \, \mathcal{N}_{\Lambda\Sigma|\Gamma\Delta} \, f^{\Gamma\Delta}_{AB} \frac{e^U}{r^2} . \tag{8.5.15}$$

In equation (8.5.15), we have set $p^{\Lambda\Sigma} = 0$ since reality of the l.h.s. and of $f^{\Gamma\Delta}_{AB}$ (see equation (8.5.29)) imply the vanishing of the magnetic charge. Furthermore, we have normalized the vielbein component of the $Usp(8)$ singlet as follows

$$\widehat{P} = 4\sqrt{3} \tag{8.5.16}$$

which corresponds to normalizing the $Usp(8)$ vielbein as

$$P^{(singlet)}_{ABCD} = \frac{1}{16}\widehat{P}C_{[AB}C_{CD]} = \frac{\sqrt{3}}{4}C_{[AB}C_{CD]}\,\mathrm{d}\Phi\,. \tag{8.5.17}$$

This choice agrees with the normalization of the scalar fields existing in the current literature. Let us now consider the gravitino equation (8.3.13). Computing the spin connection ω^a_b from equation (6.1.1), we find

$$\omega^{0i} = \frac{\mathrm{d}U}{\mathrm{d}r}\frac{x^i}{r}\,e^{U(r)}V^0$$

$$\omega^{ij} = 2\frac{\mathrm{d}U}{\mathrm{d}r}\frac{x_k}{r}\,\eta^{k[i}\,V^{j]}\,e^{U} \tag{8.5.18}$$

where $V^0 = e^U\,\mathrm{d}t$, $V^i = e^{-U}\,\mathrm{d}x^i$. Setting $\epsilon_A = e^{f(r)}\zeta_A$, where ζ_A is a constant chiral spinor, we obtain

$$\left\{\frac{\mathrm{d}f}{\mathrm{d}r}\frac{x^i}{r}\,e^{f+U}\delta^B_A V^i + \Omega^B_{A,\alpha}\partial_i\Phi^\alpha\,e^f\,V^i \right.$$

$$\left. -\frac{1}{4}\left(2\frac{\mathrm{d}U}{\mathrm{d}r}\frac{x^i}{r}\,e^U\,e^f\left(\gamma^0\gamma^i V^0 + \gamma^{ij}V_j\right)\right)\delta^B_A + \delta^B_A\,T^-_{ab}\gamma^{ab}\gamma^c\gamma^0 V_c\right\}\zeta_B = 0 \tag{8.5.19}$$

where we have used equations (8.3.13), (8.3.14), (8.5.3). This equation has two sectors; setting to zero the coefficient of V^0 or of $V^i\gamma^{ij}$ and tracing over the A, B indices we find two identical equations, namely:

$$\frac{\mathrm{d}U}{\mathrm{d}r} = -\frac{1}{8}\ell(r)^{\Lambda\Sigma}\frac{e^U}{r^2}C^{AB}\,\mathrm{Im}\,\mathcal{N}_{\Lambda\Sigma,\Gamma\Delta}f^{\Gamma\Delta}_{AB}\,. \tag{8.5.20}$$

Instead, if we set to zero the coefficient of V^i, we find a differential equation for the function $f(r)$, which is uninteresting for our purposes. Comparing now equations (8.5.15) and (8.5.20) we immediately find

$$\Phi = 2\sqrt{3}\,U\,. \tag{8.5.21}$$

8.5.1 Explicit computation of the Killing equations and of the reduced Lagrangian in the 1/2 case

In order to compute the l.h.s. of equations (8.5.15), (8.5.20) and the Lagrangian of the 1/2 model, we need the explicit form of the coset representative $\mathbb{L}$ given in equation (8.3.1). This will also enable us to compute explicitly the r.h.s. of equations (8.5.15), (8.5.20). In the present case the explicit form of $\mathbb{L}$ can be retrieved by exponentiating the $Usp(8)$ singlet generator. As stated in equation (8.3.3), the scalar vielbein in the $Usp(28, 28)$ basis is given by the off-diagonal block elements of $\mathbb{L}^{-1} d\mathbb{L}$, namely

$$\mathbb{P} = \begin{pmatrix} 0 & \overline{P}_{ABCD} \\ P_{ABCD} & 0 \end{pmatrix}. \tag{8.5.22}$$

From equation (8.5.17), we see that the $Usp(8)$ singlet corresponds to the generator

$$\mathbb{K} = \frac{\sqrt{3}}{4} \left(\begin{array}{c|c} 0 & C^{[AB}C^{HL]} \\ \hline C_{[CD}C_{RS]} & 0 \end{array} \right) \tag{8.5.23}$$

and therefore, in order to construct the coset representative of the $O(1, 1)$ subgroup of $E_{7(7)}$, we need only to exponentiate $\Phi\mathbb{K}$. Note that $\mathbb{K}$ is a $Usp(8)$ singlet in the **70** representation of $SU(8)$, but it acts non-trivially in the **28** representation of the quantized charges (q_{AB}, p^{AB}). It follows that the various powers of $\mathbb{K}$ are proportional to the projection operators onto the irreducible $Usp(8)$ representations **1** and **27** of the charges:

$$\mathbb{P}_1 = \tfrac{1}{8} C^{AB} C_{RS} \tag{8.5.24}$$

$$\mathbb{P}_{27} = (\delta^{AB}_{RS} - \tfrac{1}{8} C^{AB} C_{RS}). \tag{8.5.25}$$

Straightforward exponentiation gives

$$\exp(\Phi\mathbb{K}) = \cosh\left(\frac{1}{2\sqrt{3}}\Phi\right)\mathbb{P}_{27} + \frac{3}{2}\sinh\left(\frac{1}{2\sqrt{3}}\Phi\right)\mathbb{P}_{27}\mathbb{K}\mathbb{P}_{27}$$
$$+ \cosh\left(\frac{\sqrt{3}}{2}\Phi\right)\mathbb{P}_1 + \frac{1}{2}\sinh\left(\frac{\sqrt{3}}{2}\Phi\right)\mathbb{P}_1\mathbb{K}\mathbb{P}_1. \tag{8.5.26}$$

Since we are interested only in the singlet subspace

$$\mathbb{P}_1 \exp[\Phi\mathbb{K}]\mathbb{P}_1 = \cosh\left(\frac{\sqrt{3}}{2}\Phi\right)\mathbb{P}_1 + \frac{1}{2}\sinh\left(\frac{\sqrt{3}}{2}\Phi\right)\mathbb{P}_1\mathbb{K}\mathbb{P}_1 \tag{8.5.27}$$

$$\mathbb{L}_{singlet} = \frac{1}{8} \left(\begin{array}{c|c} \cosh(\frac{\sqrt{3}}{2}\Phi)C^{AB}C_{CD} & \sinh(\frac{\sqrt{3}}{2}\Phi)C^{AB}C^{FG} \\ \hline \sinh(\frac{\sqrt{3}}{2}\Phi)C_{CD}C_{LM} & \cosh(\frac{\sqrt{3}}{2}\Phi)C_{CM}C^{FG} \end{array} \right). \tag{8.5.28}$$

Comparing (8.5.28) with the equation (8.3.1), we find[2]:

$$f = \frac{1}{8\sqrt{2}} e^{\frac{\sqrt{3}}{2}\Phi} C^{AB} C_{CD} \tag{8.5.29}$$

$$h = -i\frac{1}{8\sqrt{2}} e^{-\frac{\sqrt{3}}{2}\Phi} C_{AB} C_{CD} \tag{8.5.30}$$

---

[2] Note that we are we are writing the coset matrix with the same pairs of indices $AB, CD, \ldots$ without distinction between the pairs $\Lambda\Sigma$ and AB as was done in section 8.3.

and hence, using $\mathcal{N} = hf^{-1}$, we find

$$\mathcal{N}_{ABCD} = -i\tfrac{1}{8}\,e^{-\sqrt{3}\Phi}C_{AB}C_{CD} \tag{8.5.31}$$

so that we can compute the r.h.s. of (8.5.15), (8.5.20). Using the relation (8.5.21) we find a single equation for the unknown functions $U(r)$, $\ell(r) = C_{\Lambda\Sigma}\ell^{\Lambda\Sigma}(r)$

$$\frac{dU}{dr} = \frac{1}{8\sqrt{2}}\frac{\ell(r)}{r^2}\exp(-2U). \tag{8.5.32}$$

At this point to solve the problem completely we have to consider also the second order field equation obtained from the Lagrangian. The bosonic supersymmetric Lagrangian of the $1/2$ preserving supersymmetry case is obtained from equation (8.3.5) by substituting the values of P_{ABCD} and $\mathcal{N}_{\Lambda\Sigma|\Gamma\Delta}$ given in equations (8.5.17) and (8.3.7) into equation (8.3.5). We find

$$\mathcal{L} = 2R - e^{-\sqrt{3}\Phi}F_{\mu\nu}F^{\mu\nu} + \tfrac{1}{2}\partial_\mu\Phi\partial^\mu\Phi. \tag{8.5.33}$$

Note that this action has the general form of 0-brane action in $D = 4$ (compare with equation (5.1.1)). Furthermore recalling equations (6.1.14), we see that the value of the parameter a is

$$a = \sqrt{3}. \tag{8.5.34}$$

According to this we expect a solution where:

$$\begin{aligned}
U &= -\tfrac{1}{4}\log H(r) \\
\Phi &= -\frac{\sqrt{3}}{2}\log H(r) \\
\ell &= 2r^3\frac{d}{dr}(H(r))^{-\frac{1}{2}} = k \times (H(r))^{-\frac{3}{2}}
\end{aligned} \tag{8.5.35}$$

where $H(r) = 1 + k/r$ denotes a harmonic function. In the next subsection, by explicit calculation we show that this is indeed the BPS solution we obtain.

8.5.2   The 1/2 solution

The resulting field equations are
Einstein equation:

$$U'' + \frac{2}{r}U' - (U)^2 = \frac{1}{4}(\Phi')^2 \tag{8.5.36}$$

Maxwell equation:

$$\frac{d}{dr}(e^{-\sqrt{3}\Phi}\ell(r)) = 0 \tag{8.5.37}$$

dilaton equation:

$$\Phi'' + \frac{2}{r}\Phi' = -e^{-\sqrt{3}\Phi+2U}\ell(r)^2\frac{1}{r^4}. \tag{8.5.38}$$

From the Maxwell equations one immediately finds

$$\ell(r) = e^{\sqrt{3}\Phi(r)}.$$ (8.5.39)

Taking into account (8.5.21), the second order field equation and the first order Killing spinor equation have the common solution

$$U = -\tfrac{1}{4}\log H(x)$$

$$\Phi = -\frac{\sqrt{3}}{2}\log H(x)$$

$$\ell = H(x)^{-\frac{3}{2}}$$ (8.5.40)

where:

$$H(x) \equiv 1 + \sum_i \frac{k_i}{\vec{x} - \vec{x}_i^0}$$ (8.5.41)

is a harmonic function describing 0-branes located at $\vec{x}_\ell^0$ for $\ell = 1, 2, \ldots$, each brane carrying a charge k_i. In particular for a single 0-brane we have:

$$H(x) = 1 + \frac{k}{r}$$ (8.5.42)

and the solution reduces to the expected form (8.5.36).

Recall also from equation (6.1.14) that for $a = \sqrt{3}$ we have $\Delta = 4$. As shown in [14] the parameter Δ is a dimensional reduction invariant so that we trace back the higher dimensional origin of the 1/2 BPS black hole from such a value. If we interpret $N = 8$ supergravity as the T^7 compactification of M-theory we can compare our present four-dimensional 0-brane solution with an 11-dimensional $M2$-brane solution. They both have $\Delta = 4$ and preserve 1/2 of the 32 supersymmetry charges. We can identify our present black hole with the wrapping of the $M2$-brane on a 2-cycle of the 7-torus.

What we have shown is that the most general BPS-saturated black hole preserving 1/2 of the $N = 8$ supersymmetry is actually described by the Lagrangian (8.5.33) with the solution given by (8.5.40), in the sense that any other solution with the same property can be obtained from the present one by an $E_{7(7)}$ (U-duality) transformation.

8.6   Detailed study of the 1/4 case

Solutions preserving $\tfrac{1}{4}$ of $N = 8$ supersymmetry have two pairs of identical skew eigenvalues in the normal frame for the central charges. In this case the stability subgroup preserving the normal form is $O(5, 5)$ with normalizer subgroup in $E_{(7,7)}$ given by $SL(2, \mathbb{R}) \times O(1, 1)$ (see [47]), according to the decomposition

$$E_{(7,7)} \supset O(5, 5) \times SL(2, \mathbb{R}) \times O(1, 1) = G_{stab} \times G_{norm}.$$ (8.6.1)

The relevant fields parametrize $\frac{SL(2,\mathbf{R})}{U(1)} \times O(1, 1)$ while the surviving charges transform in the representation $(\mathbf{2}, \mathbf{2})$ of $SL(2, \mathbf{R}) \times O(1, 1)$. The group $SL(2, \mathbf{R})$ rotates electric into electric and magnetic into magnetic charges while $O(1, 1)$ mixes them. $O(1, 1)$ is therefore a true electromagnetic duality group.

8.6.1 Killing spinor equations in the 1/4 case: surviving fields and charges

The holonomy subgroup $SU(8)$ decomposes in our case as

$$SU(8) \to Usp(4) \times SU(4) \times U(1) \qquad (8.6.2)$$

indeed in this case the Killing spinors satisfy (8.4.13) where we recall the index convention:

$$
\begin{array}{lll}
A, B = 1 \ldots 8 & SU(8) \text{ indices} & \\
a, b = 1 \ldots 4 & Usp(4) \text{ indices} & (8.6.3) \\
X, Y = 5 \ldots 8 & SU(4) \text{ indices} &
\end{array}
$$

and C_{ab} is the invariant metric of $Usp(4)$. With respect to the holonomy subgroup $SU(4) \times Usp(4)$, P_{ABCD} and T_{AB} appearing in the equations (8.3.12), (8.3.13) decompose as follows:

$$
\mathbf{70} \xrightarrow{Usp(4) \times SU(4)} (\mathbf{1}, \mathbf{1}) \oplus (\mathbf{4}, \mathbf{4}) \oplus (\mathbf{5}, \mathbf{6}) \oplus (\mathbf{1}, \mathbf{6}) \oplus \left(\mathbf{\bar{4}}, \mathbf{\bar{4}}\right) \oplus \left(\mathbf{\bar{1}}, \mathbf{\bar{1}}\right)
$$
$$
\mathbf{28} \xrightarrow{Usp(4) \times SU(4)} (\mathbf{1}, \mathbf{6}) \oplus (\mathbf{4}, \mathbf{4}) \oplus (\mathbf{5}, \mathbf{1}) \oplus (\mathbf{1}, \mathbf{1}) . \qquad (8.6.4)
$$

We decompose equation (8.3.12) according to equation (8.6.4). We obtain

$$\delta \chi_{XYZ} = 0 \qquad (8.6.5)$$
$$\delta \chi_{aXY} = 0 \qquad (8.6.6)$$
$$\delta \overset{\circ}{\chi}_{abX} = C^{ab} \delta \chi_{abX} = 0 \qquad (8.6.7)$$
$$\delta \chi_{abc} = C_{[ab} \delta \chi_{c]} = 0. \qquad (8.6.8)$$

From $\delta \chi_{XYZ} = 0$ we immediately obtain

$$P_{XYZa,\alpha} \partial_\mu \Phi^\alpha = 0 \qquad (8.6.9)$$

by means of which we recognize that 16 scalar fields are actually constant in the solution.

From the reality condition of the vielbein P_{ABCD} (equation (8.3.4)) we can also conclude

$$P_{Xabc} \equiv P_{X[a} C_{bc]} = 0 \qquad (8.6.10)$$

so that there are 16 more scalar fields set to constants.

From $\delta\chi_{aXY} = 0$ we find

$$P_{XY,i}\,\partial_\mu\Phi^i\,\gamma^\mu\gamma^0\epsilon_a = T_{XY\mu\nu}\gamma^{\mu\nu}\epsilon_a \tag{8.6.11}$$

$$\overset{\circ}{P}_{XYab,i}\,\partial_\mu\Phi^i = 0 \tag{8.6.12}$$

where we have set

$$P_{XYab} = \overset{\circ}{P}_{XYab} + \tfrac{1}{4}C_{ab}P_{XY}. \tag{8.6.13}$$

Note that equation (8.6.12) sets 30 extra scalar fields to constant.

From $\delta\chi_{Xab} = 0$, using (8.6.10), one finds that also $T_{Xa} = 0$. Finally, setting

$$P_{abcd} = C_{[ab}C_{cd]}P \tag{8.6.14}$$

$$T_{ab} = \overset{\circ}{T}_{ab} + \tfrac{1}{4}C_{ab}T \tag{8.6.15}$$

the Killing spinor equation $\delta\chi_{abc} \equiv C_{[ab}\delta\chi_{c]} = 0$ yields:

$$\overset{\circ}{T}_{ab} = 0 \tag{8.6.16}$$

$$P_{,i}\,\partial_\mu\Phi^i\,\gamma^\mu\gamma^0 - \tfrac{3}{16}T_{\mu\nu}\gamma^{\mu\nu}\epsilon_a = 0. \tag{8.6.17}$$

Performing the gamma-matrix algebra and using equation (8.5.14), the relevant evolution equations (8.6.11), (8.6.17) become

$$P_{,i}\frac{d\Phi^i}{dr} = i\frac{3}{8}\,(p+i\ell\,(r))^{\Lambda\Sigma}\,\mathrm{Im}\mathcal{N}_{\Lambda\Sigma,\Gamma\Delta}f^{\Gamma\Delta}_{AB}C^{AB}\frac{e^U}{r^2}$$

$$P_{XY,i}\frac{d\Phi^i}{dr} = 2i(p+i\ell(r))^{\Lambda\Sigma}\mathrm{Im}\mathcal{N}_{\Lambda\Sigma,\Gamma\Delta}f^{\Gamma\Delta}_{XY}\frac{e^U}{r^2}. \tag{8.6.18}$$

According to our previous discussion, $P_{XY,i}$ is the vielbein of the coset $\frac{O(1,6)}{SU(4)}$, which can be reduced to depend on six real fields Φ^i since, in force of the $SU(8)$ pseudo-reality condition (8.3.4), $P_{XY,i}$ satisfies an analogous pseudo-reality condition. On the other hand $P_{,i}$ is the vielbein of $\frac{SL(2,\mathbf{R})}{U(1)}$, and it is intrinsically complex. Indeed the $SU(8)$ pseudo-reality condition relates the $SU(4)$ singlet P_{XYZW} to the $Usp(4)$ singlet P_{abcd}. Hence $P_{,i}$ depends on a complex scalar field. In conclusion we find that equations (8.6.18) are evolution equations for eight real fields, the six on which $P_{XY,i}$ depends plus the two real fields sitting in $P_{,i}$. However, according to the discussion given above, we expect that only three scalar fields, parametrizing $\frac{SL(2,\mathbf{R})}{U(1)} \times O(1,1)$, should be physically relevant. To retrieve this number we note that $O(1,1)$ is the subgroup of $O(1,6)$ which commutes with the stability subgroup $O(5,5)$, and hence also with its maximal compact subgroup $Usp(4) \times Usp(4)$. Therefore out of the six fields of $\frac{O(1,6)}{O(6)}$ we restrict our attention to the real field parametrizing $O(1,1)$, whose corresponding vielbein is $C^{XY}P_{XY} = P_1\,d\Phi_1$. Thus the second of equations (8.6.18) can be reduced to the evolution equation for the single scalar field Φ_1, namely:

$$P_1\frac{d\Phi^1}{dr} = -2\ell\,(r)^{\Lambda\Sigma}\,\mathrm{Im}\,\mathcal{N}_{\Lambda\Sigma,\Gamma\Delta}f^{\Gamma\Delta}_{XY}\frac{e^U}{r^2}. \tag{8.6.19}$$

In this equation we have set the magnetic charge $p^{\Lambda\Sigma} = 0$ since, as we show explicitly later, the quantity $\text{Im}\,\mathcal{N}_{\Lambda\Sigma,\Gamma\Delta} f^{\Gamma\Delta}_{ab}$ is actually real. Hence, since the left hand side of equation (8.6.19) is real, we are forced to set the corresponding magnetic charge to zero. On the other hand, as we now show, inspection of the gravitino Killing spinor equation, together with the first of equations (8.6.18), further reduces the number of fields to two. Indeed, let us consider the $\delta\psi_A = 0$ Killing spinor equation. The starting equation is the same as (8.5.19), (8.3.15). In the present case, however, the indices $A, B, \ldots$ are $SU(8)$ indices, which have to be decomposed with respect to $SU(4) \times Usp(4) \times U(1)$. Then, from $\delta\Psi_X = 0$, we obtain

$$\Omega^a_X = 0 \qquad T_{Xa} = 0. \tag{8.6.20}$$

From $\delta\psi_a = 0$ we obtain an equation identical to (8.5.19) with $SU(8)$ indices replaced by $SU(4)$ indices. With the same computations performed in the $Usp(8)$ case we obtain the final equation

$$\frac{dU}{dr} = -\frac{1}{4}\ell\,(r)^{\Lambda\Sigma}\,\frac{e^U}{r^2}C^{ab}\,\text{Im}\,\mathcal{N}_{\Lambda\Sigma,\Gamma\Delta} f^{\Gamma\Delta}_{ab} \tag{8.6.21}$$

where we have taken into account that $C^{ab}\,\text{Im}\,\mathcal{N}_{\Lambda\Sigma,\Gamma\Delta} f^{\Gamma\Delta}_{ab}$ must be real, implying the vanishing of the magnetic charge corresponding to the singlet of $U(1) \times SU(4) \times Usp(4)$. Furthermore, since the right-hand side of the equation (8.6.18) is proportional to the right-hand side of the gravitino equation, it turns out that the vielbein P_i must also be real. Let us name Φ_2 the scalar field appearing in left hand side of the equation (8.6.18), and P_2 the corresponding vielbein component. Equation (8.6.18) can be rewritten as:

$$P_2\frac{d\Phi_2}{dr} = -\frac{3}{8}\ell(r)^{\Lambda\Sigma}\text{Im}\,\mathcal{N}_{\Lambda\Sigma,\Gamma\Delta} f^{\Gamma\Delta}_{XY}\,C^{XY}\frac{e^U}{r^2}. \tag{8.6.22}$$

In conclusion, we see that the most general model describing BPS-saturated solutions preserving $\frac{1}{4}$ of $N = 8$ supersymmetry is given, modulo $E_{7(7)}$ transformations, in terms of two scalar fields and two electric charges.

8.6.2 Derivation of the 1/4 reduced Lagrangian in the Young basis

Our next step is to write down the Lagrangian for this model. This implies the construction of the coset representative of $\frac{SL(2,\mathbf{R})}{U(1)} \times O(1, 1)$ in terms of which the kinetic matrix of the vector fields and the σ-model metric of the scalar fields is constructed.

Once again we begin by considering such a construction in the Young basis where the field strengths are labelled as antisymmetric tensors and the $E_{7(7)}$ generators are written as $Usp(28, 28)$ matrices.

The basic steps in order to construct the desired Lagrangian consist of

(i) embedding of the appropriate $SL(2, \mathbb{R}) \times O(1, 1)$ Lie algebra in the Usp_Y (28, 28) basis for the **56** representation of $E_{7(7)}$.

(ii) performing the explicit exponentiation of the two commuting Cartan generators of the above algebra.

(iii) calculating the restriction of the $\mathbb{L}$ coset representative to the four-dimensional space spanned by the $Usp(4) \times Usp(4)$ singlet field strengths and by their magnetic duals.

(iv) deriving the restriction of the matrix $\mathcal{N}_{\Lambda\Sigma}$ to the above four-dimensional space.

(v) calculating the explicit form of the scalar vielbein P^{ABCD} and hence of the scalar kinetic terms.

Let us begin with the first issue. To this effect we consider the following two antisymmetric 8×8 matrices:

$$\varpi_{AB} = -\varpi_{BA} = \left(\begin{array}{c|c} C & 0 \\ \hline 0 & 0 \end{array}\right) \qquad \Omega_{AB} = -\Omega_{BA} = \left(\begin{array}{c|c} 0 & 0 \\ \hline 0 & C \end{array}\right) \qquad (8.6.23)$$

where each block is 4×4 and the non-vanishing block C satisfies[3]:

$$C^T = -C \qquad C^2 = -\mathbb{1}. \qquad (8.6.24)$$

The subgroup $Usp(4) \times Usp(4) \subset SU(8)$ is defined as the set of unitary unimodular matrices that preserve simultaneously ϖ and Ω:

$$A \in Usp(4) \times Usp(4) \subset SU(8) \quad \leftrightarrow \quad A^\dagger \varpi A = \varpi \quad \text{and} \quad A^\dagger \Omega A = \Omega. \qquad (8.6.25)$$

Obviously any other linear combinations of these two matrices is also preserved by the same subgroup so that we can also consider:

$$\tau^\pm_{AB} \equiv \tfrac{1}{2}(\varpi_{AB} \pm \Omega_{AB}) = \left(\begin{array}{c|c} C & 0 \\ \hline 0 & \pm C \end{array}\right). \qquad (8.6.26)$$

Introducing also the matrices:

$$\pi_{AB} = \left(\begin{array}{c|c} \mathbb{1} & 0 \\ \hline 0 & 0 \end{array}\right) \qquad \Pi_{AB} = \left(\begin{array}{c|c} 0 & 0 \\ \hline 0 & \mathbb{1} \end{array}\right) \qquad (8.6.27)$$

we have the obvious relations:

$$\pi_{AB} = -\varpi_{AC} \varpi_{CB} \qquad \Pi_{AB} = -\Omega_{AC} \Omega_{CB}. \qquad (8.6.28)$$

---

[3] The upper and lower matrices appearing in ω_{AB} and in Ω_{AB} are actually the matrices C_{ab}, $a, b = 1, \ldots, 4$ and C_{XY} $X, Y = 1, \ldots, 4$ used in the previous section.

In terms of these matrices we can easily construct the projection operators that single out from the **28** of $SU(8)$ its $Usp(4) \times Usp(4)$ irreducible components according to:

$$\mathbf{28} \overset{Usp(4) \times Usp(4)}{\Longrightarrow} (\mathbf{1, 0}) \oplus (\mathbf{0, 1}) \oplus (\mathbf{5, 0}) \oplus (\mathbf{0, 5}) \oplus (\mathbf{4, 4}). \qquad (8.6.29)$$

These projection operators are matrices mapping antisymmetric 2-tensors into antisymmetric 2-tensors and read as follows:

$$\begin{aligned}
\mathbb{P}^{(1,0)}_{AB\ RS} &= \tfrac{1}{4} \varpi_{AB} \varpi_{RS} \\
\mathbb{P}^{(0,1)}_{AB\ RS} &= \tfrac{1}{4} \Omega_{AB} \Omega_{RS} \\
\mathbb{P}^{(5,0)}_{AB\ RS} &= \tfrac{1}{2} \left(\pi_{AR} \pi_{BS} - \pi_{AS} \pi_{BR} \right) - \tfrac{1}{4} \varpi_{AB} \varpi_{RS} \\
\mathbb{P}^{(0,5)}_{AB\ RS} &= \tfrac{1}{2} \left(\Pi_{AR} \Pi_{BS} - \Pi_{AS} \Pi_{BR} \right) - \tfrac{1}{4} \Omega_{AB} \Omega_{RS} \\
\mathbb{P}^{(4,4)}_{AB\ RS} &= \tfrac{1}{8} \left(\pi_{AR} \Pi_{BS} + \Pi_{AR} \pi_{BS} - \pi_{AS} \Pi_{BR} - \Pi_{AS} \pi_{BR} \right).
\end{aligned} \qquad (8.6.30)$$

We also introduce the following shorthand notations:

$$\begin{aligned}
\ell^{AB}_{RS} &\equiv \tfrac{1}{2} \left(\pi_{AR} \pi_{BS} - \pi_{AS} \pi_{BR} \right) \\
L^{AB}_{RS} &\equiv \tfrac{1}{2} \left(\Pi_{AR} \Pi_{BS} - \Pi_{AS} \Pi_{BR} \right) \\
U^{ABCD} &\equiv \varpi^{[AB} \varpi^{CD]} = \tfrac{1}{3} \left[\varpi^{AB} \varpi^{CD} + \varpi^{AC} \varpi^{DB} + \varpi^{AD} \varpi^{BC} \right] \\
W^{ABCD} &\equiv \Omega^{[AB} \Omega^{CD]} = \tfrac{1}{3} \left[\Omega^{AB} \Omega^{CD} + \Omega^{AC} \Omega^{DB} + \Omega^{AD} \Omega^{BC} \right] \\
Z^{ABCD} &\equiv \varpi^{[AB} \Omega^{CD]} = \tfrac{1}{6} \left[\varpi^{AB} \Omega^{CD} + \varpi^{AC} \Omega^{DB} \varpi^{AD} \Omega^{BC} \right. \\
&\qquad \left. + \Omega^{AB} \varpi^{CD} + \Omega^{AC} \varpi^{DB} + \Omega^{AD} \varpi^{BC} \right].
\end{aligned} \qquad (8.6.31)$$

Then by direct calculation we can verify the following relations:

$$\begin{aligned}
Z_{ABRS} Z_{RSUV} &= \tfrac{4}{9} \left(\mathbb{P}^{(1,0)}_{AB\ UV} + \mathbb{P}^{(0,1)}_{AB\ UV} + \mathbb{P}^{(4,4)}_{AB\ UV} \right) \\
U_{ABRS} U_{RSUV} &= \tfrac{4}{9} \ell^{AB}_{UV} \\
W_{ABRS} W_{RSUV} &= \tfrac{4}{9} L^{AB}_{UV}.
\end{aligned} \qquad (8.6.32)$$

Using the above identities we can write the explicit embedding of the relevant $SL(2, \mathbb{R}) \times O(1, 1)$ Lie algebra into the $E_{7(7)}$ Lie algebra, realized in the Young basis, namely in terms of $Usp(28, 28)$ matrices. Abstractly we have:

$$\begin{aligned}
SL(2, \mathbb{R})\ \text{algebra} &\longrightarrow \begin{cases} \left[L_+, L_- \right] = 2 L_0 \\ \left[L_0, L_\pm \right] = \pm L_\pm \end{cases} \\[1em]
O(1, 1)\ \text{algebra} &\longrightarrow \qquad \mathcal{C} \\[1em]
\text{and they commute} &\qquad \left[\mathcal{C}, L_\pm \right] = \left[\mathcal{C}, L_0 \right] = 0.
\end{aligned} \qquad (8.6.33)$$

The corresponding $E_{7(7)}$ generators in the **56** Young basis representation are:

$$L_0 = \left(\begin{array}{c|c} 0 & \frac{3}{4}(U^{ABFG} + W^{ABFG}) \\ \hline \frac{3}{4}(U_{LMCD} + W_{LMCD}) & 0 \end{array} \right)$$

$$L_\pm = \left(\begin{array}{c|c} \pm\frac{i}{2}(\ell^{AB}_{CD} - L^{AB}_{CD}) & i\frac{3}{4}(U^{ABFG} - W^{ABFG}) \\ \hline -i\frac{3}{4}(U_{LMCD} - W_{LMCD}) & \mp\frac{i}{2}(\ell^{LM}_{FG} - L^{LM}_{FG}) \end{array} \right)$$

$$C = \left(\begin{array}{c|c} 0 & \frac{3}{4}Z^{ABFG} \\ \hline \frac{3}{4}Z_{LMCD} & 0 \end{array} \right). \tag{8.6.34}$$

The non-compact Cartan subalgebra of $SL(2, \mathbb{R}) \times O(1, 1)$, spanned by L_0, C is a 2-dimensional subalgebra of the full $E_{7(7)}$ Cartan subalgebra. As such this Abelian algebra is also a subalgebra of the 70-dimensional solvable Lie algebra $Solv_7$ defined in equation (8.4.10). The scalar fields associated with L_0 and C are the two dilatons parametrizing the reduced bosonic Lagrangian we want to construct. Hence our programme is to construct the coset representative:

$$\mathbf{L}(\Phi_1, \Phi_2) \equiv \exp[\Phi_1 C + \Phi_2 L_0] \tag{8.6.35}$$

and consider its restriction to the four-dimensional space spanned by the $Usp(4) \times Usp(4)$ singlets ϖ_{AB} and Ω_{AB}. Using the definitions (8.6.31) and (8.6.32) we can easily verify that:

$$\mathbb{P}^{(1,0)}_{AB\ RS} \frac{3}{4} Z^{RSUV} \mathbb{P}^{(1,0)}_{UV\ PQ} = 0$$

$$\mathbb{P}^{(0,1)}_{AB\ RS} \frac{3}{4} Z^{RSUV} \mathbb{P}^{(0,1)}_{UV\ PQ} = 0$$

$$\mathbb{P}^{(1,0)}_{AB\ RS} \frac{3}{4} Z^{RSUV} \mathbb{P}^{(0,1)}_{UV\ PQ} = \frac{1}{8}\varpi_{AB}\,\Omega_{PQ} \tag{8.6.36}$$

$$\mathbb{P}^{(0,1)}_{AB\ RS} \frac{3}{4} Z^{RSUV} \mathbb{P}^{(1,0)}_{UV\ PQ} = \frac{1}{8}\Omega_{AB}\,\varpi_{PQ}$$

and similarly:

$$\mathbb{P}^{(1,0)}_{AB\ RS} \frac{3}{4} \left(U^{RSUV} + W^{RSUV} \right) \mathbb{P}^{(1,0)}_{UV\ PQ} = \frac{1}{2}\mathbb{P}^{(1,0)}_{AB\ PQ}$$

$$\mathbb{P}^{(1,0)}_{AB\ RS} \frac{3}{4} \left(U^{RSUV} + W^{RSUV} \right) \mathbb{P}^{(0,1)}_{UV\ PQ} = 0$$

$$\mathbb{P}^{(0,1)}_{AB\ RS} \frac{3}{4} \left(U^{RSUV} + W^{RSUV} \right) \mathbb{P}^{(0,1)}_{UV\ PQ} = \frac{1}{2}\mathbb{P}^{(0,1)}_{AB\ PQ} \tag{8.6.37}$$

$$\mathbb{P}^{(0,1)}_{AB\ RS} \frac{3}{4} Z^{RSUV} \mathbb{P}^{(1,0)}_{UV|PQ} = 0.$$

This means that in the four-dimensional space spanned by the $Usp(4) \times Usp(4)$ singlets, using also the definition (8.6.26) and the shorthand notation

$$\Phi^\pm = \frac{\Phi_2 \pm \Phi_1}{2} \tag{8.6.38}$$

the coset representative can be written as follows

$$
= \left(\begin{array}{c}
\dfrac{\exp[\Phi_1 \mathcal{C} + \Phi_2 L_0]}{\cosh \Phi^+ \frac{1}{2} \tau_{AB}^+ \tau_{CD}^+ + \cosh \Phi^- \frac{1}{2} \tau_{AB}^- \tau_{CD}^-} \\[2mm]
\sinh \Phi^+ \frac{1}{2} \tau_{AB}^+ \tau_{CD}^+ + \sinh \Phi^- \frac{1}{2} \tau_{AB}^- \tau_{CD}^-
\end{array} \right.
$$

$$
\left. \begin{array}{c}
\sinh \Phi^+ \frac{1}{2} \tau_{AB}^+ \tau_{CD}^+ + \sinh \Phi^- \frac{1}{2} \tau_{AB}^- \tau_{CD}^- \\[2mm]
\cosh \Phi^+ \frac{1}{2} \tau_{AB}^+ \tau_{CD}^+ + \cosh \Phi^- \frac{1}{2} \tau_{AB}^- \tau_{CD}^-
\end{array} \right). \tag{8.6.39}
$$

Starting from equation (8.6.39) we can easily write down the matrices $f_{AB\,CD}$, $h_{AB\,CD}$ and $\mathcal{N}_{AB\,CD}$. We immediately find:

$$
f_{AB\,CD} = \frac{1}{\sqrt{2}} \left(\exp[\Phi^+] \tfrac{1}{2} \tau_{AB}^+ \tau_{CD}^+ + \exp[\Phi^-] \tfrac{1}{2} \tau_{AB}^- \tau_{CD}^- \right)
$$

$$
h_{AB\,CD} = -\frac{i}{\sqrt{2}} \left(\exp[-\Phi^+] \tfrac{1}{2} \tau_{AB}^+ \tau_{CD}^+ + \exp[-\Phi^-] \tfrac{1}{2} \tau_{AB}^- \tau_{CD}^- \right)
$$

$$
\mathcal{N}_{AB\,CD} = -\frac{i}{4} \left(\exp[-2\,\Phi^+] \tfrac{1}{2} \tau_{AB}^+ \tau_{CD}^+ + \exp[-2\,\Phi^-] \tfrac{1}{2} \tau_{AB}^- \tau_{CD}^- \right). \tag{8.6.40}
$$

To complete our programme, the last point we have to deal with is the calculation of the scalar vielbein P^{ABCD}. We have:

$$
\mathbf{L}^{-1}(\Phi_1, \Phi_2)\, \mathrm{d}\, \mathbf{L}(\Phi_1, \Phi_2)
$$

$$
= \frac{3}{4} \left(\begin{array}{c}
0 \\[2mm]
\mathrm{d}\Phi_1\, Z^{LMCD} + \mathrm{d}\Phi_2 \left(U^{LMCD} + W^{LMCD} \right)
\end{array} \right.
$$

$$
\left. \begin{array}{c}
\mathrm{d}\Phi_1\, Z^{ABFG} + \mathrm{d}\Phi_2 \left(U^{ABFG} + W^{ABFG} \right) \\[2mm]
0
\end{array} \right) \tag{8.6.41}
$$

so that we obtain

$$
P^{ABCD} = \mathrm{d}\Phi_1\, \tfrac{3}{4}\, Z^{ABCD} + \mathrm{d}\Phi_2\, \tfrac{3}{4} \left(U^{ABCD} + W^{ABCD} \right) \tag{8.6.42}
$$

and with a straightforward calculation:

$$
P_\mu^{ABCD}\, P_{ABCD}^\mu = \tfrac{3}{2}\, \partial_\mu \Phi_1\, \partial^\mu \Phi_1 + 3\, \partial_\mu \Phi_2\, \partial^\mu \Phi_2. \tag{8.6.43}
$$

Hence recalling the normalizations of the supersymmetric $N = 8$ Lagrangian (8.3.5), and introducing the two $Usp(4) \times Usp(4)$ singlet electromagnetic fields:

$$
A_\mu^{AB} = \tau_{AB}^+ \frac{1}{2\sqrt{2}} A_\mu^1 + \tau_{AB}^- \frac{1}{2\sqrt{2}} A_\mu^2 + 26 \text{ non singlet fields} \tag{8.6.44}
$$

we obtain the following reduced Lagrangian:

$$
\mathcal{L}_{red}^{1/4} = \sqrt{-g} \left[2\, R[g] + \tfrac{1}{4}\, \partial_\mu \Phi_1\, \partial^\mu \Phi_1 + \tfrac{1}{2}\, \partial_\mu \Phi_2\, \partial^\mu \Phi_2 \right.
$$

$$
\left. - \exp[-\Phi_1 - \Phi_2] \left(F_{\mu\nu}^1 \right)^2 - \exp[\Phi_1 - \Phi_2] \left(F_{\mu\nu}^2 \right)^2 \right]. \tag{8.6.45}
$$

Redefining:

$$\Phi_1 = \sqrt{2}\,h_1 \qquad \Phi_2 = h_2 \qquad (8.6.46)$$

we obtain the final standard form for the reduced Lagrangian

$$\mathcal{L}_{red}^{1/4} = \sqrt{-g}\left[2\,R[g] + \tfrac{1}{2}\partial_\mu h_1\,\partial^\mu h_1 + \tfrac{1}{2}\partial_\mu h_2\,\partial^\mu h_2 \right.$$
$$\left. - \exp\left[-\sqrt{2}\,h_1 - h_2\right]\left(F_{\mu\nu}^1\right)^2 - \exp\left[\sqrt{2}\,h_1 - h_2\right]\left(F_{\mu\nu}^2\right)^2 \right].$$

$$(8.6.47)$$

8.6.3 Solution of the reduced field equations

We can easily solve the field equations for the reduced Lagrangian (8.6.47). The Einstein equation is

$$U'' + \frac{2}{r}U' - (U')^2 = \frac{1}{4}\left(h_1'\right)^2 + \frac{1}{4}\left(h_2'\right)^2 \qquad (8.6.48)$$

the Maxwell equations are

$$\sqrt{2}\,h_1' - h_2' = -\frac{\ell_2'}{\ell_2}$$
$$\qquad (8.6.49)$$
$$-\sqrt{2}\,h_1' - h_2' = -\frac{\ell_1'}{\ell_1}$$

the scalar equations are

$$h_1'' + \frac{2}{r}h_1' = \frac{1}{\sqrt{2}r^4}\left(e^{\sqrt{2}h_1 - h_2 + 2U}\,\ell_2^2 - e^{-\sqrt{2}h_1 - h_2 + 2U}\,\ell_1^2\right)$$
$$\qquad (8.6.50)$$
$$h_2'' + \frac{2}{r}h_2' = \frac{1}{2r^4}\left(e^{\sqrt{2}h_1 - h_2 + 2U}\,\ell_2^2 + e^{-\sqrt{2}h_1 - h_2 + 2U}\,\ell_1^2\right).$$

The solution of the Maxwell equations is

$$\ell_1(r) = \ell_1\,e^{\sqrt{2}h_1 + h_2}$$
$$\qquad (8.6.51)$$
$$\ell_2(r) = \ell_2\,e^{-\sqrt{2}h_1 + h_2}$$

where

$$\ell_1 \equiv \ell_1(\infty)$$
$$\qquad (8.6.52)$$
$$\ell_2 \equiv \ell_2(\infty)$$

$$h_1 = -\frac{1}{\sqrt{2}} \log \frac{H_1}{H_2}$$

$$h_2 = -\frac{1}{2} \log H_1 H_2 \qquad (8.6.53)$$

$$U = -\frac{1}{4} \ln H_1 H_2 \qquad (8.6.54)$$

and

$$H_1(r) = 1 + \sum_i \frac{k_i^1}{\vec{x} - \vec{x}_i^{(1)}}$$

$$\qquad (8.6.55)$$

$$H_2(r) = 1 + \sum_i \frac{l_i^2}{\vec{x} - \vec{x}_i^{(2)}}$$

are a pair of harmonic functions.

We can summarize the form of the $1/4$ solution writing:

$$ds^2 = (H_1 H_2)^{-1/2} dt^2 - (H_1 H_2)^{1/2} \left(dr^2 + r^2 d\Omega_2 \right)$$

$$h_1 = -\frac{1}{\sqrt{2}} \log \frac{H_1}{H_2} \qquad (8.6.56)$$

$$h_2 = -\frac{1}{2} \log [H_1 H_2]$$

$$F^{1,2} = -dt \wedge d\vec{x} \cdot \frac{\vec{\partial}}{\partial x} \left(H_{1,2} \right)^{-1}.$$

Let us now observe that by specializing equations (8.6.57) to the case where the two harmonic functions are equal, $H_1(r) = H_2(r) = H(r)$, we exactly match the case $a = 1$, $\Delta = 2$ of equation (6.1.14). Indeed for this choice we have $h_1 = 0$ and the two vector field strengths can be identified since they are identical. The metric, the surviving scalar field h_2 and the electric field they all have the form predicted in equation (6.1.14). Looking back at equation (8.6.47) we see that the parameter a for h_2 is indeed $a = 1$ once h_1 is suppressed. This explains why we claimed that the $a = 1$ 0-brane corresponds to the case of $1/4$-preserved supersymmetry. The generating solution for $1/4$ BPS black holes is more general since it involves two harmonic functions but the near-horizon behaviour of all such solutions is the that of $a = 1$ model. Indeed near the horizon $r = 0$ all harmonic functions are proportional.

8.6.4 Comparison with the Killing spinor equations

In this section we show that the Killing spinor equations (8.6.19), (8.6.22) are identically satisfied by the solution (8.6.57) giving no further restriction on the harmonic function H_1, H_2. Using the formalism developed in section 8.6.2 the equations (8.6.19), (8.6.22) can be combined as follows:

$$\frac{16}{3}P_2\frac{d\Phi_2}{dr} \pm P_1\frac{d\Phi_1}{dr} = -2\ell\,(r)^{\Lambda\Sigma}\,\tau^{(\pm)AB}\,\mathrm{Im}\,\mathcal{N}_{\Lambda\Sigma,\Gamma\Delta}f^{\Gamma\Delta}_{AB}\frac{e^U}{r^2} \qquad (8.6.57)$$

where

$$P_2 d\Phi_2 = \tfrac{3}{8}C^{ab}C^{cd}P_{abcd} = \tfrac{3}{8}U^{ABCD}P_{ABCD} = \tfrac{3}{4}\,d\Phi_2$$

$$P_1\,d\Phi_1 = C^{XY}C^{ab}P_{XYab} = Z^{ABCD}P_{ABCD} = 2\,d\Phi_1 \qquad (8.6.58)$$

furthermore, using equations (8.6.40), we have

$$\mathrm{Im}\,\mathcal{N}_{\Lambda\Sigma,\Gamma\Delta}f^{\Gamma\Delta}_{AB} = -\frac{1}{2\sqrt{2}}\left(e^{-\Phi_+}\tau^{(+)\Lambda\Sigma}\tau^{(+)AB} + e^{-\Phi_-}\tau^{(-)\Lambda\Sigma}\tau^{(-)AB}\right).$$
$$(8.6.59)$$

Therefore, the equation (8.6.57) becomes

$$4\frac{d\Phi_2}{dr} \pm 2\frac{d\Phi_1}{dr} = 4\sqrt{2}\tau^{(\pm)}_{\Lambda\Sigma}\ell^{\Lambda\Sigma}\frac{e^{U-\Phi_\pm}}{r^2}. \qquad (8.6.60)$$

Using equation (8.6.44) we also have

$$\tau^{(+)}_{\Lambda\Sigma}q^{\Lambda\Sigma} = \frac{1}{\sqrt{2}}\ell_1\,(r)$$

$$\tau^{(-)}_{\Lambda\Sigma}q^{\Lambda\Sigma} = \frac{1}{\sqrt{2}}\ell_2\,(r)\,. \qquad (8.6.61)$$

Comparing equation (8.6.57) with (8.5.29), (6.6.1), (8.5.12), we obtain

$$\ell_1 = -r^2\frac{H_1'}{H_1^2}\sqrt{H_1 H_2}$$

$$\ell_2 = -r^2\frac{H_2'}{H_2^2}\sqrt{H_1 H_2}. \qquad (8.6.62)$$

Using all this information, a straightforward computation shows that the Killing spinor equations are identically satisfied.

8.7 Detailed study of the 1/8 case

As previously emphasized, the most general 1/8 black hole solution of $N = 8$ supergravity is, up to U-duality transformations, a solution of an STU-model suitably embedded in the original $N = 8$ theory. Therefore, in dealing with the STU-model we would like to keep track of this embedding. To this end, we shall use, as anticipated, the mathematical tool of SLA which in general provides a suitable and simple description of the embedding of a supergravity theory in a larger one. The SLA formalism is very useful in order to give a geometrical and a quasi-easy characterization of the different dynamical scalar fields belonging to the solution. Secondly, it enables one to write down the somewhat heavy first order differential system of equations for all the fields and to compute all the geometrical quantities appearing in the effective supergravity theory in a clear and direct way. Instead of considering the STU-model embedded in the whole $N = 8$ theory with scalar manifold $\mathcal{M} = E_{7(7)}/SU(8)$, it suffices to focus on its $N = 2$ truncation with scalar manifold $\mathcal{M}_{T^6/\mathbb{Z}_3} = [SU(3, 3)/SU(3) \times U(3)] \times \mathcal{M}_{quat}$ which describes the classical limit of type IIA supergravity compactified on $T^6/\mathbb{Z}_3$, $\mathcal{M}_{quat}$ being the quaternionic manifold $SO(4, 1)/SO(4)$ describing 1 hyperscalar. Within this latter simpler model we are going to construct the $N = 2$ STU-model as a consistent truncation. Indeed the embedding of the manifold $\mathcal{M}_{T^6/\mathbb{Z}_3}$ inside $N = 8$ has already been described in equations (8.4.36) where we have shown the embedding of the corresponding solvable Lie algebra into $Solv_3 \subset Solv_7$. The embedding of the STU scalar manifold $\mathcal{M}_{STU} = (SL(2, \mathbb{R})/U(1))^3$ inside $\mathcal{M}_{T^6/\mathbb{Z}_3}$ was described as an example in section 8.2.4 (see in particular equation (8.2.9)). On the other hand in section 8.4.3 we discussed how, up to $H = SU(8)$ transformations, the $N = 8$ central charge which is an 8×8 antisymmetric complex matrix can always be brought to its *normal* form in which it is skew diagonal with complex eigenvalues Z, Z_i, $i = 1, 2, 3$ ($|Z| > |Z_i|$). In order to do this one needs to make a suitable 48-parameter $SU(8)$ transformation on the central charge. This transformation may be seen as the result of a 48-parameter $E_{7(7)}$ duality transformation on the 56-dimensional charge vector and on the 70 scalars which, in the expression of the central charge, sets to zero 48 scalars (24 vector scalars and 24 hyperscalars from the $N = 2$ point of view) and 48 charges. Taking into account that there are 16 scalars parametrizing the submanifold $SO(4, 4)/SO(4) \times SO(4)$, $SO(4, 4)$ being the centralizer of the normal form, on which the eigenvalues of the central charge do not depend at all, the central charge, in its normal form, will depend only on the six scalars and eight charges defining an STU-model. The isometry group of $\mathcal{M}_{STU}$ is $[SL(2, \mathbb{R})]^3$, which is the *normalizer* of the normal form, i.e. the residual U-duality which can still act non-trivially on the six scalars and eight charges while keeping the central charge skew diagonalized. As we show in the sequel, the six scalars of the STU-model consist of three axions a_i and three dilatons p_i, whose exponential $\exp p_i$ will be denoted by $-b_i$.

In the framework of the STU-model, the eigenvalues $Z(a_i, b_i, p^\Lambda, q_\Lambda)$ and $Z_i(a_i, b_i, p^\Lambda, q_\Lambda)$ are, respectively, the local realization on modulus space of the $N = 2$ supersymmetry algebra central charge and of the three *matter* central charges associated with the three matter vector fields (see equation (6.6.14) and (6.6.15)). The BPS condition for a 1/8 black hole is that the ADM mass should equal the modulus of the central charge:

$$M_{ADM} = |Z(a_i, b_i, p^\Lambda, q_\Lambda)|. \tag{8.7.1}$$

As explained in section 6.6.1, at the horizon the field dependent central charge $|Z|$ flows to its minimum value:

$$|Z|_{min}(p^{\Lambda}, q_{\Lambda}) = |Z(a_i^{fix}, b_i^{fix}, p^{\Lambda}, q_{\Lambda})|$$

$$0 = \frac{\partial}{\partial a_i}|Z|_{a=b=fixed} = \frac{\partial}{\partial b_i}|Z|_{a=b=fixed} \qquad (8.7.2)$$

which is obtained by extremizing it with respect to the six moduli a_i, b_i. At the horizon the other eigenvalues Z_i vanish (see chapter 7). The value $|Z|_{min}$ is related to the Bekenstein–Hawking entropy of the solution and it is expressed in terms of the quartic invariant of the 56-representation of $E_{7(7)}$, which in principle depends on all the eight charges of the STU-model (see section 7.5.1 and in particular equation (7.5.6)). Nevertheless there is a residual $[U(1)]^3 \in [SL(2, \mathbb{R})]^3$ acting on the $N = 8$ central charge matrix in its normal form. These three gauge parameters can be used to reduce the number of charges appearing in the quartic invariant (entropy) from eight to five, as we have already pointed out. We shall see how these three conditions may be implemented on the eight charges at the level of the first order BPS equations in order to obtain the five parameter generating solution for the most general 1/8 black holes in $N = 8$ supergravity. This generating solution coincides with the solution generating the orbit of 1/2 BPS black holes in the truncated $N = 2$ model describing type IIA supergravity compactified on $T^6/\mathbb{Z}_3$. Therefore, in the framework of this latter simpler model, we shall work out the STU-model and construct the set of second and first order differential equations defining our solution.

8.7.1 The STU-model in the $SU(3, 3)/SU(3) \times U(3)$ theory and solvable Lie algebras

As shown in [44] and already discussed in these lectures the hyperscalars do not contribute to the dynamics of our BPS black hole, therefore, in what follows, all hyperscalars will be set to zero and we shall forget about the quaternionic factor $\mathcal{M}_{quat}$ in $\mathcal{M}_{T_6/Z_3}$. The latter will then be the scalar manifold of an $N = 2$ supergravity describing nine vector multiplets coupled with the graviton multiplet. The 18 real scalars span the manifold $\mathcal{M}_{T_6/Z_3} = SU(3, 3)/SU(3) \times U(3)$, while the ten electric and ten magnetic charges associated with the ten vector fields transform under duality in the **20** (three times antisymmetric) of $SU(3, 3)$. As anticipated, in order to show how the STU scalar manifold $\mathcal{M}_{STU}$ is embedded in $\mathcal{M}_{T_6/Z_3}$ we just have to use the SLA description. Indeed it suffices to quote the results of section 8.2.4 where the embedding of $Solv_{STU}$ into $Solv(SU(3, 3)/SU(3) \times U(3))$ was explicitly derived.

8.7.2 First order differential equations: the algebraic approach

Now that the STU-model has been constructed from the original $SU(3, 3)/SU(3) \times U(3)$ model, we may address the problem of writing down the BPS first order equations. To this end we shall use the geometrical intrinsic approach defined in [44] and eventually compare it with the special Kähler geometry formalism.

The system of first order differential equations in the background fields is obtained from the Killing spinor equations (6.6.9), (6.6.10), with parameter as in equation (6.6.11). In our case we have that $i = 1, 2, 3$ labels the three matter vector fields, while here as in any other $N = 2$ theory $A, B = 1, 2$ are the $SU(2)$ R-symmetry indices.

Following the procedure defined in previous chapters and sections, in order to obtain a system of first order differential equations out of the Killing spinor conditions we make the *ansätze* (6.1.1) for the metric, (6.6.6), (6.6.4), (6.6.3) for the field strengths and we assume radial dependence for the scalar fields. We represent the scalars of the STU-model in terms of three complex fields $\{z^i\} \equiv \{S, T, U\}$, parametrizing each of the three factors $SL(2, \mathbb{R})/SO(2)$ in $\mathcal{M}_{STU}$.

In terms of special Kähler geometry the first order equations have already been derived for a generic $N = 2$ theory and they are given by equations (6.6.12) and (6.6.13). We just have to particularize such equations to our specific STU-model.

In order to compute the explicit form of equations (6.6.12) and (6.6.13) in a geometrical intrinsic way [44] we need to decompose the four vector fields into the graviphoton $F^0_{\mu\nu}$ and the matter vector fields $F^i_{\mu\nu}$ in the same representation of the scalars z^i with respect to the isotropy group $H = [SO(2)]^3$. This decomposition is immediately performed by computing the positive weights $\vec{v}^\Lambda$ of the $(\mathbf{2}, \mathbf{2}, \mathbf{2})$ on the three generators $\{\widetilde{g}_i\}$ of H combined in such a way as to factorize in H the automorphism group $H_{aut} = SO(2)$ of the supersymmetry algebra generated by $\lambda = \widetilde{g}_1 + \widetilde{g}_2 + \widetilde{g}_3$ from the remaining $H_{matter} = [SO(2)]^2 = \{\widetilde{g}_1 - \widetilde{g}_2, \widetilde{g}_1 - \widetilde{g}_3\}$ generators acting non-trivially only on the matter fields. The real and imaginary components of the graviphoton central charge Z will be associated with the weight, say $\vec{v}^0$ having vanishing value on the generators of H_{matter}. The remaining weights will define a representation $(\mathbf{2}, \mathbf{1}, \mathbf{1}) \oplus (\mathbf{1}, \mathbf{2}, \mathbf{1}) \oplus (\mathbf{1}, \mathbf{1}, \mathbf{2})$ of H in which the real and imaginary parts of the central charges Z^i associated with $F^i_{\mu\nu}$ transform and will be denoted by $\vec{v}^i$, $i = 1, 2, 3$. This representation is the same as the one in which the six real scalar components of $z^i = a_i + ib_i$ transform with respect to H. It is useful to define on the tangent space of $\mathcal{M}_{STU}$ curved indices α and rigid indices $\widehat{\alpha}$, both running from 1 to 6. Using the solvable parametrization of $\mathcal{M}_{STU}$, which defines real coordinates ϕ^α, the generators of $Solv_{STU} = \{T^\alpha\}$ carry curved indices since they are parametrized by the coordinates, but do not transform in a representation of the isotropy group. The compact generators $\mathbb{K} = Solv_{STU} + Solv^\dagger_{STU}$ of $[SL(2, \mathbb{R})]^3$ on the other hand transform in the $(\mathbf{2}, \mathbf{1}, \mathbf{1}) \oplus (\mathbf{1}, \mathbf{2}, \mathbf{1}) \oplus (\mathbf{1}, \mathbf{1}, \mathbf{2})$ of H and we can choose an orthonormal basis (with respect to the trace) for $\mathbb{K}$ consisting of the generators $\mathbb{K}^{\widehat{\alpha}} = T^\alpha + T^{\alpha\dagger}$. These generators now carry the rigid index and are in one to one correspondence with the real scalar fields ϕ^α. There is a one to one correspondence between the non-compact matrices $\mathbb{K}^{\widehat{\alpha}}$ and the eigenvectors $|v^i_{x,y}\rangle$ ($i = 1, 2, 3$) which are orthonormal bases (in different spaces) of the same representation of H:

$$\underbrace{\{\mathbb{K}^1, \mathbb{K}^2, \mathbb{K}^3, \mathbb{K}^4, \mathbb{K}^5, \mathbb{K}^6\}}_{\{\mathbb{K}^{\widehat{a}}\}} \leftrightarrow \underbrace{\{|v_x^1\rangle, |v_y^2\rangle, |v_y^3\rangle, |v_y^1\rangle, |v_x^2\rangle, |v_x^3\rangle\}}_{\{|v^{\widehat{a}}\rangle\}} .(8.7.3)$$

The relation between the real parameters ϕ^α of the SLA and the real and imaginary parts of the complex fields z^i is:

$$\{\phi^\alpha\} \equiv \{-2a_1, -2a_2, -2a_3, \log(-b_1), \log(-b_2), \log(-b_3)\}. \quad (8.7.4)$$

Using the $Sp(8)_D$ representation of Sol_{USTU}, we construct the coset representative $\mathbb{L}(\phi^\alpha)$ of $\mathcal{M}_{STU}$ and the vielbein $\mathbb{P}_\alpha^{\widehat{a}}$ as follows:

$$\mathbb{L}(a_i, b_i) = \exp\left(T_\alpha \phi^\alpha\right)$$

$$= (1 - 2a_1 g_1)(1 - 2a_2 g_2)(1 - 2a_3 g_3)\exp\left(\sum_i \log(-b_i)h_i\right)$$

$$(8.7.5)$$

$$\mathbb{P}^{\widehat{a}} = \frac{1}{2\sqrt{2}}\text{Tr}\left(\mathbb{K}^{\widehat{a}}\mathbb{L}^{-1}d\mathbb{L}\right) = \left\{-\frac{da_1}{2b_1}, -\frac{da_2}{2b_2}, -\frac{da_3}{2b_3}, \frac{db_1}{2b_1}, \frac{db_2}{2b_2}, \frac{db_3}{2b_3}\right\}.$$

The scalar kinetic term of the Lagrangian is $\sum_{\widehat{a}}(\mathbb{P}_{\widehat{a}})^2$. The following relations between quantities computed in the solvable approach and special Kähler formalism hold:

$$\left(\mathbb{P}_{\widehat{a}}^\alpha\langle v^{\widehat{a}}|\mathbb{L}'\mathbb{C}M\right) = \sqrt{2}\left(\begin{array}{l}\text{Re}(g^{ij^*}(\overline{h}_{j^*|\Lambda})), -\text{Re}(g^{ij^*}(\overline{f}_{j^*}^\Sigma)) \\ \text{Im}(g^{ij^*}(\overline{h}_{j^*|\Lambda})), -\text{Im}(g^{ij^*}(\overline{f}_{j^*}^\Sigma))\end{array}\right)$$

$$(8.7.6)$$

$$\left(\begin{array}{l}\langle v_y^0|\mathbb{L}'\mathbb{C}M\rangle \\ \langle v_x^0|\mathbb{L}'\mathbb{C}M\rangle\end{array}\right) = \sqrt{2}\left(\begin{array}{l}\text{Re}(M_\Lambda), -\text{Re}(L^\Sigma) \\ \text{Im}(M_\Lambda), -\text{Im}(L^\Sigma)\end{array}\right)$$

where in the first equation both sides are 6×8 matrices in which the rows are labelled by α. The first three values of α correspond to the axions a_i, the last three to the dilatons $\log(-b_i)$. The columns are to be contracted with the vector consisting of the eight electric and magnetic charges $|\vec{Q}\rangle_{sc} = 2\pi(p^\Lambda, q_\Sigma)$ in the *special coordinate* symplectic gauge of $\mathcal{M}_{STU}$. In equations (8.7.7) $\mathbb{C}$ is the symplectic invariant matrix, while M is the symplectic matrix relating the charge vectors in the $Sp(8)_D$ representation and in the *special coordinate* symplectic gauge:

$$|\vec{Q}\rangle_{Sp(8)_D} = M \cdot |\vec{Q}\rangle_{sc}$$

$$M = \begin{pmatrix} 0 & 0 & 0 & 0 & 0 & 0 & 1 & 0 \\ 1 & 0 & 0 & 0 & 0 & 0 & 0 & 0 \\ 0 & 0 & 0 & 0 & 0 & 0 & 0 & 1 \\ 0 & 0 & 0 & 0 & 0 & -1 & 0 & 0 \\ 0 & 0 & -1 & 0 & 0 & 0 & 0 & 0 \\ 0 & 0 & 0 & 0 & 1 & 0 & 0 & 0 \\ 0 & 0 & 0 & -1 & 0 & 0 & 0 & 0 \\ 0 & 1 & 0 & 0 & 0 & 0 & 0 & 0 \end{pmatrix} \in Sp(8, \mathbb{R}). \quad (8.7.7)$$

Using equations (8.7.7) it is now possible to write in a geometrically intrinsic way the first order equations:

$$\frac{d\phi^\alpha}{dr} = \left(\mp \frac{e^U}{r^2}\right) \frac{1}{2\sqrt{2}\pi} \mathbb{P}^\alpha_{\hat{a}} \langle v^{\hat{a}} | \mathbb{L}'\mathbb{C}M | \vec{Q} \rangle_{sc}$$

$$\frac{dU}{dr} = \left(\mp \frac{e^U}{r^2}\right) \frac{1}{2\sqrt{2}\pi} \langle v^0_y | \mathbb{L}'\mathbb{C}M | \vec{Q} \rangle_{sc} \qquad (8.7.8)$$

$$0 = \langle v^0_x | \mathbb{L}'\mathbb{C}M | t \rangle_{sc}.$$

The full explicit form of equations (8.7.9) can be found in the appendix , 8.7.6, where, using equation (6.6.17), everything is expressed in terms of the quantized modulus-independent charges (q_Λ, p^Σ). The fixed values of the scalars at the horizon are obtained by setting the right-hand side of the above equations to zero and the result is consistent with the literature [48]:

$$(a_1 + ib_1)_{fix} = \frac{p^\Lambda q_\Lambda - 2p^1 q_1 - i\sqrt{f(p,q)}}{2p^2 p^3 - 2p^0 q_1}$$

$$(a_2 + ib_2)_{fix} = \frac{p^\Lambda q_\Lambda - 2p^2 q_2 - i\sqrt{f(p,q)}}{2p^1 p^3 - 2p^0 q_2} \qquad (8.7.9)$$

$$(a_3 + ib_3)_{fix} = \frac{p^\Lambda q_\Lambda - 2p^3 q_3 - i\sqrt{f(p,q)}}{2p^1 p^2 - 2p^0 q_3}$$

where $f(p,q)$ is the $E_{7(7)}$ quartic invariant $S^2(p,q)$ (see equations (7.5.6) and (7.5.14)) expressed as a function of all the eight charges (and whose square root is the entropy of the solution):

$$S^2(p,q)0 = f(p,q)$$
$$= -(p^0 q_0 - p^1 q_1 + p^2 q_2 + p^3 q_3)^2$$
$$+ 4(p^2 p^3 - p^0 q_1)(p^1 q_0 + q_2 q_3). \qquad (8.7.10)$$

The last of equations (8.7.9) expresses the reality condition for $Z(\phi, p, q)$ and it amounts to fixing one of the three $SO(2)$ gauge symmetries of H giving therefore a condition on the eight charges. Without spoiling the generality (up to U-duality) of the black hole solution it is still possible to fix the remaining $[SO(2)]^2$ gauges in H by imposing two conditions on the phases of the $Z^i(\phi, p, q)$. For instance we could require two of the $Z^i(\phi, p, q)$ to be imaginary. This would imply two more conditions on the charges, leading to a generating solution depending only on five parameters as we expect it to be [52]. Hence we can conclude with the following:

Statement 8.7.1. Since the radial evolution of the axion fields a_i is related to the real part of the corresponding central charge $Z^i(\phi, p, q)$ (see (6.6.12)), up to U duality transformations, the *five parameter generating solution* will have *three dilatons* and *one axion* evolving from their fixed value at the horizon to the boundary value at infinity, and two constant axions whose value is the corresponding fixed one at the horizon (*double fixed*).

8.7.3   The first order equations: the special geometry approach

The first order BPS equations may be equivalently formulated within a special
Kähler description of the manifold $\mathcal{M}_{STU}$. In the *special coordinate* symplectic
gauge, all the geometrical quantities defined on $\mathcal{M}_{STU}$ may be deduced form a
cubic *prepotential* $F(X)$:

$$\{z^i\} = \{S, T, U\} \qquad \Omega(z) = \begin{pmatrix} X^\Lambda(z) \\ F_\Sigma(z) \end{pmatrix}$$

$$X^\Lambda(z) = \begin{pmatrix} 1 \\ S \\ T \\ U \end{pmatrix}$$

$$F_\Sigma(z) = \partial_\Sigma F(X)$$

$$\mathcal{K}(z, \bar{z}) = -\log(8|\mathrm{Im}\, S\, \mathrm{Im}\, T\, \mathrm{Im}\, U|) g_{ij^*}(z, \bar{z}) = \partial_i \partial_{j^*} \mathcal{K}(z, \bar{z})$$

$$= \mathrm{diag}\{-(\bar{S} - S)^{-2}, -(\bar{T} - T)^{-2}, -(\bar{U} - U)^{-2}\} \quad (8.7.11)$$

$$\mathcal{N}_{\Lambda\Sigma} = \overline{F}_{\Lambda\Sigma} + 2i \frac{\mathrm{Im}\, F_{\Lambda\Omega}\, \mathrm{Im}\, F_{\Sigma\Pi} L^\Omega L^\Pi}{L^\Omega L^\Pi\, \mathrm{Im}\, F_{\Omega\Pi}}$$

$$F_{\Lambda\Sigma}(z) = \partial_\Lambda \partial_\Sigma F(X)$$

$$F(X) = \frac{X^1 X^2 X^3}{X^0}.$$

The covariantly holomorphic symplectic section $V(z, \bar{z})$ and its covariant derivative $U_i(z, \bar{z})$ are:

$$V(z, \bar{z}) = \begin{pmatrix} L^\Lambda(z, \bar{z}) \\ M_\Sigma(z, \bar{z}) \end{pmatrix} = e^{\mathcal{K}(z, \bar{z})/2} \Omega(z, \bar{z})$$

$$U_i(z, \bar{z}) = \begin{pmatrix} f_i^\Lambda(z, \bar{z}) \\ h_{i|\Sigma}(z, \bar{z}) \end{pmatrix} = \nabla_i V(z, \bar{z}) = \left(\partial_i + \frac{\partial_i \mathcal{K}}{2}\right) V(z, \bar{z})$$

$$\overline{U}_{i^*}(z, \bar{z}) = \begin{pmatrix} \overline{f}_{i^*}^\Lambda(z, \bar{z}) \\ \overline{h}_{i^*|\Sigma}(z, \bar{z}) \end{pmatrix} = \nabla_{i^*} \overline{V}(z, \bar{z}) = \left(\partial_{i^*} + \frac{\partial_{i^*} \mathcal{K}}{2}\right) \overline{V}(z, \bar{z})$$

$$\hspace{8cm} (8.7.12)$$

$$M_\Sigma(z, \bar{z}) = \mathcal{N}_{\Sigma\Lambda}(z, \bar{z}) L^\Lambda(z, \bar{z})$$
$$h_{i|\Sigma}(z, \bar{z}) = \overline{\mathcal{N}}_{\Sigma\Lambda}(z, \bar{z}) f_i^\Lambda(z, \bar{z}).$$

The real and imaginary part of $\mathcal{N}$ in terms of the real part a_i and imaginary part
b_i of the complex scalars z^i are:

$$\mathrm{Re}\mathcal{N} = \begin{pmatrix} 2\,a1\,a2\,a3 & -(a2\,a3) & -(a1\,a3) & -(a1\,a2) \\ -(a2\,a3) & 0 & a3 & a2 \\ -(a1\,a3) & a3 & 0 & a1 \\ -(a1\,a2) & a2 & a1 & 0 \end{pmatrix}$$

$$
\operatorname{Im} \mathcal{N} =
\begin{pmatrix}
\frac{a1^2\,b2\,b3}{b1} + \frac{b1\left(a3^2\,b2^2 + \left(a2^2 + b2^2\right)b3^2\right)}{b2\,b3} & -\frac{a1\,b2\,b3}{b1} & -\frac{a2\,b1\,b3}{b2} & -\frac{a3\,b1\,b2}{b3} \\
-\frac{a1\,b2\,b3}{b1} & \frac{b2\,b3}{b1} & 0 & 0 \\
-\frac{a2\,b1\,b3}{b2} & 0 & \frac{b1\,b3}{b2} & 0 \\
-\frac{a3\,b1\,b2}{b3} & 0 & 0 & \frac{b1\,b2}{b3}
\end{pmatrix}.
$$

$$(8.7.13)$$

Using the above defined quantities, the first order BPS equations can be written in a complex notation as in equation (6.6.12):

$$
\frac{dS}{dr} = \pm \left(\frac{e^{\mathcal{U}(r)}}{r^2}\right) \sqrt{\left|\frac{\operatorname{Im}(S)}{2\operatorname{Im}(T)\operatorname{Im}(U)}\right|}
$$

$$
\times [q_0 + \overline{U}\,q_3 - \overline{U}\,p^2\,S + q_1\,S + \overline{T}\,(-(\overline{U}\,p1) + q_2 + \overline{U}\,p^0\,S - p^3\,S)]
$$

$$
\frac{dT}{dr} = \pm \left(\frac{e^{\mathcal{U}(r)}}{r^2}\right) \sqrt{\left|\frac{\operatorname{Im}(T)}{2\operatorname{Im}(S)\operatorname{Im}(U)}\right|}
$$

$$
\times [q_0 + \overline{U}\,q_3 - \overline{U}\,p^1\,T + q_2\,T + \overline{S}\,(-(\overline{U}\,p^2) + q_1 + \overline{U}\,p^0\,T - p^3\,T)]
$$

$$
\frac{dU}{dr} = \pm \left(\frac{e^{\mathcal{U}(r)}}{r^2}\right) \sqrt{\left|\frac{\operatorname{Im}(U)}{2\operatorname{Im}(S)\operatorname{Im}(T)}\right|}
$$

$$
\times [q_0 + \overline{t}\,q_2 - \overline{T}\,p^1\,U + q_3\,U + \overline{S}\,(-(\overline{T}\,p^3) + q_1 + \overline{T}\,p^0\,U - p^2\,U)]
$$

$$
\frac{d\mathcal{U}}{dr} = \pm \left(\frac{e^{\mathcal{U}(r)}}{r^2}\right) \left(\frac{1}{2\sqrt{2}(|\operatorname{Im}(S)\operatorname{Im}(T)\operatorname{Im}(U)|)^{1/2}}\right)
$$

$$
[q_0 + S(T\,U\,p^0 - U\,p^2 - T\,p^3 + q_1) + T(-(U\,p^1) + q_2) + U\,q_3]. \quad (8.7.14)
$$

Note that to avoid confusion in the above equations the real function $U(r)$ appearing in the metric has been renamed $\mathcal{U}(r)$ so that it should not be confused with the complex scalar field U. The naming STU for the three complex scalar fields of the $SL(2, \mathbf{R})^3$ model has by now become so traditional that it could not be avoided: in the literature such an $N = 2$ theory is universally referred to as the STU-model. On the other hand the name U for the scalar function in the *ansatz* (6.1.1) for the extremum 4D black hole is also universally adopted so that the conflict of notation limited to equation (8.7.14) could be removed only at the price of using unfamiliar notations throughout all the lectures.

The central charge $Z(z, \overline{z}, p, q)$ being given by:

$$
Z(z, \overline{z}, p, q) = -\left(\frac{1}{2\sqrt{2}(|\operatorname{Im}(S)\operatorname{Im}(T)\operatorname{Im}(U)|)^{1/2}}\right)
$$

$$
\times [q_0 + S(T\,U\,p^0 - U\,p^2 - T\,p^3 + q_1)
$$

$$
+ T(-(U\,p^1) + q_2) + U\,q_3]. \quad (8.7.15)
$$

8.7.4 The solution: preliminaries and comments on the most general one

In order to find the solution of the STU-model we need also the equations of motion that must be satisfied together with the first order ones. We go on using the special Kähler formalism in order to let the comparison with various papers being more immediate. Let us first compute the field equations for the scalar fields z_i, which can be obtained from an $N = 2$ pure supergravity action coupled to three vector multiplets. From the action of equation (6.5.1) we obtain it Maxwell's equations.

The field equations for the vector fields and the Bianchi identities read:

$$\partial_\mu(\sqrt{-g}\,\widetilde{G}^{\mu\nu}) = 0$$
$$\partial_\mu(\sqrt{-g}\,\widetilde{F}^{\mu\nu}) = 0. \tag{8.7.16}$$

Using the *ansatz* (6.6.4) the second equation is automatically fulfilled while the first equation, as anticipated before, requires the quantized electric charges q_Λ defined by equation (6.6.17) to be r-independent (equation (6.6.18)).

Scalar equations: varying with respect to z^i one obtains

$$-\frac{1}{\sqrt{-g}}\partial_\mu\left(\sqrt{-g}\,g^{\mu\nu}g_{ij^\star}\partial_\nu\bar{z}^{j^\star}\right) + \partial_i(g_{kj^\star})\partial_\mu z^k \partial_\nu\bar{z}^{j^\star}g^{\mu\nu}$$
$$+ (\partial_i\,\mathrm{Im}\mathcal{N}_{\Lambda\Sigma})F^\Lambda_{..}F^{\Sigma|..} + (\partial_i\,\mathrm{Re}\mathcal{N}_{\Lambda\Sigma})F^\Lambda_{..}\widetilde{F}^{\Sigma|..} = 0 \tag{8.7.17}$$

which, once projected onto the real and imaginary parts of both sides, read:

$$\frac{e^{2\mathcal{U}}}{4b_i^2}\left(a_i'' + 2\frac{a_i'}{r} - 2\frac{a_i'b_i'}{b_i}\right) = -\frac{1}{2}\left((\partial_{a_i}\,\mathrm{Im}\mathcal{N}_{\Lambda\Sigma})F^\Lambda_{..}F^{\Sigma|..}\right.$$
$$\left. + (\partial_{a_i}\,\mathrm{Re}\mathcal{N}_{\Lambda\Sigma})F^\Lambda_{..}\widetilde{F}^{\Sigma|..}\right)$$
$$\frac{e^{2\mathcal{U}}}{4b_i^2}\left(b_i'' + 2\frac{b_i'}{r} + \frac{(a_i'^2 - b_i'^2)}{b_i}\right) = -\frac{1}{2}\left((\partial_{b_i}\,\mathrm{Im}\mathcal{N}_{\Lambda\Sigma})F^\Lambda_{..}F^{\Sigma|..}\right.$$
$$\left. + (\partial_{b_i}\,\mathrm{Re}\mathcal{N}_{\Lambda\Sigma})F^\Lambda_{..}\widetilde{F}^{\Sigma|..}\right). \tag{8.7.18}$$

Einstein equations: varying the action (6.5.1) with respect to the metric we obtain the following equations:

$$R_{MN} = -g_{ij^\star}\partial_M z^i \partial_N \bar{z}^{j^\star} + S_{MN}$$
$$S_{MN} = -2\,\mathrm{Im}\,\mathcal{N}_{\Lambda\Sigma}\left(F^\Lambda_M.F^{\Sigma|.}_N - \frac{1}{4}g_{MN}F^\Lambda_{..}F^{\Sigma|..}\right) \tag{8.7.19}$$
$$-2\,\mathrm{Re}\,\mathcal{N}_{\Lambda\Sigma}\left(F^\Lambda_M.\widetilde{F}^{\Sigma|.}_N - \frac{1}{4}g_{MN}F^\Lambda_{..}\widetilde{F}^{\Sigma|..}\right).$$

Projecting on the components $(M, N) = (\underline{0}, \underline{0})$ and $(M, N) = (\underline{a}, \underline{b})$, respectively, these equations can be written in the following way:

$$\mathcal{U}'' + \frac{2}{r}\mathcal{U}' = -2\,e^{-2\mathcal{U}}\,S_{\underline{00}}$$

$$(\mathcal{U}')^2 + \sum_i \frac{1}{4b_i^2}\left((b_i')^2 + (a_i')^2\right) = -2\,e^{-2\mathcal{U}}\,S_{\underline{00}} \qquad (8.7.20)$$

where:

$$S_{\underline{00}} = -\frac{2e^{4\mathcal{U}}}{(8\pi)^2 r^4}\,\mathrm{Im}\mathcal{N}_{\Lambda\Sigma}(p^\Lambda p^\Sigma + \ell(r)^\Lambda \ell(r)^\Sigma). \qquad (8.7.21)$$

In order to solve these equations one would need to make explicit the right-hand side expression in terms of scalar fields a_i, b_i and quantized charges (p^Λ, q_Σ). In order to do that, one has to consider the *ansatz* for the field strengths (6.6.4) substituting for the modulus-dependent charges $\ell_\Lambda(r)$ appearing in the previous equations their expression in terms of the quantized charges obtained by inverting equation (6.6.17):

$$\ell^\Lambda(r) = \mathrm{Im}\mathcal{N}^{-1|\Lambda\Sigma}\left(q_\Sigma - \mathrm{Re}\mathcal{N}_{\Sigma\Omega}p^\Omega\right). \qquad (8.7.22)$$

Using now the expression for the matrix $\mathcal{N}$ in equation (8.7.13), one can find the explicit expression of the scalar field equations of motion written in terms of the quantized r-independent charges. They can be found in the appendix. In the appendix we report the full explicit expression of the equations of motion for both the scalars and the metric. Let us stress that, in order to find the five-parameter generating solution of the STU-model, it is not sufficient to substitute for each charge, in the scalar fixed values of equation (8.7.9), a corresponding harmonic function $(q_i \rightarrow H_i = 1 + q_i/r)$. As already explained, the generating solution should depend on five parameters and four harmonic functions, as in [53]. In particular, as explained above, two of the six scalar fields parametrizing the STU-model, namely two axion fields, should be taken to be constant. Therefore, in order to find the generating solution one has to solve the two systems of equations (8.7.37) (first order) and (8.7.38) (second order) explicitly putting as an external input the information on the constant nature of two of the three axion fields. As is evident from the above quoted system of equations, it is quite difficult to give a not double extreme solution of the combined system that is both explicit and manageable. It is, however, work in progress for a forthcoming paper [55].

8.7.5 The solution: a simplified case, namely $S = T = U$

In order to find a fully explicit solution we can deal with, let us consider the particular case where $S = T = U$. Although simpler, this solution encodes all non-trivial aspects of the most general one: it is regular, i.e. has non-zero entropy, and the scalars do evolve, i.e. it is an extreme but *not* double extreme solution.

First of all let us notice that equations (8.7.37) remain invariant if the same set of permutations are performed on the triplet of subscripts $(1, 2, 3)$ in both the fields and the charges. Therefore the solution $S = T = U$ implies the positions $q_1 = q_2 = q_3 \equiv q$ and $p^1 = p^2 = p^3 \equiv p$ on the charges and therefore it will correspond to a solution with (apparently only) four independent charges (p^0, p, q_0, q). According to this identification, what we do expect now is to find a solution which depends on (apparently) only three charges and two harmonic functions. Notice that this is not simply an axion–dilaton black hole: such a solution would have a vanishing entropy differently from our case. The fact that we have just one complex field in our solution is because the three complex fields are taken to be equal in value. The equations (8.7.37) simplify in the following way:

$$\frac{da}{dr} = \pm \left(\frac{e^{\mathcal{U}(r)}}{r^2}\right) \frac{1}{\sqrt{-2b}} \left(bq - 2ab\,p + \left(a^2 b + b^3\right) p^0\right)$$

$$\frac{db}{dr} = \pm \left(\frac{e^{\mathcal{U}(r)}}{r^2}\right) \frac{1}{\sqrt{-2b}} \left(3aq - \left(3a^2 + b^2\right) p + \left(a^3 + a b^2\right) p^0 + q_0\right)$$

$$\text{(8.7.23)}$$

$$\frac{d\mathcal{U}}{dr} = \pm \left(\frac{e^{\mathcal{U}(r)}}{r^2}\right) \left(\frac{1}{2\sqrt{2}(-b)^{3/2}}\right) \left(3aq - \left(3a^2 - 3b^2\right) p \right.$$
$$\left. + \left(a^3 - 3ab^2\right) p^0 + q_0\right)$$

$$0 = 3bq - 6ab\,p + \left(3a^2 b - b^3\right) p^0$$

where $a \equiv a_i$, $b \equiv b_i$ $(i = 1, 2, 3)$. In this case the fixed values for the scalars a, b are:

$$a_{fix} = \frac{pq + p^0 q_0}{2\,p^2 - 2\,p^0 q}$$

$$\text{(8.7.24)}$$

$$b_{fix} = -\frac{\sqrt{f(p, q, p^0, q_0)}}{2(p^2 - p^0 q)}$$

where

$$f(p, q, p^0, q_0) = 3p^2 q^2 + 4p^3 q_0 - 6\,p p^0 q\,q_0 - p^0(4\,q^3 + p^0 q_0^2).$$

Computing the central charge at the fixed point $Z_{fix}(p, q, p^0, q_0) = Z(a_{fix}, b_{fix}, p, q, p^0, q_0)$ one finds:

$$Z_{fix}(p, q, p^0, q_0) = |Z_{fix}| e^\theta$$

$$|Z_{fix}(p, q, p^0, q_0)| = f(p, q, p^0, q_0)^{1/4}$$

$$\text{(8.7.25)}$$

$$\sin\theta = \frac{p^0 f(p, q, p^0, q_0)^{1/2}}{2(p^2 - qp^0)^{3/2}}$$

$$\cos\theta = \frac{-2\,p^3 + 3\,p\,p^0 q + p^{0^2} q_0}{2\left(p^2 - p^0 q\right)^{3/2}}.$$

The value of the U-duality group quartic invariant (whose square root is the entropy) is:

$$S^2(p, q, p^0, q_0) = |Z_{fix}(p, q, p^0, q_0)|^4 = f(p, q, p^0, q_0). \quad (8.7.26)$$

We see from equations (8.7.26) that in order for Z_{fix} to be real and the entropy to be non-vanishing the only possibility is $p^0 = 0$ corresponding to $\theta = \pi$. It is in fact necessary that $\sin\theta = 0$ while keeping $f \neq 0$. We are therefore left with three independent charges (q, p, q_0), as anticipated.

8.7.5.1 Solution of the first order equations

Setting $p^0 = 0$ the fixed values of the scalars and the quartic invariant become:

$$a_{fix} = \frac{q}{2p}$$

$$b_{fix} = -\frac{\sqrt{3q^2 + 4q_0 p}}{2p} \quad (8.7.27)$$

$$I_4 = (3q^2 p^2 + 4q_0 p^3).$$

From the last of equations (8.7.23) we see that in this case the axion is double fixed, namely does not evolve, $a \equiv a_{fix}$ and the reality condition for the central charge is fulfilled for any r. Of course, also the axion equation is fulfilled and therefore we are left with two axion-invariant equations for b and $\mathcal{U}$:

$$\frac{db}{dr} = \pm \frac{e^{\mathcal{U}}}{r^2 \sqrt{2b}} \left(q_0 + \frac{3q^2}{4p} - b^2 p \right)$$

$$\frac{d\mathcal{U}}{dr} = \pm \frac{e^{\mathcal{U}}}{r^2 (2b)^{3/2}} \left(q_0 + \frac{3q^2}{4p} + 3b^2 p \right) \quad (8.7.28)$$

which admit the following solution:

$$b(r) = -\sqrt{\frac{(A_1 + k_1/r)}{(A_2 + k_2/r)}}$$

$$e^{\mathcal{U}} = \left(\left(A_2 + \frac{k_2}{r} \right)^3 (A_1 + k_1/r) \right)^{-1/4}$$

$$k_1 = \pm \frac{\sqrt{2}(3q^2 + 4q_0 p)}{4p} \quad (8.7.29)$$

$$k_2 = \pm \sqrt{2}p.$$

In the limit $r \to 0$:

$$b(r) \to -\left(\frac{k_1}{k_2} \right)^{1/2} = b_{fix}$$

$$e^{\mathcal{U}(r)} \to r\, (k_1 k_2^3)^{-1/4} = r f^{-1/4}$$

as expected, and the only undetermined constants are A_1, A_2. In order for the solution to be asymptotically Minkowskian it is necessary that $(A_1 A_2^3)^{-1/4} = 1$. There is then just one undetermined parameter which is fixed by the asymptotic value of the dilaton b. We choose for simplicity it to be -1, therefore $A_1 = 1$, $A_2 = 1$. This choice is arbitrary in the sense that the different value of b at infinity the different universe ($\equiv$ black hole solution), but with the same entropy. Summarizing, before considering the equations of motion, the solution is:

$$a = a_{fix} = \frac{q}{2p}$$

$$b = -\sqrt{\frac{(1 + k_1/r)}{(1 + k_2/r)}} \tag{8.7.30}$$

$$e^{\mathcal{U}} = \left[(1 + k_1/r)(1 + k_2/r)^3 \right]^{-1/4}$$

with k_1 and k_2 given in (8.7.29).

8.7.5.2 Solution of the second order equations

In the case $S = T = U$ the structure of the $\mathcal{N}$ matrix (8.7.13) and of the field strengths reduces considerably. For the period matrices one simply obtains:

$$\mathrm{Re}\mathcal{N} = \begin{pmatrix} 2a^3 & -a^2 & -a^2 & -a^2 \\ -a^2 & 0 & a & a \\ -a^2 & a & 0 & a \\ -a^2 & a & a & 0 \end{pmatrix}$$

$$\mathrm{Im}\mathcal{N} = \begin{pmatrix} 3a^2 b + b^3 & -(ab) & -(ab) & -(ab) \\ -(ab) & b & 0 & 0 \\ -(ab) & 0 & b & 0 \\ -(ab) & 0 & 0 & b \end{pmatrix} \tag{8.7.31}$$

while the dependence of $\ell^\Lambda(r)$ from the quantized charges simplifies to:

$$\ell^\Lambda(r) = \begin{pmatrix} \dfrac{-3a^2\,p + 3\,a\,q + q_0}{b^3} \\[4pt] \dfrac{-3a^3\,p + b^2\,q + 3\,a^2\,q + a\,(-2\,b^2\,p + q_0)}{b^3} \\[4pt] \dfrac{-3a^3\,p + b^2\,q + 3\,a^2\,q + a\,(-2\,b^2\,p + q_0)}{b^3} \\[4pt] \dfrac{-3a^3\,p + a^4\,p^0 + b^2\,q + 3\,a^2\,q + a\,(-2\,b^2\,p + q_0)}{b^3} \end{pmatrix}. \tag{8.7.32}$$

Inserting (8.7.32) in the expressions (6.6.4) and substituting the result in the equations of motion (8.7.19) one finds:

$$\left(a'' - 2\frac{a'b'}{b} + 2\frac{a'}{r} \right) = 0$$

$$\left(b'' + 2\frac{b'}{r} + \frac{(a'^2 - b'^2)}{b}\right) = -\frac{b^2 e^{2U}\left(p^2 - \frac{(-3a^2 p + 3aq + q_0)^2}{b^6}\right)}{r^4}.$$

$$(8.7.33)$$

The equation for a is automatically fulfilled by our solution (8.7.30). The equation for b is fulfilled as well and both sides are equal to:

$$\frac{(k_2 - k_1)\, e^{4U}\left(k_1 + k_2 + \frac{2k_1 k_2}{r}\right)}{2b r^4}.$$

If $(k_2 - k_1) = 0$ both sides are separately equal to 0 which corresponds to the double fixed solution already found in [48]. Let us now consider the Einstein equations. From equations (8.7.21) we obtain in our simpler case the following ones:

$$U'' + \frac{2}{r}U' = (U')^2 + \frac{3}{4b^2}\left((b')^2 + (a')^2\right)$$

$$U'' + \frac{2}{r}U' = -2e^{-2U} S_{\underline{00}}.$$

$$(8.7.34)$$

The first of equations (8.7.35) is indeed fulfilled by our *ansatz*. Both sides are equal to:

$$\frac{3(k_2 - k_1)^2}{16 r^4 (H_1)^2 (H_2)^2}.$$

$$(8.7.35)$$

Again, both sides are separately zero in the double extreme case $(k_2 - k_1) = 0$. The second equation is fulfilled, too, by our *ansatz* and again both sides are zero in the double extreme case. Therefore we can conclude with the

Statement 8.7.2. Equation (8.7.31) yields a $\frac{1}{8}$ supersymmetry preserving solution of $N = 8$ supergravity that is *not double extreme* and has a *finite entropy*:

$$S_{BH} = \frac{1}{4\pi}\left(3 p^2 q^2 + 4 q_0 p^3\right)^{1/2}$$

$$(8.7.36)$$

depending on three of the five truly independent charges.

On the other hand if in equation (8.7.31) we set $k_1 = k_2$ we see that both the dilaton and the axion are set to a constant value equal to the fixed value. At the same time the form of the metric becomes that of an $a = 0$ 0-brane solution as described in equation (6.1.14). As the critical values $a = \sqrt{3}$ and $a = 1$ are respectively associated with 1/2 and 1/4 supersymmetry preserving black holes, in the same way the last critical value $a = 0$ is associated with 1/8 preserving black holes. This case is the only one where scalar fields are finite at the horizon and correspondingly the entropy can be non-vanishing and finite.

8.7.6   Appendix: The full set of first and second order differential equations for the STU-model

Setting $z^i = a_i + ib_i$ equations (8.7.14) can be rewritten in the form:

$$\frac{da_1}{dr} = \pm \frac{e^{\mathcal{U}(r)}}{r^2} \sqrt{-\frac{b_1}{2b_2b_3}} [-b_1q_1 + b_2q_2 + b_3q_3 + (-(a_2\,a_3\,b_1)$$

$$+ a_1\,a_3\,b_2 + a_1\,a_2\,b_3 + b_1\,b_2\,b_3)\,p^0$$
$$+ (-(a_3\,b_2) - a_2\,b_3)\,p^1 + (a_3\,b_1 - a_1\,b_3)\,p^2$$
$$+ (a_2\,b_1 - a_1\,b_2)\,p^3]$$

$$\frac{db_1}{dr} = \pm \frac{e^{\mathcal{U}(r)}}{r^2} \sqrt{-\frac{b_1}{2b_2b_3}} [a_1q_1 + a_2q_2 + a_3q_3$$

$$+ (a_1\,a_2\,a_3 + a_3\,b_1\,b_2 + a_2\,b_1\,b_3 - a_1\,b_2\,b_3)\,p^0$$
$$+ (-(a_2\,a_3) + b_2\,b_3)\,p^1 - (a_1\,a_3 + b_1\,b_3)\,p^2$$
$$- (a_1\,a_2 + b_1\,b_2)\,p^3 + q_0]$$

$$\frac{da_2}{dr} = (1, 2, 3) \rightarrow (2, 1, 3) \tag{8.7.37}$$

$$\frac{db_2}{dr} = (1, 2, 3) \rightarrow (2, 1, 3)$$

$$\frac{da_3}{dr} = (1, 2, 3) \rightarrow (3, 2, 1)$$

$$\frac{db_3}{dr} = (1, 2, 3) \rightarrow (3, 2, 1)$$

$$\frac{d\mathcal{U}}{dr} = \pm \frac{e^{\mathcal{U}(r)}}{r^2} \frac{1}{2\sqrt{2}(-b_1b_2b_3)^{1/2}} [a_1q_1 + a_2q_2 + a_3q_3$$

$$+ (a_1\,a_2\,a_3 - a_3\,b_1\,b_2 - a_2\,b_1\,b_3 - a_1\,b_2\,b_3)\,p^0$$
$$- (a_2\,a_3 - b_2\,b_3)\,p^1 - (a_1\,a_3 - b_1\,b_3)\,p^2 - (a_1\,a_2 - b_1\,b_2)\,p^3 + q_0]$$
$$0 = b_1q_1 + b_2q_2 + b_3q_3 + (a_2\,a_3\,b_1 + a_1\,a_3\,b_2 + a_1\,a_2\,b_3 - b_1\,b_2\,b_3)\,p^0$$
$$- (a_3\,b_2 + a_2\,b_3)\,p^1$$
$$- (a_3\,b_1 + a_1\,b_3)\,p^2 - (a_2\,b_1 + a_1\,b_2)\,p^3.$$

The explicit form of the equations of motion for the most general case is:

Scalar equations:

$$\left(a_1'' - 2\frac{a_1'b_1'}{b_1} + 2\frac{a_1'}{r}\right) = \frac{-2\,b_1\,e^{2\mathcal{U}}}{r^4} [a_1\,b_2\,b_3\,(p^{1^2} - \ell(r)_1{}^2)$$

$$+ b_2\,(-(b_3\,p^1\,p^2) + b_3\,\ell(r)_1\,\ell(r)_2)$$
$$+ b_1\,(-2\,a_2\,a_3\,p^1\,\ell(r)_1 + a_3\,p^3\,\ell(r)_1$$
$$+ a_2\,p^4\,\ell(r)_1 + a_3\,p^1\,\ell(r)_3$$
$$- p^0\,\ell(r)_3 + a_2\,p^1\,\ell(r)_0 - p^3\,\ell(r)_0)]$$

$$\left(b_1'' + 2\frac{b_1'}{r} + \frac{(a_1'^2 - b_1'^2)}{b_1}\right) = -\frac{e^{2\mathcal{U}}}{b_2 b_3 r^4}[-(a_1{}^2 b_2{}^2 b_3{}^2 p^{1^2}) + b_1{}^2 b_2{}^2 b_3{}^2 p^{1^2}$$

$$+ 2 a_1 b_2{}^2 b_3{}^2 p^1 p^2$$
$$- b_2{}^2 b_3{}^2 p^{2^2} + b_1{}^2 b_3{}^2 p^{3^2} + b_1{}^2 b_2{}^2 p^{0^2}$$
$$+ a_1{}^2 b_2{}^2 b_3{}^2 \ell(r)_1{}^2$$
$$- b_1{}^2 b_2{}^2 b_3{}^2 \ell(r)_1{}^2 + a_3{}^2 b_1{}^2 b_2{}^2 (p^{1^2} - \ell(r)_1{}^2)$$
$$+ a_2{}^2 b_1{}^2 b_3{}^2 (p^{1^2} - \ell(r)_1{}^2)$$
$$- 2 a_1 b_2{}^2 b_3{}^2 \ell(r)_1 \ell(r)_2 + b_2{}^2 b_3{}^2 \ell(r)_2{}^2$$
$$- b_1{}^2 b_3{}^2 \ell(r)_3{}^2$$
$$+ 2 a_2 b_1{}^2 b_3{}^2 (-(p^1 p^3) + \ell(r)_1 \ell(r)_3)$$
$$- b_1{}^2 b_2{}^2 \ell(r)_0{}^2$$
$$+ 2 a_3 b_1{}^2 b_2{}^2 (-(p^1 p^0) + \ell(r)_1 \ell(r)_0)]$$

$$\left(a_2'' - 2\frac{a_2' b_2'}{b_2} + 2\frac{a_2'}{r}\right) = (1, 2, 3) \rightarrow (2, 1, 3)$$

$$\left(b_2'' + 2\frac{b_2'}{r} + \frac{(a_2'^2 - b_2'^2)}{b_2}\right) = (1, 2, 3) \rightarrow (2, 1, 3)$$

$$\left(a_3'' - 2\frac{a_3' b_3'}{b_3} + 2\frac{a_3'}{r}\right) = (1, 2, 3) \rightarrow (3, 2, 1)$$

$$\left(b_3'' + 2\frac{b_3'}{r} + \frac{(a_3'^2 - b_3'^2)}{b_3}\right) = (1, 2, 3) \rightarrow (3, 2, 1).$$

Einstein equations:

$$\mathcal{U}'' + \frac{2}{r}\mathcal{U}' = -2 e^{-2\mathcal{U}} S_{00}$$

$$(\mathcal{U}')^2 + \sum_i \frac{1}{4b_i^2}\left((b_i')^2 + (a_i')^2\right) = 2 e^{-2\mathcal{U}} S_{00} \qquad (8.7.38)$$

where the quantity S_{00} on the right hand side of the Einstein equations has the following form:

$$S_{00} = \frac{e^{4\mathcal{U}}}{4 b_1 b_2 b_3 r^4}(a_1{}^2 b_2{}^2 b_3{}^2 p_1{}^2 + b_1{}^2 b_2{}^2 b_3{}^2 p_1{}^2$$
$$- 2 a_1 b_2{}^2 b_3{}^2 p_1 p_2 + b_2{}^2 b_3{}^2 p_2{}^2 + b_1{}^2 b_3{}^2 p_3{}^2$$
$$+ b_1{}^2 b_2{}^2 p_0{}^2 + a_1{}^2 b_2{}^2 b_3{}^2 \ell(r)_1{}^2 + b_1{}^2 b_2{}^2 b_3{}^2 \ell(r)_1{}^2$$
$$+ a_3{}^2 b_1{}^2 b_2{}^2 (p_1{}^2 + \ell(r)_1{}^2)$$
$$+ a_2{}^2 b_1{}^2 b_3{}^2 (p_1{}^2 + \ell(r)_1{}^2) - 2 a_1 b_2{}^2 b_3{}^2 \ell(r)_1 \ell(r)_2$$
$$+ b_2{}^2 b_3{}^2 \ell(r)_2{}^2 + b_1{}^2 b_3{}^2 \ell(r)_3{}^2$$
$$- 2 a_2 b_1{}^2 b_3{}^2 (p_1 p_3 + \ell(r)_1 \ell(r)_3) + b_1{}^2 b_2{}^2 \ell(r)_0{}^2$$
$$- 2 a_3 b_1{}^2 b_2{}^2 (p_1 p_0 + \ell(r)_1 \ell(r)_0)). \qquad (8.7.39)$$

The explicit expression of the $\ell_\Lambda(r)$ charges in terms of the quantized ones is computed from equation (8.7.22):

$$\ell_\Lambda(r) = \frac{1}{b_1 b_2 b_3}$$

$$\left(\begin{array}{c} \left[q_1 + a_1 \left(a_2 a_3 p^1 - a_3 p^3 - a_2 p^0 + q_2 \right) \right. \\ \left. + a_2 \left(- \left(a_3 p^2 \right) + q_3 \right) + a_3 q_0 \right] \\ \hline \left[a_1{}^2 \left(a_2 a_3 p^1 - a_3 p^3 - a_2 p^0 + q_2 \right) \right. \\ + b_1{}^2 \left(a_2 a_3 p^1 - a_3 p^3 - a_2 p^0 + q_2 \right) \\ \left. + a_1 \left(q_1 + a_2 \left(- \left(a_3 p^2 \right) + q_3 \right) + a_3 q_0 \right) \right] \\ \hline \left[a_1 \left(a_2{}^2 \left(a_3 p^1 - p^0 \right) + b_2{}^2 \left(a_3 p^1 - p^0 \right) + a_2 \left(- \left(a_3 p^3 \right) + q_2 \right) \right) \right. \\ \left. + a_2{}^2 \left(- \left(a_3 p^2 \right) + q_3 \right) + b_2{}^2 \left(- \left(a_3 p^2 \right) + q_3 \right) + a_2 \left(q_1 + a_3 q_0 \right) \right] \\ \hline \left[a_3 q_1 + a_1 \left(- \left(a_3{}^2 p^3 \right) \right) - b_3{}^2 p^3 \right. \\ + a_2 \left(a_3{}^2 p^1 + b_3{}^2 p^1 - a_3 p^0 \right) + a_3 q_2 \right) \\ \left. - a_2 \left(a_3{}^2 p^2 + b_3{}^2 p^2 - a_3 q_3 \right) + a_3{}^2 q_0 + b_3{}^2 q_0 \right] \end{array} \right) \times \quad .$$

$$(8.7.40)$$

Acknowledgments

The authors are particularly grateful to their friends, collaborators and former students Matteo Bertolini and Mario Trigiante for their invaluable help throughout the preparation of these lecture notes. In particular they want to stress that the material presented in section 8.7 concerning the 1/8 supersymmetry preserving $N = 8$ black holes just coincides with the results of a research article by M Bertolini, M Trigiante and P Fré quoted in the references.

References

[1] Strominger A and Vafa C 1986 *Phys. Lett.* B **379** 99

[2] For a recent update see: Maldacena J M 1996 Black holes in string theory, hep-th/9607235

[3] References on SUSY black holes: Gibbons G 1982 Unified theories of elementary particles. Critical assessment and prospects *Proc. Heisenberg Symp. (München, 1981) (Lecture Notes in Physics 160)* ed P Breitenlohner and H P Dürr (Berlin: Springer)

Gibbons G W and Hull C M 1982 *Phys. Lett.* B **109** 190

Gibbons G W 1984 Supersymmetry, supergravity and related topics *Proc. XVth GIFT Int. Phys. (Girona, 1984)* ed F del Aguila, J de Azcárraga and L Ibáñez (Singapore: World Scientific) p 147

Kallosh R, Linde A, Ortin T, Peet A and Van Proeyen A 1992 *Phys. Rev.* D **46** 5278

Kallosh R, Ortin T and Peet A 1993 *Phys. Rev.* D **47** 5400

Kallosh R 1992 *Phys. Lett.* B **282** 80

Kallosh R and Peet A 1992 *Phys. Rev.* D **46** 5223

Sen A 1995 *Nucl. Phys.* B **440** 421

Sen A 1995 *Phys. Lett.* B **303** 221

Sen A 1995 *Mod. Phys. Lett.* A **10** 2081

Schwarz J and Sen A 1993 *Phys. Lett.* B **312** 105

[4] Kallosh R and Ortin T 1993 *Phys. Rev.* D **48** 742

Bergshoeff E, Kallosh R and Ortin T 1996 *Nucl. Phys.* B **478** 156

[5] Duff M J, Khuri R R and Lu J X 1995 *Phys. Rep.* **259** 213–326

[6] For recent reviews see: Schwarz J 1996 *Preprint* CALTECH-68-2065, hep-th/9607201

Duff M 1996 *Preprint* CPT-TAMU-33-96, hep-th/9608117

Sen A 1996 *Preprint* MRI-PHY-96-28, hep-th/9609176

[7] Sen A and Schwarz J 1993 *Phys. Lett.* B **312** 105

Sen A and Schwarz J 1993 *Nucl. Phys.* B **411** 35

[8] Hull C M and Townsend P K 1995 *Nucl. Phys.* B **451** 525

Witten E 1995 *Nucl. Phys.* B **443** 85

[9] Hull C M and Townsend P K 1995 *Nucl. Phys.* B **438** 109

[10] Schwarz J H 1996 M–theory extension of *T*–duality, hep-th/9601077

Vafa C 1996 Evidence for F–theory, hep-th/9602022

[11] Polchinski J 1995 *Phys. Rev. Lett* **75** 4724

[12] For a recent review see: Polchinski J 1996 TASI lectures on D–branes and Polchinski J, Chaudhuri S and Johnson C 1996 Notes on D–branes, hep-th/9602052

[13] Lu H, Pope C N, Sezgin E and Stelle K S 1995 hep-th/9508042
Lu H and Pope C N 1996 hep-th/9605082
Duff M J, Lu H and Pope C N 1996 hep-th/9604052
Lu H, Pope C N and Stelle K S 1996 hep-th/9602140

[14] The literature on this topic is quite extended. As a general review, see the lecture notes: Stelle K 1997 *Lectures on Supergravity p–Branes, 1996 ICTP Summer School (Trieste)* hep-th/9701088

[15] For a recent comprehensive updating on M-brane solutions see also: Townsend P K 1997 *M–Theory from its Superalgebra, NATO Advanced Study Institute on Strings, Branes and Dualities (Cargese, 26 May–14 June 1997)* hep-th/9712004

[16] For a review on supergravity and superstring theory the reader can see the book by Castellani L, D'Auria R and Fré P 1990 *Supergravity and Superstring Theory: a Geometric Perspective* (Singapore: World Scientific)

[17] Castellani L, Ceresole A, D'Auria R, Ferrara S, Fré P and Trigiante M 1998 G/H M-branes and AdS$_{p+2}$ geometries, hep-th/9803039

[18] Freund P G O and Rubin M A 1980 *Phys. Lett* B **97** 233

[19] Castellani L, Romans L J and Warner N P 1984 *Nucl. Phys.* B **241** 429

[20] Duff M J, Lü H, Pope C N and Sezgin E 1996 *Phys. Lett.* B **371** 206

[21] Cremmer E, Ferrara S, Girardello L and Van Proeyen A 1983 *Nucl Phys.* B **212** 413

[22] Fré P and Soriani P 1995 *The N = 2 Wonderland: from Calabi–Yau manifolds to Topological Field Theories* (Singapore: World Scientific)

[23] Ceresole A, D'Auria R and Ferrara S 1996 *Nucl. Phys.* B 46 67

[24] Cremmer E and Van Proeyen A 1985 *Class. Quantum Grav.* **2** 445

[25] Gaillard M K and Zumino B 1981 *Nucl. Phys.* B **193** 221

[26] Fré P 1996 *Nucl. Phys. Proc. Suppl.* B, C **45** 59–114

[27] Andrianopoli L, Bertolini M, Ceresole A, D'Auria R, Ferrara S, Fré P and Magri T 1997 *J. Geom. Phys.* **23** 111

[28] Ferrara S and Van Proeyen A 1989 *Class. Quantum Grav.* **2** 124

[29] Ferrara S, Girardello L and Porrati M 1995 Minimal Higgs Branch for the breaking of half of the supersymmetry in $N = 2$ supergravity *Preprint* CERN-TH/95-268, hep-th/9510074

[30] Bertotti B 1959 *Phys. Rev.* **116** 1331
Robinson I 1959 *Bull. Acad. Pol.* **7** 351

[31] Calabi E and Visentini E 1960 *Ann. Math.* **71** 472

[32] Ceresole A, D'Auria R, Ferrara S and Van Proeyen A 1995 *Nucl. Phys.* B **444** 92
Ceresole A, D'Auria R, Ferrara S and Van Proeyen A 1995 On electromagnetic duality in locally supersymmetric $N = 2$ Yang–Mills theory *Proc. Workshop on Physics from the Planck Scale to Electromagnetic Scale (Warsaw, 1994)*

[33] Ferrara S, Kallosh R and Strominger A 1995 $N = 2$ extremal black holes, hep-th/9508072

[34] Strominger A 1996 Macroscopic entropy of $N = 2$ extremal black holes, hep-th/9602111

[35] Ferrara S and Kallosh R 1996 Supersymmetry and attractors, hep-th/9602136

[36] Castellani L, Ceresole A, D'Auria R, Ferrara S, Fré P and Maina E 1986 *Nucl. Phys.* B **286** 317

[37] Bergshoeff E, Koh I G and Sezgin E 1985 *Phys. Lett.* B **155** 71
de Roo M and Wagemans F 1985 *Nucl. Phys.* B **262** 644

[38]   Salam A and Sezgin E 1989 *Supergravities in Diverse Dimensions* vol 1 ed A Salam and E Sezgin (North-Holland, World Scientific)

[39]   Cremmer E *Supergravity '81* ed S Ferrara and J G Taylor, p 313

      Julia B 1981 *Superspace & Supergravity* ed S W Hawking and M Rocek (Cambridge: Cambridge University Press) p 331

[40]   See for example: Helgason S (ed) 1962 *Differential Geometry and Symmetric Spaces* (New York: Academic)

      Gilmore R 1974 *Lie Groups, Lie Algebras and Some of Their Applications* (New York: Wiley)

[41]   Andrianopoli L, D'Auria R, Ferrara S, Fré P and Trigiante M 1997 *Nucl. Phys.* B **496** 617

[42]   Andrianopoli L, D'Auria R, Ferrara S, Fré P, Minasian R and Trigiante M 1997 *Nucl. Phys.* B **493** 249

[43]   Alekseevskii D V 1975 *Math. USSR Izv.* **9**

[44]   Andrianopoli L, D'Auria R, Ferrara S, Fré P and Trigiante M 1998 *Nucl. Phys.* B **509** 463

[45]   Arcioni G, Ceresole A, Cordaro F, D'Auria R, Fré P, Gualtieri L and Trigiante M 1998 $N = 8$ BPS black holes with $1/2$ or $1/4$ supersymmetry and solvable Lie algebra decompositions, hep-th/9807136

[46]   Bertolini M, Fré P and Trigiante M 1998 $N = 8$ BPS black holes preserving $1/8$ supersymmetry, hep-th 9811251

[47]   Ferrara S and Gunaydin M 1998 Orbits of exceptional groups, duality and BPS states in string theory *Int. J. Mod. Phys.* A **13** 2075

[48]   Duff M J, Liu J T and Rahmfeld J 1996 *Nucl. Phys.* B **459** 125

      Kallosh R, Shmakova M and Wong W K 1996 *Phys. Rev.* D **54** 6284

      Behrndt K, Kallosh R, Rahmfeld J, Smachova M and Wond W K 1996 STU black holes and string triality, hep-th/9608059

      Lopes Cardoso G, Lust D and Mohaupt T 1996 *Phys. Lett.* B **388** 266

      Behrndt K, Lust D and Sabra W A 1997 Stationary solutions of $N = 2$ supergravity, hep-th/9705169

[49]   Ferrara S and Maldacena J 1998 *Class. Quant. Grav.* **15** 749

[50]   Ferrara S, Gunaydin M and Savoy C A, unpublished

[51]   Cvetic M and Hull C M 1996 *Nucl. Phys.* B **480** 296

[52]   Balasubramanian V 1997 How to count the states of extremal black holes in $N = 8$ supergravity, hep-th/9712215

[53]   Cvetic M and Youm D 1996 *Phys. Rev.* D **53** 584

      Cvetic M and Youm D 1996 *Nucl. Phys.* B **472** 249

      Cvetic M and Youm D 1996 *Phys. Rev.* D **54** 2612

      Cvetic M and Tseytlin A 1996 *Phys. Rev.* D **53** 5619 (Erratum 1997 *Phys. Rev.* D **55** 3907)

[54]   Behrndt K, Lust D and Sabra W A 1998 *Nucl. Phys.* B **510** 264

      Behrndt K, Cvetic M and Sabra W A 1998 *Phys. Rev.* D **58** 84 018

[55]   Bertolini M, Fré P and Trigiante M, work in progress

PART 4

Aldo Treves and Francesco Haardt

*Dipartimento di Scienze Chimiche, Fisiche e Matematiche,
via Lucini 3, 22100 Como, Italy*

Chapter 9

Astrophysics of stellar mass black holes

9.1 Introduction

The conception of black holes (BHs) of astrophysicists is different from that of mathematicians, or mathematical physicists. For the latter a BH is essentially a solution of Einstein equations, which is endowed with certain properties of regularity, and satisfactory asymptotic behaviour. For the astronomer BHs are rather a possible end point of stellar evolution, and therefore they are related to white dwarfs (WDs) and neutron stars (NSs).

Since, as we will see later, there is no upper limitation to the mass of BHs, they can be the end of the evolution of a condensation of matter the mass of which is billions of solar masses, comparable to the mass of a galaxy $10^{11} M_\odot$. In this case one speaks of supermassive BHs, and they are a basic ingredient of active galactic nuclei (AGNs), i.e. objects like quasars, Seyfert galaxies, BL Lacs etc.

Much of the progress in the last 30 years derived from the effort of filling the gap between the two conceptions of BHs, the astronomical one which is largely responsible for the collection of relevant observations and the mathematical–physical one which is interpreting the observational results on the basis of classical general relativity (GR).

In the last decade a new conception of BHs has been acquiring an extraordinary popularity. This started with the remarkable suggestion by Hawking (1975) and Bekenstein (1974) of evaporation of mini black holes, deriving in turn from the so called BH thermodynamics, and specifically on the definition of BH entropy. These very concepts are being reconsidered in the framework of the main stream of theoretical physics, the string theory, which could be the precursor of a 'fundamental theory'. The gap between theory and observations has widened again, and possibly it has never been as large as it is now. It seems to us that the danger of reducing the theory to solipsism, and of not making use in observations of the tremendous progress of the theory, dictate a brave effort on both sides, at least to start a scientific dialogue. We think that this is the main scope and motivation of the school.

From our side we will illustrate some results on 'astrophysics of BHs', making a special effort in stressing how the gap with classical general relativity was successfully bridged. It is not clear to us whether ours is the main way of approaching 'observationally' the 'new BH physics' or rather it should be through cosmology or experiments with elementary particles.

It has often happened that subtle but fundamental links have appeared in physical disciplines which apparently were moving in divergent directions. The basic step is to create an interaction between the communities, following the progress and making an effort to understand the fundamental goals.

9.1.1   Outline of the lectures

Our lectures will concentrate on solar mass black holes, though some references to AGNs are made. The reason is that, apart from the desire to not widen too much the scope of the course, we wish to stress the affinity with NS astrophysics.

First we recall the concept of a collapsed star, introducing the Chandrasekhar mass, and NS critical mass. We describe the search for NSs and BHs, and their essentially serendipitous discovery as pulsars and x-ray binaries. The basic process which makes a BH visible is its accretion of surrounding matter, and we report the fundamentals of 'accretion theory'.

BH candidates in x-ray binaries are then discussed, distinguishing the persistent and the transient ones, and the newly discovered 'superluminals'. We focus then on a number of 'observational probes' of BH physics, based on variability and spectroscopic observations.

The last chapter covers a topic on which present research efforts are concentrating, and, while the results are still controversial, the opportunities of sensational progress are indeed substantial. It follows from the recent discovery of millisecond 'quasiperiodic oscillations' (QPOs), and the proposal that part of the phenomenon should be interpreted in terms of the Lense–Thirring effect for a rapidly spinning NS. A further and even more controversial proposal is that one can extrapolate to rotating BHs.

9.2   Prehistory and history of BHs

9.2.1   Collapsed objects

Michell and Laplace (see Hawking and Ellis 1973, Gribbin and Rees 1990) noted that the escape velocity of a test particle from an object the radius of which is

$$r < r_G \equiv 2GM/c^2 \qquad (9.1)$$

would exceed c. Therefore, within a corpuscular theory, light should not be emitted by such objects, which should be black. This illustrates that the gravitational radius r_G can be introduced on a quasi-Newtonian basis. For a solar mass $M_\odot = 2 \times 10^{33}$ g, $r_G = 3 \times 10^5$ cm.

9.2.2   End points of stellar evolution

9.2.2.1   *White dwarfs*

In the 1920s it was clear that the equilibrium of stars was hydrostatic, with gravity being compensated by a pressure gradient. This in turn requires that energy is generated in the star core. The mechanism of energy generation (e.g. fusion of H into He) was not clear at the time, but the theory of star equilibrium could be discussed in detail (see, e.g., Eddington 1926).

A class of stars was defying this picture. These were white dwarfs, discovered at the beginning of the century by Adams (1914). They exhibited very low luminosities compared to the Sun, indicating radii $\sim 10-2R_\odot$ ($R_\odot = 7 \times 10^{10}$ cm). Their mass, as demonstrated by the observation of binary systems (see, e.g., the companion of Sirius), was close to $M_\odot$. No hydrostatic model appeared applicable. The solution was heavily rooted in quantum mechanics and came immediately after the introduction of Fermi–Dirac statistics. Gravitation was compensated by the degeneracy pressure of an electron Fermi gas. Chandrasekhar noted that stable solutions were possible only as long as the gas was non-relativistic, a condition which imposes an upper bound to the mass. The critical mass for WDs (the so called Chandrasekhar mass) can be written in the form

$$M_{\text{Ch}} = \left(\frac{hc}{G}\right)^{3/2} \frac{1}{m_{\text{p}}^2} \frac{1}{8^{3/2}} \frac{3}{2h} \frac{1}{\mu_{\text{E}}} \simeq 1.4 \, M_\odot \tag{9.2}$$

where $(hc/G)^{1/2} = 5.5 \times 10^{-5}$ g is the Planck mass, m_{p} the proton mass and $\mu_{\text{E}} \simeq 2$ the mean molecular weight. This illustrates how the solar mass can essentially be expressed through fundamental constants. The radii of a WD deduced from the model agree well with observations, and are typically $\simeq 10^{2-3} r_{\text{G}}$. For $M < M_{\text{Ch}}$, white dwarfs are a stable end point of stellar evolution. WDs shine because of their internal heat, and they fade as they slowly cool off, becoming increasingly difficult to observe.

9.2.2.2   *Neutron stars*

Shortly after the clarification of the nature of WD another possible stable end point of stellar evolution was envisaged. Suppose that one starts with a WD and increases the matter density. Through beta reactions electrons and protons will yield neutrons, which will become the dominant particles.

No equilibrium configuration can be reached until nuclear densities are approached ($\simeq 10^{14}$ g cm^3). Here one should consider the appearance of a repulsive component of the nuclear potential, and at the same time the degeneracy of neutrons. The star behaves like an enormous nucleus of $\sim 10^{57}$ nucleons.

The possibility of an equilibrium configuration of this kind was proposed in a seminal paper by Landau (1932) before the actual discovery of neutrons. As in the case of WDs, a critical mass, close to $M_\odot$, should exist. From the condition that the density has to be close to nuclear, one obtains a radius $a \simeq 10^6$ cm, which

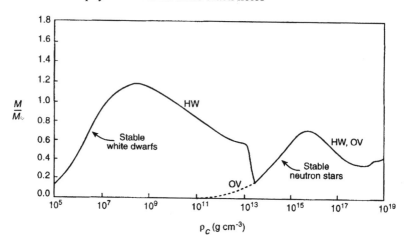

Figure 9.1. Gravitational radius versus central density for two different equations of state (HW and OV). The stable WD and NS branches are designated by a heavy solid line (from Shapiro and Teukolsky 1983).

is a factor of 3 larger than r_G. Realistic models of NS must take into account the details of nuclear interaction, and need to be treated fully in the framework of GR.

The nuclear interaction is directly known up to energies (densities) which can be explored with accelerators; beyond one should rely on models. This reflects in a certain degree of uncertainty on the NS critical mass, which is generally evaluated as $M_c^{NS} \simeq 2\text{--}3\ M_\odot$.

In figure 9.1 we report a mass/central density plot for a given nuclear potential. Stable configurations turn out to correspond to positive derivatives. Critical masses for WDs and NSs are apparent.

NSs can be stable end points of stellar evolution. Note that, in contrast to the case of WDs, NSs were theoretically proposed before being observed.

Just after the first formulation of NS theory, Baade and Zwicky (1934) pointed out that the binding energy of an NS should be

$$W \sim \frac{GM^2}{a} \sim 10^{54}\ \text{erg.} \tag{9.3}$$

This is comparable with the energy released in a supernova event. NSs and supernovae appeared closely linked, both as the product of gravitational collapse.

9.2.2.3 Black holes

Above the critical mass M_c^{NS} no equilibrium solution is possible. Therefore for stars which exhaust their nuclear fuel maintaining a mass above M_c^{NS}, complete gravitational collapse becomes inescapable. This was already clear in the papers of the 1930s of Landau and Chandrasekhar. Following the line of Tolman (1934), Oppenheimer and Volkoff (1939) gave a first general relativistic description of the gravitational collapse and of the formation of what we call now a BH. The gap between astronomy and mathematical physics was starting to be filled.

9.3   Search for NSs and BHs

In the 1940s and 50s NSs and BHs were not much more than the product of the fertile fantasy of theoretical physicists, perhaps analogous to antielectrons and magnetic monopoles after the Dirac proposal. In the early 1960s the challenge of actually searching for NSs and BHs was faced. The attitude at the time is well illustrated by a review paper by Wheeler (1966), where possible strategies for finding objects of both classes were considered, and the conclusion was that the most promising technique was detection of gravitational waves. Were this true we would still be searching for BHs and NSs. Nature has been merciful, and in the 1960s two major discoveries, essentially serendipitous, have completely changed the observational situation.

9.3.1   Pulsars

These objects were discovered in 1968, as pointlike radio-sources characterized by extremely stable periodicities. The first case CP 1919+22 yielded a period $P \simeq 1$ s. Later it was shown that the period was increasing with $\dot{P} = 10^{-15}$ s s^{-1}.

The identification of pulsars with rotating NSs was immediately proposed and became robust with the discovery of two objects associated with the young supernova remnants Vela and Crab, in view of the aforementioned proposal by Baade and Zwicky (1934).

Pulsars are powered by the neutron star rotation energy

$$W = \frac{1}{2} I \left(\frac{2\pi}{P} \right)^2 \tag{9.4}$$

where I is the moment of inertia $I = (2/5)Ma^2 \sim 10^{45}$ g cm^2 ($M = 1$ $M_\odot$, $a \simeq 10^6$ cm is the NS radius). The energy loss is mediated through the magnetic field endowed to the NS, which is, in general, misaligned with respect to the spin axis. This is the ultimate reason for periodic emission. In the simplest models, the pulsar energy loss is given by

$$\frac{dW}{dt} = \frac{2}{3} \left(\frac{2\pi}{P} \right)^4 \frac{B_0^2 a^6}{c^3} \tag{9.5}$$

where B_0 is the magnetic field at the pole. This formula immediately follows considering the pulsar as a spinning magnet rotating *in vacuo*. Equation (9.5) can be used to evaluate B_0. For CP 1919+22 and for the vast majority of pulsars, which have similar values of P and $\dot{P}$, $B_0 \simeq 10^{12}$ G. The ages of pulsars can then be taken as

$$\tau = \frac{1}{2} \frac{P}{\dot{P}}. \qquad (9.6)$$

For a typical pulsar, this turns out to be $\sim 10^{6-7}$ years. Pulsars are therefore very young NSs.

A second class of pulsars, the detection of which was more difficult, was later discovered, the so called 'millisecond' pulsars. For such objects, $P \sim 10$ ms, $\dot{P} \simeq 10^{-18}$ s s^{-1}, corresponding to a magnetic field $B_0 \simeq 10^8$ G and to an age $\sim 10^8$ years. A large fraction of ms pulsars are in close binary systems ($P_{orb} \simeq 1$ day), which allows a direct estimate of the masses of the components (see the following sections). The best studied case is that of PSR 1913-16, which yielded very stringent constrains on the NS mass, and, through the decay of P_{orb}, allowed an estimate of the energy loss (e.g., Taylor and Weisberg 1982).

From the basic role played by the magnetic field in pulsar emission one can exclude that pulsars are related to BHs. In fact the so called 'no hair theorem' (e.g., Carter 1973) guarantees that the magnetic field of a BH should be proportional to its net charge (Kerr–Newman solution), which is negligible in any realistic steady astrophysical situation. Moreover the magnetic moment must be aligned with the spin axis, and therefore no spin periodicity should be observable.

9.3.2   X-ray binaries: the case of Cyg X-1

Extra-solar x-ray astronomy started in 1962 (Giacconi *et al* 1962). The spectral band which was first explored was the 2–10 keV interval, which could be looked at by means of proportional counters. The high absorption of the atmosphere requires a spaceborne x-ray detector. The first results derived from suborbital flights allowed us to discover a population of x-ray sources associated with the Galactic plane, with typical luminosities $L_X = 10^{37}$ erg s^{-1}. Because of the difficulty of optical identification its nature remained mysterious, until the beginning of the 1970s. A fundamental contribution came from the first satellite completely dedicated to x-ray astronomy, Uhuru, launched in 1970.

A basic step forward came from the study of the bright x-ray source Cyg X-1, which is of particular importance, since it is the first x-ray source whose binary nature was demonstrated, and the first, and probably the best studied, BH candidate.

In 1971 the source entered a new x-ray state, characterized by a significant increase in soft x-rays. This is indicated as the high state of Cyg X-1, which repeats sporadically every 2–5 years, and has a duration of 2–3 months. During the transition to high state, a radio source appeared in the x-ray error box, which had a size of $1°$, and was full of stars. The radio-source could be located with the precision of $1'$, and supposing that it was related to Cyg X-1, it was possible to pinpoint the optical candidate. It turned out to be a rather normal B supergiant.

The key discovery was that the star is a spectroscopic binary. In such systems from the Doppler shift of the lines, one can reconstruct the radial velocity curve

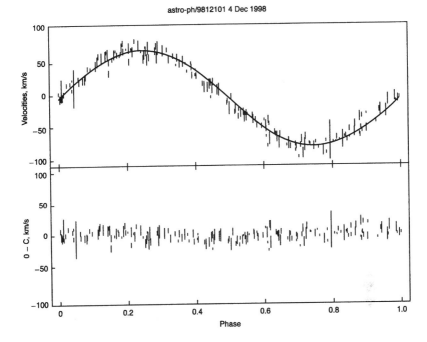

astro-ph/9812101 4 Dec 1998

Figure 9.2. Best fitting fixed period, circular orbit to helium radial velocity data of Cyg X-1. The lower panel shows the residuals to the fit (from La Sala *et al* 1998).

(see figure 9.2). From its examination the following quantities can be measured: the period, the systemic velocity and the radial velocity amplitude (here we consider non-eccentric orbit, an excellent approximation for Cyg X-1). If only one star is visible, and this is obviously the case of a binary system containing a BH, from the Kepler 3D law, the following quantity, which is called the mass function $f(M)$, can be measured:

$$f(M_X) = \frac{M_X^3 \sin^3 i}{(M_X + M_1)^2} = K_1^3 \frac{P}{2\pi G}. \tag{9.7}$$

'1' labels the observed star, 'X' the unseen one, K_1 is the velocity amplitude, P the period and i is the inclination angle. From $f(M)$ and an estimate of M_1 and $\sin i$, one can infer a value of M_X, which is found to be $M_X \simeq 10 \, M_\odot$. Since there was no trace of a second normal star in the optical spectra and photometry, the consequence was that the star was collapsed and, since M_X greatly exceeds the NS critical mass, it was proposed that the object was a BH. Note that from equation (9.7), one obtains

$$M_X > f(M_X) \tag{9.8}$$

independently of the mass of the normal star and of the inclination angle. In the case of Cyg X-1, $f(M) = 0.25\,M_\odot$. Therefore the presence of a BH relies heavily on the estimates of M_1 and $\sin i$.

9.4   Why x-ray binaries?

The discovery of the binary nature of Cyg X-1 was a fundamental step in understanding the nature of bright x-ray sources populating the Galaxy. The terms of the problem were essentially how to explain a temperature of few keV, and a luminosity $\sim 10^4\,L_\odot$. The solution was in fact proposed before the optical identification of Cyg X-1, but it was not recognized as such. At the time the majority of the astronomical community propended for models involving rapidly spinning NSs, in analogy with pulsars.

9.4.1   Some elements of the theory of accretion

Take a star of mass M and radius a immersed in a gas cloud at temperature T. Under the action of the gravitational field the cloud material will flow onto the star, and the process is referred to as 'accretion'. If the cloud material could be treated as a fluid, Bondi (1952) showed that there are steady solutions describing the accretion process. A most significant length-scale entering in the theory is the so called accretion radius r_A, i.e. the radius at which the thermal energy density equates the gravitational energy density. Note that in its simplest form the theory is purely Newtonian. One has

$$\frac{\rho G M}{r_A} \simeq \frac{\rho k T}{m_p} \tag{9.9}$$

where ρ is the fluid density, and we suppose that the medium is pure hydrogen. The dynamics of the fluid is dominated by the gravitational field within r_A, while the field is negligible in the outer region. Introducing a sound (thermal) velocity $v_s^2 \simeq (kT/m_p)^{1/2}$, one finds

$$r_A \sim \frac{G M}{v_s^2}. \tag{9.10}$$

The accretion radius is therefore approximately given by the radius at which the sound velocity equates the free fall velocity. The accretion rate can be written as

$$\frac{dM}{dt} \simeq 4\pi r_A^2 \rho v_s \simeq 4\pi \rho G^2 M^2 / v_s^3. \tag{9.11}$$

The energy released by the accretion process is obviously the gravitational binding energy multiplied by the accretion rate, i.e.

$$\frac{dW}{dt} \simeq \frac{G M}{a} \frac{dM}{dt}. \tag{9.12}$$

If one measures the star radius in units of the gravitational radius, i.e. $a = \tilde{r} r_G$, obtains

$$\frac{dW}{dt} \simeq \frac{c^2}{2\tilde{r}} \frac{dM}{dt}. \tag{9.13}$$

$\epsilon = 1/(2\tilde{r})$ is therefore a measure of the absolute efficiency of transformation of mass into energy. ϵ is a parameter which increases with compactness of the star. For instance for the Sun $\epsilon = 2 \times 10^{-6}$, orders of magnitude below the efficiency of nuclear reaction ($\epsilon \simeq 0.007$). For a WD $\epsilon \simeq 0.001$, and for an NS $\epsilon \simeq 0.1$, more than an order of magnitude above the nuclear reaction case. Therefore accretion of a neutron star is indeed an efficient mechanism of energy production. The case of BHs should be treated separately, since it cannot be discussed in terms of a Newtonian model: this is done in the following sections. As it will be seen, also for a BH an efficiency $\epsilon = 0.1$, close to that of an NS, can be expected.

Suppose that an NS is moving in a medium of density $n = 1$ cm^{-3} with $v_s = 10^7$ cm s^{-1}. These are realistic parameters for the interstellar medium. From equation (9.12), the emitted luminosity turns out to be

$$\frac{dW}{dt} \simeq 10^{30} \text{ erg s}^{-1} \sim 0.001 \, L_\odot. \tag{9.14}$$

Since this luminosity is a mere 10^{-7} times that observed in x-ray sources, it is clear that accretion of an NS (or a BH) is an effective way of producing energy, but the density of the accreted medium must be much larger than the interstellar one, i.e. accretion must occur only in the vicinity of a companion star. This should clarify the basic point that, in order to explain the high luminosity of galactic x-ray sources, the accretion process should occur on a collapsed object in a binary system. The 'normal' star provides the required accretion material. In the following we will concentrate on binaries containing BHs or NSs, while we will not treat the case of WDs, which corresponds to so called 'cataclysmic variables'.

As mentioned above the typical luminosity of x-ray binaries is $L_X = 10^{37}$ erg s^{-1}. This luminosity is presumably connected with the so called Eddington luminosity L_E, the luminosity at which the push due to radiation pressure compensates the gravitational force. For pure hydrogen and for photon–electron interaction described by the Thomson cross section, σ_T, $L_E = 4\pi G M m_p c/\sigma_T$. For $M = M_\odot$, $L_E \simeq 10^{38}$ erg s^{-1}.

The spectrum of an accreting NS is hard to calculate especially if the NS is endowed with a magnetic field B. This will be considered in the next section. Now we give a simple argument neglecting B. We simply suppose that the NS radiates its luminosity as a black body. Using the Stefan–Boltzmann law and the luminosity and dimension given above, one finds a radiation temperature

$$T \simeq \left(\frac{L}{4\pi a^2 \sigma} \right)^{1/4} \sim 1 \text{ keV} \tag{9.15}$$

which is indeed in (soft) x-rays.

9.4.2 Role of the magnetic field

Suppose that the magnetic field of the collapsed object (NS or WD) is characterized by a polar magnetic field B_0. The effect of the field becomes dominant if the magnetic energy density overcomes the gravitational energy density. The two quantities are equal at a radius r_{Alf} (Alfven radius), which is given by

$$r_{\text{Alf}} = \left(\frac{B_0^2 a^6}{\sqrt{2GM\dot{M}}} \right)^{2/7}. \tag{9.16}$$

In the simplest models, within r_{Alf} the accreting matter is supposed to follow the field lines (magnetic surfaces), which produce a pair of magnetic funnels in correspondence of the poles. The matter interacts with the star only on two caps, the dimension of which is much smaller than the star radius, which increases the value of the effective temperature (see equation (9.15)).

9.5 Accretion onto BHs

9.5.1 Spherical case

First we will consider the spherical case. The material which is accreted will be heated up because of compression. If compression is adiabatic with a polytropic index $\gamma = 5/3$, a Newtonian model gives the temperature profile:

$$T \sim \frac{3}{20} \frac{m_{\text{p}}}{k} \frac{GM}{r}. \tag{9.17}$$

For r approaching r_{G}, the proton temperature is a few MeV. The energy released is only that radiated by the gas before crossing the horizon. The accreting matter heats up, but in general radiates only a fraction of its heat. The calculation of the efficiency is not trivial. It depends on the electron–proton coupling and on the details of the radiation process. In particular the role of the magnetic field entangled in the accreting material and compressed by the flow becomes critical. If the accretion rate is $\dot{M} \ll \dot{M}_{\text{E}} \equiv L_{\text{E}}/c^2$, the efficiency turns out to be extremely small. Only for $\dot{M} \simeq \dot{M}_{\text{E}}$ values of $\epsilon = 0.1$ are found, similar to those of NSs. Note that, in general, the smaller is the efficiency, the higher are the temperatures.

9.5.2 Accretion discs

If the accreting matter owns sufficient angular momentum, accretion takes place in a dis-like fashion. Such a configuration is far more interesting than spherical accretion, since a large fraction of the rest-mass energy can be converted into radiation. The standard way to proceed is to assume a steady state, axisymmetric geometry, and to average out the gas properties along the vertical direction. Three main equations specify the gas dynamics, i.e. conservation of mass, of angular momentum and of energy. Our ignorance about the actual viscous mechanism at work is by passed using a viscous prescription of the form $\nu \equiv \alpha v_s H$, where ν is the coefficient of kinematic viscosity, α a dimensionless parameter, v_s the sound speed, H the disc height-scale, leaving as input parameters the accretion rate $\dot{M}$, together with the BH mass M. What is implicitly assumed is that, for every value of $\dot{M}$, the viscosity is exactly what needed to sustain such an accretion rate. The energy conservation equation has the form

$$Q^+ = Q^-_{\text{rad}} + Q^-_{\text{adv}} \qquad (9.18)$$

where $Q^+ \sim GM\dot{M}/R^3$ is the local heat generation rate (per unit surface). The two cooling terms represent the radiative cooling and the advective cooling, i.e. the part of the heating rate which goes into internal energy of the gas. Three time-scales govern the gas dynamics and thermal properties, i.e. the viscous time-scale $t_{\text{visc}} \sim R^2/\nu$, the dynamic time-scale $t_{\text{dyn}} \sim 1/\Omega_K$, where Ω_K is the Keplerian angular velocity, and the thermal time-scale t_{th}, which is defined as the heat (per unit area) divided by the dissipation rate (per unit area). The three time-scales are such that $t_{\text{dyn}} \sim t_{\text{th}} \ll t_{\text{visc}}$.

9.5.2.1 Shakura–Sunyaev accretion discs

Roughly, there exist four branches of the accretion disc solution. For a large value of the density, the disc is geometrically thin, optically thick and the pressure is gas dominated. In this regime, $Q^+ = Q^-_{\text{rad}}$, and the disc emits locally as a black body, with temperature $T \propto r^{-3/4} M^{-1/4}$. For stellar mass BHs, $T \sim 1$ keV, while for $10^8 \, M_\odot$ objects, $T \sim 5$ eV. This solution, given by Shakura and Sunyaev (1973) (SS), can explain the basic properties of the UV spectra of QSO, and accounts for the thermal component observed in galactic BHs in their 'high state'.

For lower values of the density the disc can be optically thin. In this case the radiation spectrum is probably dominated by Comptonized free–free radiation (Shapiro *et al* 1976). Such a branch is, however, thermally unstable, and hence not relevant for realistic flows.

Finally, there exist two branches where $Q^+ = Q^-_{\text{adv}}$, and that are then called ADAFs, after advection dominated accretion flows.

9.5.2.2 ADAFs

On the high density branch, at very high accretion rates, the diffusion time-scale, which is the time-scale needed by radiation to diffuse through the disc and reach the photosphere, is $\gg t_{\text{visc}}$, and hence photons are trapped into the flow. Under such circumstances the disc body puffs up, supported by radiation pressure. The so called 'slim-disc' solution is relevant for super-Eddington flows with large densities (Abramowicz *et al* 1988). Such a solution can be considered the high accretion rate limit of the SS disc.

In the opposite limit of very low densities, there exists another branch of stable solutions. Such solutions extend from extremely low to medium accretion rates, and can be considered the low density counterpart of the SS disc. The key point is that when the density is very low Coulomb collisions are not fast enough to transfer energy from ions (supposed to be directly energized by the accretion process) to electrons. If the interaction between the two species is slow, then ions 'advect in' part of the energy gained as they fall into the hole. Also in this case the whole process can be understood in terms of time-scales. The three competing time-scales are t_{visc}, the Coulomb time-scale t_{p-e}, and the cooling time-scale for electrons t_{cool}. Viscous processes provide the heating term for the ions, while Coulomb collisions are at the same time the heating term for the electrons, and the cooling term for the ions. In the proper regime for ADAFs, in the inner part of the flow we have $t_{cool} \ll t_{visc} \ll t_{p-e}$. The above relations immediately show that: (a) the energy is advected in by protons (or more generally ions); and (b) the plasma is at two temperatures, with $T_e < T_p < (m_p/m_e)T_e$. Once heated, electrons cool down very rapidly via Comptonization of free-free and cyclo-synchrotron radiation.

The key feature of the solution described above is that a low density ADAF is extremely hot, with temperatures of few hundred keV, and underluminous. They have been proposed to explain the x-ray emission from underluminous AGNs, and are thought to be relevant for the so called 'low state' of galactic BHs (see section 1.3.2, and, for a review, Narayan *et al* 1998, and references therein).

9.5.3   Disc inner edge

The issue of the choice of the inner boundary condition in the solution of the disc equations needs some further comments. In the simplest formulations, the disc inner edge is taken as the radius $\tilde{r}$ of the last stable orbit of a test particle rotating around a BH. In this way one neglects the energy production between $\tilde{r}$ and the horizon.

The value of $\tilde{r}$ depends on the angular momentum of the BHs. The fractional binding energy ϵ corresponding to $\tilde{r}$ yields the fraction of energy which can be radiated, and therefore strictly compares with the accretion efficiency introduced in section 1.4.1.

For a Schwarzschild BH $\tilde{r} = 6GM/c^2$, $\epsilon = 0.057$. Increasing the angular momentum, $\tilde{r}$ decreases if the particle is corotating with the hole; it increases with respect to the spherical case if it counterrotates. For an extreme Kerr BH, $\tilde{r} = GM/c^2$, $\epsilon = 0.42$ for corotation, $\tilde{r} = 9GM/c^2$, $\epsilon = 0.038$ for counterrotation. On these issues we refer to Misner *et al* (1973). For a fixed accretion rate the shrinking of $\tilde{r}$ corresponds to an increase of the disc luminosity and temperature.

9.6   LMC X-3

As discussed above luminous x-ray binaries host accreting collapsed objects. The distinction between NSs and BHs is mainly based on the mass function. The argument pointing to a BH is essentially the exclusion of the NS possibility, and therefore it is rather indirect. In particular one can always envisage a situation which mimics the presence of a BH. Take for instance a triple system, with a primary of high mass and the other two components close together, being an

NS and a normal star, the sum of the masses exceeding M_c^{NS}. Though from evolutionary considerations the system would appear rather unlikely to form, and its lifetime may be short, one cannot exclude this possibility, which would imitate the appearance of a BH.

Astronomical observations can only lead to BH candidates. Astronomical evidence is obviously different from that of the laboratory where an experiment is performed under controlled conditions. The 'discovery' of BHs required to be substained by other candidates, so that it could be established on a statistical basis. For almost a decade Cyg X-1 remained the only BH candidate, which provoked some uneasiness in the BH fans.

A new BH candidate was announced in 1983 by Cowley *et al.* The procedure was similar to that followed for Cyg X-1. The persistent x-ray source LMC X-3 (in the Large Magellanic Cloud) was optically identified with a spectroscopic binary ($P = 1.7$ d). The mass function was then measured as $f(M_X) = 2.3\ M_\odot$. From the estimates of the mass of the non-collapsed component and of the orbital inclination, it was found that $M_X = 10\ M_\odot$, pointing to a BH. Note that in this case the evidence is stronger than for Cyg X-1. In fact here the BHs is more massive than its companion, which, as discussed in section 3.2, eliminates some ambiguity. As from equation (9.7), the sole measurement of the mass function yields a lower limit to M_X, which is close to, but not above, M_c^{NS}.

Among persistent x-ray sources there is only one more robust BH candidate, LMC X-1. In fact optical identifications can be performed only in the Galaxy and in the Magellanic clouds. The present instrumentation has already allowed the detection of all persistent sources above 10^{37} erg s^{-1}, and optical identification has been obtained for most of them. Therefore unless there is a class of BH candidates much below the previous luminosity limit, we do not expect a substantial increase in the number of persistent x-ray sources hosting a BH.

9.7   BH in transient x-ray binaries

In addition to persistent x-ray binaries, since the time of the Uhuru satellite, a class of sources which brighten transiently has been discovered. In x-ray transients the accretion process occurs sporadically. The classification of transients is complex. The most important ones for BH candidates are the 'x-ray novae'. In this case in a week the x-ray luminosity increases from an undetectable level, to values close to the Eddington limit for an $M = 1\ M_\odot$; the decay is exponential and occurs on month time-scales. Each year some of these sources are detected. The optical counterparts are easy to discover, since they are typically stars of mass similar to that of the Sun, and during the x-ray active state the optical luminosity increases by four to five magnitudes (as in classical novae). During the bright phase the optical spectrum is extremely complicated, but on a year time-scale it is dominated by the normal star, and the mass function is easy to measure.

Most BH candidates known thus far were found in x-ray transients. In contrast to the case of persistent sources, their number increases steadily, and there are now more than a dozen robust candidates, two of which we discuss briefly below.

9.7.1   A0620–20 (Nova Monoceros)

The x-ray transient appeared in 1975. The luminosity at maximum was 10^{38} erg s^{-1}. The nature of the system was clarified only years later, when no x-ray flux remained detectable. The source was found to be a binary of period $P = 0.31$ d. The non-collapsed component is a K3V star ($M = 0.7\,M_\odot$). The mass function is $f(M) = 3.2 \pm 0.2$ and the deduced value of the unseen component $M > 7.3\,M_\odot$. Note that in this case, from equation (9.7), the evidence of a BH is independent of the estimates of the mass of the normal star and of the inclination angle.

9.7.2   GS 2023+33 = V404 Cyg

This is the most extreme BH candidate. The x-ray behaviour is similar to that of the source discussed above. From the study of the optical counterpart one derives: $P = 6.47$ d, a companion of mass $0.6\,M_\odot$, $L_X = 10^{38}$ erg s^{-1}, $f(M) = 6.3 \pm 0.3$, $M_X > 8\,M_\odot$. The mass function exceeds by a factor of 2 the critical mass of an NS!

9.7.3   Galactic superluminals

These are a rather remarkable subclass of x-ray transients hosting BHs. They have various peculiar characteristics; the most notable is associated with the radio counterpart. During the active phases, radio blobs with apparent velocities exceeding the velocity of light are emitted. This is a phenomenon rather common in active galactic nuclei, and it is associated with relativistic effects appearing when the bulk velocity of the emitting plasma (jet) is close to c and directed towards the observer. These transients are therefore like quasars in miniature.

The best studied objects are 1915+105 and 1655–40. Only for the latter are there spectroscopic measurements, which point to a BH of 5 $M_\odot$. The former is most likely very similar. The jet velocity is in both cases about 0.9 c. The luminosities are close to the Eddington limit, but the dominant energy loss is probably that associated with the relativistic outflow, which could be as large as 10^{41} erg s^{-1}.

The two sources are ideal laboratories for special and general relativistic effects, superior to AGNs, since the existence of the companion allows us to constrain directly the system parameters.

9.8   Probes of BH physics

A direct way of probing BHs is through the measurement of photon and particle trajectories. Deviations from the Newtonian case will become more and more important as the trajectory approaches the gravitational radius. In particular, within few r_G, differences between the Newtonian case, Schwarzschild and Kerr BH become significant. This is an issue of great importance, since, in practice, it may be a way of measuring the angular momentum of BHs.

9.8.1 Emission from the inner accretion disc

Models taking into account classical GR effects in the disc emission (such as Doppler boosting, gravitational redshift and lensing properties) have been developed with a high degree of sophistication, expanding the SS and ADAF models (see sections 5.2.1, 5.2.2). As an example, consider the case of a Schwarzschild BH. The disc inner edge is then just 3 r_G (see section 5.3). In principle a spectral fit to the data can constrain r_G and therefore M. In the cases where this procedure was carried out, there is a good agreement with the value obtained from the mass function. One can go a step further. Suppose that M is well known from $f(M)$, and leave the BH angular momentum free to vary. A basic difference between the Schwarzschild and Kerr BHs is the radius of the last stable orbit, which is generally assumed to coincide with the inner edge of the disc. Since the smaller is the inner edge, the hotter is its emission, one expects, for a fixed mass, that the hardness of the spectrum increases with the angular momentum. The hardness of the spectrum becomes therefore a measure of the BH angular momentum. This recipe has been followed for instance in the case of LMC X-3 (Ebisawa 1991).

9.8.2 Fe lines

As discussed above, the basic picture of emission of an accreting BH is via an accretion disc. However, this is a an oversimplification, and in order to account for the high energy spectrum of Cyg X-1, it was proposed that the disc is surrounded by a hot corona ($T \sim 100$ keV), which may be heated by the dissipation of the magnetic field (see in this regard the discussion on spherical accretion, section 5.1). This general picture has been extended to AGNs and in particular to Seyfert galaxy spectra (Haardt and Maraschi 1991). The hot corona produces hot photons via Compton scattering, which in part leave the system, and in part interact with the cooler accretion disc. The spectrum reflected by the disc appears hotter than the original one, since the soft part is absorbed. Moreover, it is characterized by fluorescence lines, the most important ones being the Fe K-lines in the 6–9 keV region. The shape of the reflected component has been studied observationally in great detail, since it is a probe of the coronal spectrum. The lines have gathered a particular interest. In fact they are produced within few gravitational radii of the hole. They will be distorted by Doppler shift due to the disc rotation, by gravitational redshift and gravitational focusing effects (see figure 9.3).

One can calculate the theoretical shape of the line, which has two characteristic peaks, due to redshifted and blueshifted components, the latter being magnified with respect to the former. Obviously the line is a wonderful tracer of the disc inner part and of GR effects (Fabian *et al* 1989, Stella 1990). Observations of Cyg X-1 demonstrated the validity of the procedure, yielding a black hole mass close to that deduced from the mass function. For AGNs this appears the most promising way of measuring the BH mass, since in this case there is no binary companion. The technique will yield substantial progress, in view of the spectral resolution achievable with future space missions.

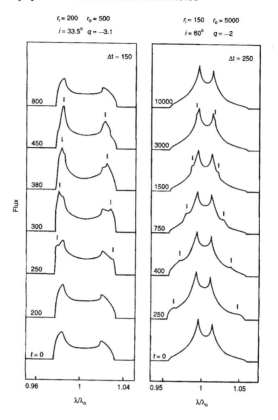

Figure 9.3. Response of the line profile from a relativistic Keplerian disc to flux variations in the central source (from Stella 1990, see the original paper for details).

9.8.3   Power spectrum

The accreting mechanism is probably not smooth, and even if it can be described by a disc, the disc itself will not be homogeneous, with structure like turbulence cells, magnetic spots or vortices, which we can refer to as 'bubbles'. Suppose that bubbles are x-ray emitters, either because of magnetic field reconnection as in solar flares, or because they reflect x-rays. Their presence produces some variability in the x-ray emission, the characteristics of which depend on the orbital properties of the bubbles themselves. The idea is to use x-ray variability structure, as obtained from a power density spectrum, for producing some basic time-scale, like the inner orbit period. This quantity, as we have already discussed, is related to the mass and angular momentum of the BH. The interesting time-scales are of the order of the light travel time of r_{BG}. This is a fraction of ms for a stellar mass BH, and hundreds of seconds for an AGN. First results with this procedure are encouraging (e.g. McHardy 1999).

9.9 Quasiperiodic oscillations

The recent discovery of kHz QPOs from x-ray binaries obtained by the Rossi X-Ray Timing Explorer launched in December 1995 (see, e.g., van der Klis 1998) represents a very effective probe of GR effects around NSs. In particular, it may have yielded a direct observation of the Lense–Thirring precession. It has been proposed that a similar picture can be extended to BH candidates. This would result in a new way of measuring the BH angular momentum. Even though the results are still preliminary, a rather important perspective was opened, with substantial progress potentialities. The present picture is based on the study of short time-scale variability in x-ray binaries, and specifically on the so called QPOs. The subject is somewhat intricate, and requires the description of a rather complex phenomenology.

9.9.1 X-ray pulsators (HMXRBs)

If the neutron star accreting in a binary system is endowed with a strong magnetic field, as discussed in section 4.2, the accreted matter flows onto the magnetic caps, which are the sources of the x-ray luminosity. The rotation of the NS induces a modulation of the emission and one has what is called an x-ray pulsator. Several dozens of such objects are known in persistent and transient x-ray binaries. The rotation periods range from fractions of a second, to days. The energy source is the accretion, and in most cases x-ray pulsators are spinning up (in contrast to pulsars), since in general the accreted matter increases the angular momentum. From the spin-up one can deduce the value of the magnetic field, which is typically $\sim 10^{12}$ G, as in pulsars (see 3.1). In some cases a cyclotron line at energy

$$h\nu = \frac{1}{2\pi} h \frac{eB}{m_e c} \simeq 50 \text{ keV} \tag{9.19}$$

has been measured, which is a direct probe of the magnetic field.

Most x-ray pulsators are in binaries where the non-collapsed component is a massive star (i.e., $M \gtrsim 1 \ M_\odot$), which dominates the optical emission. They are therefore indicated as high mass x-ray binaries (HMXRBs). The orbital periods are typically larger than one day, and the mass transfer occurs through a wind, rather than through Roche lobe overflow which is instead the case of closer systems.

The power density spectrum is characterized by very narrow peaks, corresponding to the rotation period and its harmonics.

9.9.2 Low mass x-ray binaries (LMXRBs)

These systems in most cases are rather close binaries ($P \lesssim 1$ d). The optical emission is dominated by the accretion disc. It is difficult to detect eclipses and to reconstruct a radial velocity curve and the mass function. The very binary nature of such objects may be hard to demonstrate. No regular x-ray pulsation is detected, indicating that the accreting neutron star is probably endowed with a magnetic field much more modest than that of x-ray pulsators. In fact, whenever the Alfven radius is comparable to the NS radius, a strong modulation is not expected. From equation (9.16) one can deduce a magnetic field $\lesssim 10^9$ G. The absence of regular pulsations and of cyclotron lines makes impossible a direct measurement of the field.

In the two prototype LMXRBs, namely Sco X-1 and Cyg X-2, it has been noted since the 1970s that there is no well defined dependence of the x-ray spectrum on luminosity. In particular, the spectrum does not harden with increasing luminosity, as would be expected, for instance, in the case of black body emission (see equation (9.15)).

9.9.3   QPOs

It was in the beginning of the 1980s, especially through the satellite EXOSAT, that power spectrum analysis was systematically performed for a number of LMXRBs (see, e.g., van der Klis 1989, 1995). The motivation of the search was not only for narrow peaks like those characteristic of HMXRBs, but also to study the broad power spectrum, similarly to what was performed for BHs (see section 8.3). The key discovery was that of broad peaks with centroid frequencies of 5–50 Hz (see figure (9.4)). The peak widths are comparable with the central frequency. The structure persists for 10^5 cycles. It disappears and later it may recur. One is clearly facing a quasiperiodic phenomenon which is indicated as quasiperiodic oscillations (QPOs).

The power spectrum properties are connected with those of the energy spectrum. This is apparent in the x-ray colour (hardness ratio) intensity plot, or even better the hard colour/soft colour plots. It is found that a given source moves in these planes along a very regular pattern (see figure 9.5). Rarely was the astronomical evidence so clean, but this does not imply a simple explanation. The most luminous sources, those which are close to the Eddington limit, exhibit a very characteristic pattern which looks like a Z (figure 9.5). Sco X-1 and Cyg X-2 belong to this class. In the horizontal branch (top of Z), the hard colour is constant and the soft one increases. In the normal branch (diagonal of Z), both colours decrease, and in the flaring branch (bottom of Z), both colours increase. The power spectra change drastically depending on the branch. The higher frequency QPOs show up in the horizontal branch, while lower frequencies in the normal branch. It is obviously a rather complex phenomenon, which, however, seems to be regulated by a single parameter: the accretion rate, which is increasing from top to bottom.

In less luminous sources the shape in the colour/colour plane is closer to a circle (atoll sources).

QPOs have been observed not only in sources which supposedly contain an NS (both Z and atoll sources), but also in BH candidates. For instance, in Cyg X-1 the QPO frequency is $\simeq 9$ Hz, and in 1655-40 it is $\simeq 300$ Hz.

Probably one has to invoke different scenarios for explaining various types of QPO. The phenomenon is only partly understood, and in particular the relation between power and energy spectrum properties remains unclear. An idea which

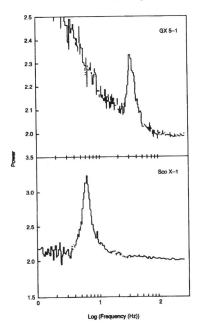

Figure 9.4. Power spectra of the x-ray intensity variations of two bright LMXRBs (from van der Klis 1989).

has been invoked essentially since the times of the discovery of QPOs asserts that there are two basic periodicities, which regulate the accretion rate, the NS spin and a Keplerian periodicity, corresponding in the simplest model to the Alfven radius. Note that the latter is not strictly periodic, since, as apparent from equation (9.16), the Alfven radius depends on the accretion rate. Accretion should occur at a frequency which is a beat of the two basic frequencies.

9.9.4 KHz QPOs

After a decade of little progress in the field, an important development came through the RXTE satellite. The instrument consists of very large area proportional counters, which allow us to study variability of the brightest sources down to fractions of milliseconds. This is a very significant time scale since the light time of the gravitational radius of a solar mass is 10^{-2} ms, and the Keplerian frequency at 15 km of a 1.4 $M_\odot$ NS is 1 kHz.

RXTE detected kHz QPO from a number of LMXRBs, both Z and atoll sources (see figure 9.6). A rather unexpected and significant discovery is that in various cases, there are twin peaks QPO. The two peaks vary their centroid frequency, in the same manner as lower frequency QPOs do, but the difference in frequencies remains stable within the uncertainties. For instance in GX 5-1,

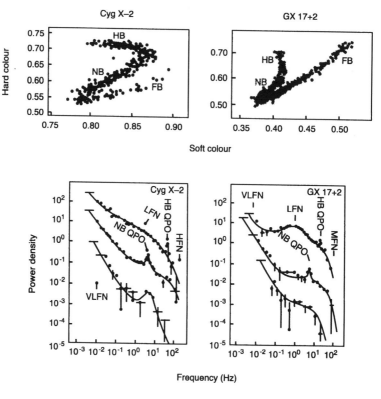

Figure 9.5. Z source behaviour. x-ray colour diagrams (top panels), and power spectra (bottom panels) (from van der Klis 1989, see the original paper for details).

$325 < \nu_1 < 448, 652 < \nu_2 < 746$ and $\nu_2 - \nu_1 = \nu_0 = 327 \pm 11$ (Hz). The stability of ν_0 is noticeable, and it is also noticeable that for a dozen sources $\nu_0 \simeq 3$ ms. It is widely accepted that ν_0 is the rotation frequency of the NS, a quantity which was searched for unsuccessfully since the identification of LMXRBs as a class. LMXRBs appear therefore to be rapidly accreting NSs, with rotation frequencies close to those of ms pulsars (see section 3.1). The picture which is coming out, and is an evolution of the beat frequency model described in the previous section, is that there are two basic frequencies, the spin and a Keplerian one. While the spin is not observable, the beat frequency and the Keplerian are apparent. What remains to be understood is the role of the low frequency QPOs.

9.9.5 Lense–Thirring precession

Since the infancy of GR it was known that the rotation of a body induces a frame dragging (Lense and Thirring 1918, see also Wilkins 1972). In particular, if one

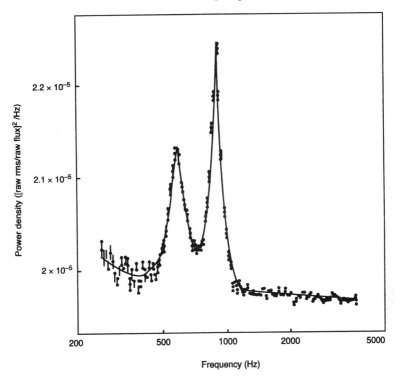

Figure 9.6. Twin peaks in the power spectrum of Sco X-1 (from van der Klis 1998).

considers a test particle moving in a non-equatorial plane around a rotating body, the spin of the body induces a prograde precession of the orbital plane. Stella and Vietri (1998) considered the beat frequency model for kHz QPOs and proposed to interpret the low frequency QPOs at 10 KHz as the Lense–Thirring (LT) precession of the inner edge of an accreting disc.

In a weak field limit the LT precession frequency is given by

$$\nu_{\text{LT}} \simeq \frac{GM\widetilde{a}}{\pi c^2 r^3} \tag{9.20}$$

where $\widetilde{a}$ is the specific angular momentum, and r is the orbital radius.

Introducing the Keplerian frequency ν_{K} and the spin frequency ν_{s}, the equation translates into the form

$$\nu_{\text{LT}} \simeq \frac{8\pi I \nu_{\text{K}} \nu_{\text{s}}}{c^2 M} \tag{9.21}$$

where I is the moment of inertia. For $\nu_{\text{K}} = 100$ Hz, $\nu_{\text{s}} = 300$ Hz, one has $\nu_{\text{LT}} \simeq 10$ kHz.

The proposal of Stella and Vietri (1998) is still in a phase of verification. The complexity of the astrophysical picture has been explored by Markovic and Lamb (1998). Here we would like to stress the potential importance of the result, which would constitute a first astrophysical application of a basic effect of GR.

For a strange conspiracy, the astrophysical evidence would be very close in time with the first observation of the effect related to the Earth's rotation (Ciufolini *et al* 1998), some 80 years after the theoretical prediction.

9.9.6   Extension of LT precession to BHs

If one has a rotating BH, an LT frame dragging is expected in complete analogy to the NS case. The role of the Keplerian frequency may be played by the frequency corresponding to the last stable orbit. This was discussed in section 5.3 and it was shown to depend on the mass and angular momentum of the BH. Cui *et al* (1998) suggested interpreting the QPOs observed in BH candidates as LT precession. In such a picture, once the BH mass is known, the angular momentum can be measured. Cui *et al* (1998) showed that the values obtained in this way are fully consistent with those deriving from the spectral fitting procedure (see section 8.1).

While the results of Cui *et al* (1998) in our opinion should be taken with caution, since they are based on an oversimplified disc model, they certainly illustrate the potentialities of a direct observation of LT precession in BH.

9.10   Concluding remarks

In the last three decades the effort in the astrophysical study of BH candidates has been enormous. However, in our opinion, only one robust result of importance for GR was obtained: the demonstration of the very existence of BHs through the measurement of their mass. In the last decade the efforts have focused on the determination of BH angular momentum. The results are encouraging but still preliminary. Since the no hair theorem (section 3.1) states that BHs are characterized solely by their mass and angular momentum, in a way the astrophysical programme is close to completion.

Obviously from the point of view of testing the gravitational theories, there is still a long way to go. Up to now astrophysics has explored the weak field limit only. The recent results on kHz QPOs seem to open our search to regions comparable in size to the gravitational radius, where strong field effects may really show up.

References

As a general bibliography to the arguments treated in this paper we quote:

Cowley A 1992 *Annu. Rev. Astron. Astrophys.* **30** 287
Frank J, King A R and Raine D J 1986 *Accretion Power in Astrophysics* (Cambridge: Cambridge University Press)
Lewin W, van Paradijs J and van den Heuvel E (eds) 1995 *X-Ray Binaries* (Cambridge: Cambridge University Press)
Shapiro S L and Teulkoski S A 1983 *Black Holes, White Dwarfs and Neutron Stars: the Physics of Compact Objects* (New York: Wiley)
Srinivasan G 1997 *Stellar Remnants (Saas Fee Advanced Course 25)* (Berlin: Springer)
Tanaka Y and Shibazaki N 1996 *Annu. Rev. Astron. Astrophys.* **34** 607
Treves A, Maraschi L and Abramowicz M 1988 *Pubb. Astr. Soc. Pac.* **100** 3

References to papers quoted in the text:

Abramowicz M A, Czerny R, Lasota J-P and Szuszkiewicz E 1988 *Astrophys. J.* **332** 646
Adams W S 1914 *Pubb. Astr. Soc. Pac.* **26** 198
Baade W and Zwiky F 1934 *Proc. Natl Acad. Sci.* **20** 259
Bekenstein J D 1974 *Phys. Rev.* D **9** 3292
Bondi H 1952 *Mon. Not. R. Astron. Soc.* **112** 195
Carter B 1973 *Black Holes* ed B DeWitt and C DeWitt (New York: Gordon and Breach)
Ciufolini I, Pavlis E, Chieppa F, Fernandez-Vieira E and Perez-Marcader J 1998 *Science* **279** 2100
Cowley A P, Crampton D and Hutchings J 1983 *Astrophys. J.* **272** 118
Cui W, Zhang S N and Chen W 1998 *Astrophys. J. Lett.* **492** 53
Ebisawa K 1991 *PhD Thesis* ISAS (Tokyo: Sagamhana)
Eddington A 1926 *The Internal Constitution of Stars* (Cambridge: Cambridge University Press)
Fabian A C, Rees M J, Stella L and White N E 1989 *Mon. Not. R. Astron. Soc.* **238** 729
Giacconi R, Gursky H, Paolini F R and Rossi B B 1962 *Phys. Rev. Lett.* **9** 439
Gribbin J and Rees M J 1990 *The Stuff of the Universe* (London: Heinemann)
Haardt F and Maraschi L 1991 *Astrophys. J. Lett.* **380** 51
Hawking S W 1975 *Commun. Math. Phys.* **43** 199
Hawking S W and Ellis G F R 1973 *The Large Scale Structure of Space Time* (Cambridge: Cambridge University Press)
Landau L D 1932 *Phys. Z. Sow* **1** 285

La Sala J, Charles P A, Smith R A D, Balucinska-Church M and Church M J 1998 *Mon. Not. R. Astron. Soc.* **301** 285

Lense J and Thirring H 1918 *Phys. Z.* **19** 156

Markovic D and Lamb F K 1998 *Astrophys. J.* **507** 316

McHardy I 1999 *Pubb. Astr. Soc. Pac. (Conf. Ser. 159)* 155

Misner C W, Thorne K S and Wheeler J A 1973 *Gravitation* (San Francisco: Freeman) section 338

Narayan R, Mahadevan R and Quatert E 1998 *The Theory of Black Hole Accretion Disks* ed M A Abramowicz, G Bjiornsson and J R Pringle (Cambridge: Cambridge University Press) p 148

Oppenheimer J R and Volkoff G M 1939 *Phys. Rev.* **55** 374

Shakura N I and Sunyaev R A 1973 *Astron. Astrophys.* **24** 33

Shapiro S L, Lightman A P and Eardley D M 1976 *Astrophys. J.* **204** 187

Stella L 1990 *Nature* **344** 747

Stella L and Vietri M 1998 *Astrophys. J. Lett.* **492** 59

Taylor J H and Weisberg J M 1982 *Astrophys. J.* **253** 908

Tolman R C 1934 *Relativity, Thermodynamics and Cosmology* (Oxford: Oxford University Press)

van der Klis M 1989 *Annu. Rev. Astron. Astrophys.* **27** 51

——1995 *X-ray Binaries* ed W Lewin, J van Paradijs and E van den Heuvel (Cambridge: Cambridge University Press)

——1998 *Nucl. Phys.* B **69** 103

Wheeler J A 1966 *Annu. Rev. Astron. Astrophys.* **4** 393

Wilkins D C 1972 *Phys. Rev* D **5** 814

PART 5

Richard A Matzner

Center for Relativity and Physics Department,
University of Texas at Austin, TX 78712, USA

Chapter 10

Computational black holes

10.1 Introduction

We outline some of the basic methods and concepts in treating the Einstein field equations as a computational system. We present discretization methods and describe the theoretical framework behind current approaches to the study of black hole dynamics. Finally, we present some results from one of the extant attempts computationally to handle the binary black hole problem, the Binary Black Hole Grand Challenge Alliance code.

10.2 Overview, analytical background

The problem is to solve the Einstein equations. Einstein equations are hyperbolic: they naturally lend themselves to (require) a stepwise integration for the solution. We will consider *black hole* solutions, which means we are solving the vacuum Einstein equations:

$$G_{\mu\nu} = 0 \tag{10.1}$$

where $G_{\mu\nu}$ are the components of the Einstein tensor, and μ, ν are indices 0, 1, 2, 3 referring to the coordinates ('0' typically refers to time) [2]. The black holes are specified, and evolved, as classically stable structures in the gravitational field configuration.

To achieve a decomposition to a form which allows stepwise evolution, different approaches are available, depending on the *slicing* used to express the separation between space and time.

These three approaches utilize

- spacelike 3-surfaces evolving in time [1]
- characteristic 3-surfaces evolving in time [3]
- hyperbolic 3-surfaces evolving in time [4, 15].

The *ADM method* [1] method sets 3-spaces evolving in time. This means that the gravitational field is described by the dynamics of three-dimensional (tensor)

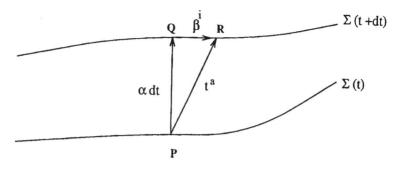

Figure 10.1.

quantities: the spatial 3-metric, and the extrinsic curvature, a spatial tensor which is effectively the momentum of the 3-metric. See figure 10.1.

The characteristic approach pioneered by Bondi *et al* [8], Sachs [9] and Penrose [10] has been developed into a computational method in more recent papers, including work by Friedrich and Stewart [14], and by Winicour and collaborators [11–13]. In this case, the constant 'time' surfaces are in fact null hypersurfaces of constant advanced time, or retarded time. For instance one may coordinatize the spacetime based on a sequence of outgoing null hypersurfaces ('nested null cores'). The 'time' then labels what would be called the retarded time of a distant timelike observer. This time is constant on each such outgoing hypersurface. These hypersurfaces are generated by outgoing null rays. Angle coordinates (θ, φ) can be defined to be constant along the rays; a radial coordinate can be defined along the rays (affine distance for instance, or an areal coordinate), and the retarded time gives the fourth coordinate.

Once the set of hypersurfaces has been chosen, the coordinate set is quite rigid. This means that there is less coordinate, or 'gauge ', freedom in the characteristic set than in the ADM formulation. Besides this advantage, the characteristic formulation also allows *compactification* of infinity, which in some cases means we can compute *to infinity* with a *finite* amount of computational effort. The applicability of such techniques in fairly general situations requires that the *incoming* wave field becomes quiescent as $r \rightarrow \mathcal{I}^+$ (future null infinity). If this is not the case, then the compactification (which is carried out by rescaling the radial variable r, e.g. $r \rightarrow u = $ arctanh r or $r \rightarrow u = 1/r$) will lead to arbitrarily short wavelength oscillations in terms of the new variable, near the outer boundary $(r = \infty)$. On the other hand, outgoing radiation travels near the $r = $ constant surfaces. If the characteristics are precisely the characteristics for the radiating system, then since the characteristic surfaces are surfaces of constant phase, the outgoing radiating fields can change only slowly, even when compactified. Then the fields have only a finite, typically polynomial, dependence on the compactified coordinates, and so the whole field out to infinity can be calculated or computed. See figure 10.2.

COMPACTIFICATION

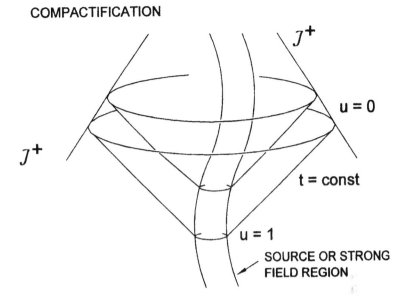

$\mathcal{J}^+$

$u = 0$

$\mathcal{J}^+$

$t = \text{const}$

$u = 1$

SOURCE OR STRONG
FIELD REGION

Figure 10.2.

Because of this feature, it might seem that a characteristic formulation is the perfect choice for black hole dynamics, where one expects small amounts of ingoing radiation and dominantly outgoing radiation near null infinity (also known as radiative infinity) $\mathcal{I}^+$. Unfortunately the physical process of gravitational lensing near black holes makes construction of a characteristic coordinate system near a pair of black holes very difficult. If a reasonable congruence of outgoing rays is set up around one black hole, the focusing and caustics due to the second black hole spoils it (figure 10.3). While there has been some work to integrate through caustics in simple spacetime situations [14], at present no general method to do this is known.

The third approach, developed by Friedrich [4] and implemented in spherically symmetric computations by Hübner [15, 16], slices spacetime into spacelike hyperboloidal surfaces (which become null at infinity). Because they become null at infinity, these surfaces share the compactification at infinity (and the attendant wave extraction capabilities) with the characteristic surfaces. In fact, Friedrich [5–7] has written a *conformal equation set* which subsumes the Einstein equations, and in which the hypersurface $\mathcal{I}^+$, for instance, is just another null surface through which one can integrate. Because this formulation is based on slices that are spacelike in the interior, but become null as they reach to null infinity, it apparently has built-in capability to handle strong-field dynamics (no caustic problems) and extract them smoothly to infinity. However, this approach to the evolution equations has not yet been the subject of substantial development (three-dimensional

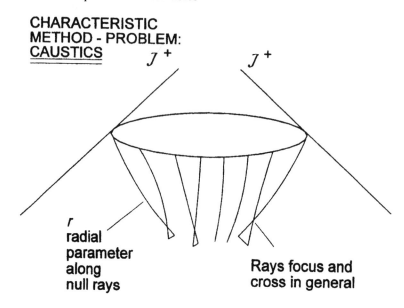

CHARACTERISTIC METHOD - PROBLEM: CAUSTICS

J^+ J^+

r
radial parameter along null rays

Rays focus and cross in general

Figure 10.3.

examples have not yet been demonstrated, and black hole data have not been constructed), so will not be the subject of closest interest in these lectures.

By relinquishing Friedrich's conformal formulation as an option, we are left with the problem of correctly describing the two regions of the computational domain: the strong-field region where the characteristic method fails, and the weaker-field region where rays can be found connecting to null infinity without caustics. Efficiency in extracting the radiation at infinity demands that the strong-field Cauchy evolution not be extended to infinity, because the number of interior points at which the fields have to be computed will scale as the volume, hence tends to infinity as the Cauchy volume is taken larger.

Even without developed hyperboloidal-slicing techniques, such extremely large computational domains are unnecessary because two alternatives exist, both of which assume there is no ingoing radiation at late times. These are the *matching alternatives*: (a) match to a perturbation extraction to infinity, or (b) match to a compactified characteristic evolution to infinity.

The success of the perturbative solutions in providing an extraction to infinity lies in the fact that the perturbation equations can be analytically (or quasi-analytically) solved. (By quasi-analytically I mean that typically only *one* decoupled ordinary differential equation has to be solved. The current situation is that solvers for such equations are very fast, stable and accurate, so they return a result which is very close to the analytic result). Perturbation theory can be a linearization around flat space, or around a black hole space, Kerr (rotating) or Schwarzschild. In all cases what the linearization removes is wave–wave interactions, but the interaction of the wave with the large-scale background induced by the mass (e.g. backscattering from the curvature of the background spacetime) *is* supported. We will further discuss the perturbative and the match to characteristic, below.

10.3   Computational relativity and new physics

Most of the tone of this manuscript is oriented toward the question of the behaviour of astrophysical black hole sources and the gravitational radiation produced. However, computational relativity also provides new arenas of research (e.g. the implementation of adaptive methods, which insert refinements where more accuracy is found to be needed). Computational relativity has also produced surprises. One particular surprise is Choptuik's discovery of critical phenomena in collapse of scalar radiation [55].

Choptuik considered slicing in terms of spacelike sections with spherical symmetry, solving the linear wave equation for a scalar field ϕ.

$$\Box \phi = 0.$$

The operator $\Box$ involves covariant derivatives due to the gravitational curvature of the background. The gravitational effect of the scalar field modifies the background from flat space. For large amplitudes of an initially ingoing shell, a black hole forms. For weak initial waves, the ingoing pulse just passes through the origin and reexplodes. Exactly at the dividing amplitude between black hole formation and non-formation, Choptuik discovered critical behaviour in his computational simulation.

This critical behaviour consists of repeated bursts of scalar radiation, most of which escapes to infinity while a small part (about 3%) continues to fall in, on a self-similar but smaller scale. Because the scale is smaller, this remaining infalling matter goes through a dynamical cycle more quickly (about a factor of 30 more quickly), and the same pattern repeats. See figure 10.4 which illustrates the phenomenon. (We take the speed of light $c = 1$, so an ingoing pulse of radiation is a 45° ingoing cone.)

If data are set up near the critical solution, then typically this repeated behaviour eventually terminates. Either the last amount of radiation escapes to infinity, or the last amount collapses into a black hole. Exactly at the critical solution, however, the process continues for an infinite number of cycles, but in a finite time. The dynamics occur on a shorter and shorter scale, and Choptuik could compute these small features only because he had implemented an adaptive mesh refinement, which triggered on a measure of the error in the computation.

Remarkably, Choptuik's computational evolution showed the development of a naked singularity. The peak curvature tensor associated with the collapsing solution has magnitude $\sim M/r^3$, where M and r refer to the small amount of ingoing radiation. Since the repeated dynamics reaches $M/r \sim 0.5$, the value of the curvature scales as $\sim M^{-2}$ and diverges just as the mass is all radiated away.

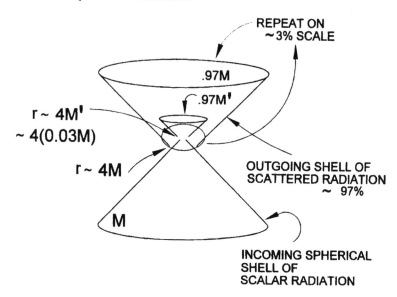

Figure 10.4.

Because of the exponential speedup of this behaviour, it is visible at infinity at a finite time, a naked singularity. See figure 10.5.

The Choptuik solution is perhaps the most dramatic example of new physics obtained from a computational approach. This is a modern development, which complements and extends the more traditional application of computation to modelling, e.g. in astrophysics, where one starts with a general idea of the desired results.

10.4   Example codes

I will give descriptions of three three-dimensional black hole evolution codes, each representing developments from one of the analytical descriptions above.

10.4.1   The Binary Black Hole Alliance code

This code uses the Arnowitt–Deser–Misner [1] formulation of the Einstein equations. The equations break up into four *constraint equations*,

$$H = K^2 - K_{ij}K^{ij} + R = 0 \tag{10.2}$$

and

$$D_j(K^{ij} - g^{ij}K) = 0 \tag{10.3}$$

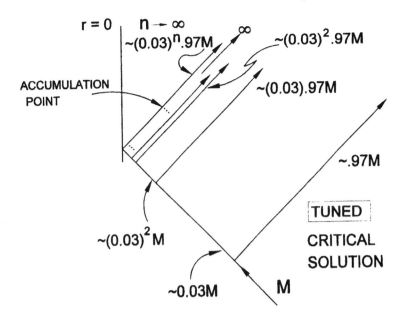

Figure 10.5.

and the dynamical Einstein equations (12 first-order equations).

$$\widehat{\partial}_0 K_{ij} = -D_i D_j \alpha + \alpha [R_{ij} + K K_{ij} - 2K_{il} K_j^l] \tag{10.4}$$

$$\widehat{\partial}_0 g_{ij} = -2\alpha K_{ij} \tag{10.5}$$

where

$$\widehat{\partial}_0 = (\partial_t - \mathcal{L}_\beta) \tag{10.6}$$

is a differentiation operator that moves along the normal to the hypersurface. Here quantities without a superscript are three-dimensional quantities, g_{ij} is the 3-spatial metric, and g^{ij} is the three-dimensional inverse. In the equations, g_{ij} and K_{ij} are the dynamical variables; K_{ij} is a momentum conjugate to g_{ij} and $K \equiv K_i^i$. Also, R_{ij} is the three-dimensional Ricci tensor and $R = R_i^i$ the Ricci scalar. D_j is the three-dimensional covariant derivative. See also [2].

In this formulation, one has available the full coordinate freedom of general relativity, both a freedom (to choose any way of defining simultaneity and any way of imposing spatial coordinates on the 3-spaces of simultaneity) and a curse (because choices *must* be made, and made artfully). The coordinate freedom is expressed in terms of a scalar function $\alpha(x,^i t)$ called the lapse, and a vector function $\beta^i(x^j, t)$ called the shift. These are viewed as functions of spatial coordinates at any one time. They are explained by figure 10.1. The lapse function α gives the relationship between coordinate time difference and the normally measured proper time interval separating spaces of simultaneity. The shift vector $\beta^i(x^l,t)\,dt$ describes a recoordinatization in the 3-space of constant t. Because the action of $\mathcal{L}_\beta$ is only a recoordinization of the surface, we are allowed to introduce $\widehat{\partial_0}$ above leading to a substantial simplification of notation. It can be seen that α is the more important in determining how the 'shape' of the constant-t space evolves. The delicacy of setting α can be seen even in Minkowski space. In particular, if α is briefly (in time) substantially longer in a finite region than elsewhere, the evolution into the future will produce a constant-t space that bulges 'forward in time'. The result is that the later 3-space is not flat and the timelike normals necessarily converge to the future. If subsequent evolution follows $\alpha = 1$, then in a finite time, supposedly distinct points in the 3-space will evolve to be atop one another, a singularity, but one totally dependent on our poor choice of slicing conditions. The shift vector can ameliorate the development of this singularity, but cannot prevent it. The recoordinatization of the 3-space will move apart the coordinate labels of points to counter the induced convergence, but near the point of the singularity this would require an infinite coordinate transformation. We will return often to this question of coordinatization.

10.4.2   The Bona–Masso formulation

The Bona–Masso [17, 18] extension of the ADM approach uses an additional set of variables, which are the spatial derivatives of variables entering the ADM formulation:

$$A_k = \partial_k \ln \alpha \qquad (10.7)$$
$$B_k^i = \tfrac{1}{2}\partial_k \beta^i \qquad (10.8)$$
$$D_{kij} = \tfrac{1}{2}\partial_k g_{ij}. \qquad (10.9)$$

With the aid of these definitions, Bona *et al* are able to write the evolution equations in the form of a first-order conservation law:

$$\partial_t U + \partial_i F^i(U) = S(U) \qquad (10.10)$$

where U is a 'vector' of the basic field variables, and F^i is a 3-vector formed from these field variables. S is a source 'vector' formed from the vector U. Notice that spatial gradients of the field variables occur only in the $\partial_i F^i$ terms. This conservative form is a standard tool in computational hydrodynamics, so this formulation allows application of many known techniques from that discipline to computational relativity. Bona *et al* further show that by defining

$$Q = -\alpha^{-1}[\partial_t \ln \alpha - \beta^i \partial_i \ln \alpha] \qquad (10.11)$$

and specifying a type of gauge condition

$$Q - f(\alpha)\, tr\, K = 0 \qquad (10.12)$$

a number of standard gauges are included:

geodesic: special case of $f = 0$

maximal: limit $f = \infty$

harmonic: $f = 1$

algebraic '1 + log': $f = 1/\alpha$.

These techniques are included in the Cactus code.

10.4.2.1 *Cactus code*

The Cactus code is maintained at the Max Planck Albert-Einstein-Institut in Potsdam. It is a modular code written to carry out parallel simulations in 3D relativity. It is designed to be collaborative in the sense that the basic infrastructure is provided and users can add routines ('thorns') that carry out specific tasks. As delivered, Cactus can handle Bona–Masso, and regular 3 + 1 (ADM) styles of general relativity. The language MPI is used to handle parallelization, via either of two structures called PUGH and DAGH. (DAGH also handles adaptive refinement.) The relativity module contains a variety of options: some kinds of initial datum (Schwarzschild black holes and wave perturbation at present), ADM or Bona–Masso formulation, initial lapse (e.g. unit lapse, or Schwarzschild lapse), choices for slicing: geodesic, external, harmonic, '1 + log', maximal. The Cactus code can automatically run convergence tests on the code. It contains a substantial data handling ability, and is oriented toward using *IDL* (a graphics system) to viewing resulting outputs: 1–2, 3D and movie output.

For more information, contact cactus-maintali-potsdam.mpg.de.

10.4.3 Hyperbolic formulations

The formulation of Bona *et al* [17, 18] achieves most of the advantages of a symmetric hyperbolic formulation of the Einstein equations. Here we now describe a somewhat different hyperbolic approach, which introduces extra time derivatives into the formulation. The study of hyperbolic schemes in handling the Einstein equations has a long history, going back to early work by Lichnerowicz [19] and Choquet [20]. A number of hyperbolic developments to the present are reviewed by Friedrich [4–6].

Choquet, York and collaborators have developed a form based on what might be called the 'standard' 3+1 equations. There is an unfortunate shift of notation; for instance, a recent summary [21] uses N to represent the lapse (and α to represent a closely related quantity).

The fundamental system considered by [21] is the Einstein–Ricci system, where the basic equation is the time derivative of the Ricci tensor (for vacuum spacetimes)

$$\widehat{\partial}_0 R_{ij} = 2D_{(i}R_{j)0} \tag{10.13}$$

which leads to

$$-N\widehat{\square}K_{ij} = J_{ij} - S_{ij} \tag{10.14}$$

where J_{ij} is symmetric and contains second spatial derivatives in the metric, first spatial derivatives in K_{ij} and second spatial derivatives in N. (See [22], [24] for the explicit form of J_{ij}.) The operator $\widehat{\Box}$ is a simple wave operator:

$$\widehat{\Box} \equiv -N^{-1}\widehat{\partial_0}N^{-1}\widehat{\partial_0} + D^k D_k. \tag{10.15}$$

From (10.14), and the noted derivatives in J_{ij}, it can be seen that except for possible 'infection' from S_{ij}, this is a wave equation for K_{ij}. (Recall also that $\widehat{\partial_0}g_{ij} = -2\alpha K_{ij}$, so g_{ij} can be straightforwardly evolved as K_{ij} is evolved.) However, we must still consider

$$-S_{ij} = -N^{-1}D_i D_j(\widehat{\partial_0}N + N^2 K_l^l). \tag{10.16}$$

This S_{ij} *does* contain second derivatives of the trace of K_{ij}, and these additional terms can act to spoil the hyperbolicity of the wave operators on K_{ij}. However, note this term has been also written to contain $\partial_0 N$. Thus, a wise choice of the lapse can, for instance, set S_{ij} to zero; this is called a 'harmonic' time slice. Or one may introduce a specified 'gauge source' function $f(t, x^i)$ [4], and equate

$$\partial_0 N + N^2 K_i^i = N f(t, x^i). \tag{10.17}$$

This system, then which is one higher derivative on the usual form of the Einstein equations, is a hyperbolic system. This is the first form to be described by York and Choquet-Bruhat, though they have developed a number of variants. Their developments have had substantial influence in computation and theoretical relativity, and in field theory. This approach has been embodied in a Binary Black Hole Grand Challenge Alliance code, the *Empire code*. I will briefly describe this code below. Many of the features developed for it reside now in the current Alliance code, which, however, is not of this third time derivative type.

The Empire code implements one of the Choquet-Bruhat and York versions of the Einstein equation of general relativity. This version of the equations is manifestly simply hyperbolic, first-order and in flux-conservative form. The line element is the standard ADM line element:

$$ds^2 = -N^2 dt^2 + g_{ij}(dx^i + \beta^i dt)(dx^j + \beta^j dt). \tag{10.18}$$

The fundamental variables are:

spatial metric:	g_{ij}	(10.19a)
lapse function:	N	(10.19b)
shift vector:	β^i	(10.19c)
extrinsic curvature:	$K_{ij} = -\frac{1}{2}N^{-1}\widehat{\partial_0}g_{ij}$	(10.19d)
time derivative of extrinsic curvature:	$L_{ij} \equiv N^{-1}\widehat{\partial_0}K_{ij}$	(10.19e)

spatial covariant derivative

of extrinsic curvature: $\qquad M^k_{ij} \equiv D^k K_{ij}$ (10.19f)

acceleration: $\qquad a_i \equiv D_i \ln N$ (10.19g)

time derivative of

the acceleration: $\qquad a_{0i} \equiv N^{-1}\widehat{\partial}_0 a_i$ (10.19h)

spatial covariant derivative

of the acceleration: $\qquad a^j_i \equiv D^j a_i.$ (10.19i)

The lapse is set as a solution to the harmonic–time condition:

$$\widehat{\partial}_0 \alpha = -\alpha^2 K. \tag{10.20}$$

In terms of these variables, the equations evolved in the Empire code are

$$\widehat{\partial}_0 g_{ij} = -2N K_{ij} \tag{10.21a}$$

$$\widehat{\partial}_0 K_{ij} = N L_{ij} \tag{10.21b}$$

$$\widehat{\partial}_0 a_i = N a_{0i} \tag{10.21c}$$

$$\widehat{\partial}_0 \Gamma^k_{ij} = N(M^k_{ij} - a_i K^k_j$$
$$-a_j K^k_i + a^k K_{ij} - M_{ij}{}^k - M_{jk}{}^k) \tag{10.21d}$$

$$\widehat{\partial}_0 L_{ij} - N\partial_k M^{k1}{}_{ij} = N[\text{function of variables defined in}$$
$$\text{equation (3.19); no derivatives}] \tag{10.21e}$$

$$\widehat{\partial}_0 M^k{}_{ij} - Ng^{kl}\partial_l L_{ij} = N[\text{function of variables defined in}$$
$$\text{equation (3.19); no derivatives}] \tag{10.21f}$$

$$\widehat{\partial}_0 a_{0i} - N\partial_j a^j{}_i = N[\text{function of variables defined in}$$
$$\text{equation (3.19); no derivatives}] \tag{10.21g}$$

$$\widehat{\partial}_0 a^j_i - Ng^{kj}\partial_k a_{0i} = N[\text{function of variables defined in}$$
$$\text{equation (3.19); no derivatives}]. \tag{10.21h}$$

The Empire code uses standard second-order, cell-centred finite-difference techniques. The evolution scheme is a variation of the Macormack predictor/corrector algorithm which minimizes the required storage and still maintains second-order convergence in space and time.

The code is written in a mixture of C++ and FORTRAN-90 and uses the Alliance developed DAGH library to implement parallelism and (in principle) adaptivity.

A fourth-order system also exists [23]. To construct it, one constructs [21]

$$\widehat{\partial}_0(\widehat{\partial}_0 R_{ij} - 2D_{(i}R_{j)0}) + D_i D_j R_{00} = 0 \tag{10.22}$$

(for vacuum).

It should be noted that although the presence of the shift vector β^i has been suppressed in these expressions, β^i must be defined for the time evolution operator to make sense. The choice of β^i has an important role to play in controlling the coordinatization of the evolution. An especially important situation can be demonstrated in terms of a single black hole. Freely falling observers fall into the black hole, but it is possible, by shifting the coordinate labels outward on the freely falling curves to obtain a static description of the exterior of the black hole. More on this later. In any case it is necessary to 'know', or to compute via some algorithm, the correct choice of shift vector β^i.

10.5   Differencing schemes and convergence

The Einstein equations are almost universally treated computationally by discretizing the differential equations form to a finite-mathematics representation. Instead of partial differential equations, we have a large number of coupled algebraic equations. (An alternative approach, Regge calculus [25, 26], discretizes the action and derives coupled algebraic equations from this discretization of the action. However, the Regge calculus approach has been less well developed than that of discretizing the differential equations, and so we do not discuss it further here.)

There are two standard approaches to discretizing differential equations. They go by the names 'finite differences' and 'finite elements', [34]. Of the two, finite-element methods have been less used in computational relativity, though there are some example applications [32] to the solution of elliptic equations that arise (e.g. in the solution of the constraints, or in specification of gauge choices via elliptic equations). Finite-element techniques express the balance of equations acting on solid (three-dimensional) volumes. The fundamental approach is again to discretize the integral defining the action and give a certain simple rule for how functions change (e.g. linearly) across the volume of the element. The coupled algebraic equations are then derived from a variation; they show a close correspondence to the differential equation. Because of this correspondence, it is also typically possible to extend the finite-element approach to new similar elliptic equations without actually going through the discretization of an action. Although a direct application of finite elements could be the discretization of the spatial derivatives in relativistic simulations, I know of no case in which the *time* discretization is handled this way. A major advantage of finite-element approaches is that they can be refined [32] and so handling holes (e.g. black holes) is easily accomplished. Further, applications using the finite-element method are the preferred method in mechanical and aeronautical engineering, to compute stress patterns in complicated geometrics such as those describing vehicles. Despite this, finite-element methods have not been widely used in computational relativity. Instead, the field has been dominated by finite-*difference* techniques.

Finite-difference techniques are based an more-or-less straightforward replacement of derivatives by difference equations. For instance, if we consider a one-dimensional problem with a uniform grid of points spaced by $\Delta x > 0$, derivative $\frac{df}{dx}\big|_{x=x_0}$ may be approximated by

$$\frac{\Delta f}{\Delta x}\bigg|_{+1} = \frac{f(x_0 + \Delta x) - f(x_0)}{\Delta x} \tag{10.23}$$

or

$$\frac{\Delta f}{\Delta x}\bigg|_{-1} = \frac{f(x_0) - f(x_0 - \Delta x)}{\Delta x} \tag{10.24}$$

or

$$\frac{\Delta f}{\Delta x}\bigg|_{2} = \frac{f(x_0 + \Delta x) - f(x_0 - \Delta x)}{2\Delta x}; \tag{10.25}$$

many other (more complicated) approximate expressions are possible.

Since the basic idea is to consider refinement of the approximation ($\Delta x \to 0$), we shall be constantly concerned with the accuracy of such expressions. This can be investigated by considering a Taylor expansion of the function f around $f(x_0)$:

$$\begin{aligned}
f(x_0 + \Delta x) &= f(x_0) + \frac{\partial f}{\partial x}\bigg|_{x_0} \Delta x + \frac{1}{2}\frac{\partial^2 f}{\partial x^2}\bigg|_{x_0} (\Delta x)^2 \\
&\quad + \frac{1}{3!}\frac{\partial^3 f}{\partial x^3}\bigg|_{x_0} (\Delta x)^3 + O((\Delta x)^4) \tag{10.26a}
\end{aligned}$$

$$\begin{aligned}
f(x_0 - \Delta x) &= f(x_0) - \frac{\partial f}{\partial x}\bigg|_{x_0} \Delta x + \frac{1}{2}\frac{\partial^2 f}{\partial x^2}\bigg|_{(\Delta x)^2} \\
&\quad - \frac{1}{3!}\frac{\partial^3 f}{\partial x^3}\bigg|_{x_0} (\Delta x)^3 + O((\Delta x)^4) \cdots .
\end{aligned} \tag{10.26b}$$

Then from (10.23)–(10.25)

$$\begin{aligned}
\frac{\Delta f}{\Delta x}\bigg|_{1} &= \frac{f(x_0) + \frac{\partial f}{\partial x}\big|_{x_0} \Delta x + \frac{1}{2}\frac{\partial^2 f}{\partial x^2}\big|_{x_0} (\Delta x)^2 + O(\Delta x)^3 - f(x_0)}{\Delta x} \\
&= \frac{\partial f}{\partial x}\bigg|_{x_0} + O(\Delta x) \tag{10.26c}
\end{aligned}$$

$$\begin{aligned}
\frac{\Delta f}{\Delta x}\bigg|_{-} &= \frac{f(x_0) - f(x_0) + \frac{\partial f}{\partial x}\big|_{x_0} \Delta x - \frac{1}{2}\frac{\partial^2 f}{\partial x^2}\big|_{x_0} (\Delta x)^2 + O(\Delta x)^3}{\Delta x} \\
&= \frac{\partial f}{\partial x_0}\bigg|_{x_0} + O(\Delta x) \tag{10.26d}
\end{aligned}$$

$$(2\Delta x) \left.\frac{\Delta f}{\Delta x}\right|_2 = f(x_0) + \left.\frac{\partial f}{\partial x}\right|_{x_0} \Delta x + \left.\frac{1}{2}\frac{\partial^2 f}{\partial x^2}\right|_{x_0} (\Delta x)^2$$

$$+ \frac{1}{3!}\frac{\partial^3 f}{\partial x^3}(\Delta x)^3 + O((\Delta x)^4)$$

$$- f(x_0) + \left.\frac{\partial f}{\partial x}\right|_{x_0} \Delta x - \left.\frac{1}{2}\frac{\partial^2 f}{\partial x^2}\right|_{x_0} (\Delta x)^2$$

$$+ \frac{1}{3}\frac{\partial^3 f}{\partial x^3}(\Delta x)^3 - O(\Delta x^4)$$

$$= 2 \left.\frac{\partial f}{\partial x}\right|_{x_0} \Delta x + O((\Delta x)^3). \tag{10.26e}$$

Hence

$$\left.\frac{\Delta f}{\Delta x}\right|_1 \text{ and } \left.\frac{\Delta f}{\Delta x}\right|_{-1} \text{ have first-order error;}$$

$$\left.\frac{\Delta f}{\Delta x}\right|_2 \text{ has second-order error;}$$

it is a second-order method. Clearly, higher-order methods are desirable because the approximation converges to the differential expression more rapidly as the discretization is refined. To achieve higher-order approximations to the derivative requires that we use more than two points to define the approximation to the derivative. The correct coefficients of the terms at each point can be found by expanding a Taylor series based on x_0 to each point. An important consideration is that there is an advantage to eliminating odd-order terms in the error. This imposes extra restrictions on the coefficients. It is generally not profitable (particularly in scientific *development* code) to go beyond a fourth-order method because the discretized expressions become extremely complicated as the order of accuracy is increased. However, automatic coding methods are beginning to relieve some of the difficulty of such development, and definitely allow easier implementation of complicated schemes.

As a simple example, let us consider a very simple problem, the linear wave equation in $1 + 1$ dimensions. First, we treat it as a Cauchy problem. The wave equation is:

$$\partial_t^2 f - \partial_x^2 f = 0.$$

Analytically we require data set at some time $t = 0$; we set $f(t = 0)$, $\partial_x f(t = 0)$.

There are different ways that we could discretize this expression. First of all, how do we write a second derivative? The second-order accurate second derivative is obtained from:

$$f(x_0 - h) = f(x_0) - h\left.\frac{\partial f}{\partial x}\right|_{x_0} + \left.\frac{h^2}{2}\frac{\partial^2 f}{\partial x^2}\right|_{x_0} - \left.\frac{h^3}{3!}\frac{\partial^3 f}{\partial x^3}\right|_{x_0} + \left.\frac{h^4}{4!}\frac{\partial^4 f}{\partial x^4}\right|_{x_0} - O(h^5). \tag{10.27}$$

Where I write $h > 0$ for $|\Delta x|$, the last term indicates the sign of the next power of h in the expansion, and

$$f(x_0 + h) = f(x_0) + h\frac{\partial f}{\partial x}\bigg|_{x_0} + \frac{h^2}{2}\frac{\partial^2 f}{\partial x^2}\bigg|_{x_0} + \frac{h^3}{3!}\frac{\partial^3 f}{\partial x^2}\bigg|_{x_0} + \frac{h^4}{4!}\frac{\partial^4 f}{\partial x^3}\bigg|_{x_0} + O(h^5).$$
(10.28)

The sum of these two equations contains no odd powers of h and gives at a general point x_0:

$$\frac{\partial^2 f}{\partial x^2}\bigg|_{x_0} = \frac{f(x_0 + h) - 2f(x_0) + f(x_0 - h)}{h^2} - \frac{h^2}{12}\frac{\partial^4 f}{\partial x^4}\bigg|_{x_0} + O(h^4). \quad (10.29)$$

Note: This is a second-order accurate expression, and we can compute the co-efficient of the dominant error term. Also, because of the symmetry, there is no third-order error.

We now have different possibilities for constructing differenced equivalents to the wave equation. These are best defined by drawing diagrams of the stencils; three possibilities are:

For instance, (a) means:

$$\frac{f(t_0 + 2h_t, x_0) - 2f(t_0 + h_t, x_0) + f(t_0, x_0)}{h_t^2}$$
$$= \frac{f(t_0, x_0 + h_x) - 2f(t_0, x_0) + f(t_0, x_0 - h_x)}{h_x^2} \quad (10.30)$$

and similar expressions for each of (b), (c). Firstly, notice that one needs the value of f at two time intervals (t_0, and $t_0 + h_t$) to obtain f at the latest time, $t_0 + 2h_t$. (Of course, the *labels* can be shifted by multiples of h_x, h_t.) This should not be surprising; the wave equation is second order in time, and we need some way to convey both the value of f and its time derivative.

Stencils (a) and (b) lead to completely *explicit* schemes to obtain f at the latest time. For instance, from stencil (b):

$$f(t + 2h_t, x_0) = \left(\frac{h_t}{h_x}\right)^2 [f(t_0 + h_t, x_0 + h_x) + f(t_0 + h_t, x_0 - h_x)]$$

$$+ 2\left(1 - \left(\frac{h_t}{h_x}\right)^2\right) f(t_0 + h_t, x_0) - f(t_0, x_0). \quad (10.31)$$

The quantity $|h_t/h_x|$ is called the Courant–Friedrichs–Lewy factor. This equation allows immediate explicit statement of f at the latest time level (we defer for a while the question of boundary conditions). The main requirement for an explicit solution is that the latest time step involve the field variable at only *one* spatial point. Stencil (c), on the other hand, leads to:

$$-f(t_0 + 2h_t, x_0 + h_x)\left(\frac{h_t}{h_x}\right)^2 + f(t_0 + 2h_t, x_0)\left(1 + 2\left(\frac{h_t}{h_x}\right)^2\right)$$

$$- f(t_0 + 2h_t, x_0 - h_x)\left(\frac{h_t}{h_x}\right)^2$$

$$= 2f(t_0 + h_t, x_0) - f(t_0, x_0).$$

$$(10.32)$$

It can be seen that the values of the field on the latest level are related (coupled) to one another. If we change notation to write $f(t', x_0 + kh_x) = f_k(t')$ we have

$$M_{kl} f_l(t_0 + 2h_t) = 2 f_k(t_0 + h_t) - f_k(t_0) \quad (10.33)$$

or

$$f_l(t_0 + 2h_t) = (M^{-1})_{lk}[2 f_k(t_0 + h_t) - f_k(t_0)] \quad (10.34)$$

with, in this case

$$M_{ab} = \mathbb{1}_{ab} + \left(\frac{h_t}{h_x}\right)^2 \begin{bmatrix} \ddots & & & & \\ -1 & 2 & -1 & & \\ & -1 & 2 & -1 & \\ & & -1 & 2 & -1 \\ & & & & \ddots \end{bmatrix}. \quad (10.35)$$

Thus to solve this problem requires inverting a matrix. In this one-dimensional case there are straightforward schemes to do so, since the matrix is tridiagonal. In higher-dimensional cases, one wonders why anyone would prefer to handle the equations by implicit methods, since the extra work of inverting a matrix complicates the algorithms.

There are two answers. First, *efficient* computational matrix inversion methods (multigrid methods) exist, which allow solution of matrix problems in $O(n)$

effort, where n is the number of quantities to be determined. Their implementation does require coding effort, of course, but they do not overly burden the computational efficiency. Much more important is an area we have not yet addressed: the stability of the evolution.

10.6 Simple stability analysis

To analyse the stability of the equations just written, we in principle want a proof that if we set numerical data which is close to the initial analytical situation, the computational evolution will remain close to the actual analytical solution. For the *linear wave equation*, we can concentrate on the propagation of small deviations from zero; the linearity guarantees that we will then have the behaviour of small deviations from any particular solution.

The von Neumann analysis proceeds by assuming a particular spatial Fourier decomposition ($e^{ikx} \equiv e^{iklh_x}$), where l counts the number of discretization steps, here assumed uniform. The time behaviour is similarly expressed in terms of $e^{i\omega t} = e^{i\omega m h_t}$. However, we allow for the possibility that ω may be complex. Thus we more generically write

$$\xi \equiv e^{i\omega m h_t}. \tag{10.36}$$

Note that here ξ is a complex number.

Our small disturbance is then expressed as

$$f(nh_t, lh_x) = \xi^n e^{ilkh_x}. \tag{10.37}$$

When we insert this expression into the stencil

assume the earliest time is labelled $n = 0$, and the central x is labelled $l = 0$, a little algebra leads to

$$\xi^2 - 2\xi + 1 = -a^2 \tag{10.38}$$

where

$$a^2 = 4\left(\frac{h_t}{h_x}\right)^2 \sin^2\left(\tfrac{1}{2}kh_x\right). \tag{10.39}$$

Similarly

leads to

$$\xi^2 - 2\xi + 1 = -a^2\xi \tag{10.40}$$

and

$$\times \qquad \otimes \qquad \times \qquad\qquad (c)$$

leads to

$$\xi^2 - 2\xi + 1 = -a^2\xi^2. \tag{10.41}$$

Stability requires $|\xi| \leqslant 1$. Solve for ξ for stencil (a):

$$\xi = 1 \pm \sqrt{1 - (1 + a^2)}$$
$$= 1 \pm i|a|. \tag{10.42}$$

This method is *unstable* for all spatially varying modes, because $|\xi|^2 = 1 + a^2 > 1$. Similarly solve for ξ for stencil (b)

$$\xi = 1 - \tfrac{1}{2}a^2 \pm \sqrt{\left(1 - \tfrac{1}{2}a^2\right)^2 - 1}.$$

The radical is imaginary, equal to $i| - a^2 + \tfrac{1}{4}a^4|^{1/2} = i\left(a^2 - \tfrac{1}{4}a^4\right)^{1/2}$ when $a < 2$. Hence

$$|\xi|^2 = \left(1 - \frac{a^2}{2}\right)^2 + a^2 - \frac{a^4}{4}$$
$$= 1 - a^2 + \frac{a^4}{4} + a^2 - a\frac{4}{4}$$
$$= 1. \tag{10.43}$$

It can be easily seen that $|\xi|^2 = 1$ also when $a = 2$, but $|\xi|^2 > 1$ when $a > 2$. Thus this method is stable for all spatial wave modes so long as $a = 2|h_t/h_x \sin(\tfrac{1}{2}kh_x)| \leqslant 2$.

Since the sine can contribute a factor of unity for k such that $kh_x = \pi$, we must require $|h_t/h_x| \leqslant 1$ for stability. This *Courant limit* means that the computational 'light cone' must not be narrower than the physical light cone. (The Courant limit arises often in discussing the stability of computational methods.) If we take $h_t/h_x \leqslant 1$, this method gives stable evolution for the wave equation.

Note on size of the wave number k: On a discrete grid, the smallest wavelength that can be represented is of wavelength $2h_x$. This allows zone-to-zone oscillations to represent this wavelength. Hence in fact it is meaningless to state a k larger than (π/h_x). Thus the shortest wavelength that can be achieved makes $\tfrac{1}{2}kh_x = \tfrac{1}{2}\pi$, and makes the sine appearing in a take the value unity.

Our final example, stencil (c), is an *implicit method*, but the stability analysis proceeds as before. Here

$$\xi = \frac{1 \pm \sqrt{1 - (1 + a^2)}}{1 + a^2}$$
$$= \frac{1 \pm i|a|}{1 + a^2}. \tag{10.44}$$

Thus

$$|\xi| = \sqrt{1 + a^2}/(1 + a^2) = \frac{1}{\sqrt{1 + a^2}} < 1. \tag{10.45}$$

Thus $|\xi| \to a^{-1} \to 0$ as $a \to \infty$. This method is unconditionally stable because $|\xi| < 1$ for all $|h_t/h_x|$, and for all kh_x. This is the reason that implicit methods are preferred: they have a larger domain of stability in terms of simple parameters describing the simulation. Also, notice for a given h_t/h_x it is the large-k (short-spatial-wavelength) variations which are most suppressed; this also is characteristic of implicit methods in contrast for example to the case of stencil (b), where $|\xi| = 1$ at the shortest wavelength in the problem. With implicit methods there is the promise that large-scale features can be evolved with very large timesteps, and smaller-scale features will be 'smoothed'; *explicit* methods always have a version of the Courant condition and *require* small timesteps for stability.

One must also note that these simple analyses are correct only for this linear wave problem. For more complicated systems with non-linear behaviour, the results from these simple calculations must be viewed as optimistic estimates (at best) of the domain of stability, and computational experimentation is the only recourse in those cases.

10.7   Accuracy

Besides stability, we must also consider the accuracy of the solutions we have obtained. We have already demonstrated that our second-derivative formula is second-order accurate. However, that second-order behaviour depended to a large extent on the symmetry of the stencil. Of the three stencils we wrote for the one-dimensional wave equation, only stencil (b) is symmetric in time.

We can extract accuracy information from the results of the stability analysis. First of all, we will consider small wavenumber (long wavelength) $k \ll 1$, because this is the domain in which we expect best representation of the continuous solution in the discretized one. Because we have $f \sim e^{iklh_x}$, and we are solving a simple 1D wave equation, we expect an analytic time dependence of the form $\xi_a^n = e^{iknh_t}$.

For short time ($n = 2$) this is

$$\xi_a = 1 \pm 2ikh_t - \tfrac{1}{2}(2kh_t)^2. \tag{10.46}$$

The subscript 'a' on ξ_a indicates that this is the analytically expected value of ξ, in contrast to the previously determined computational ξ, to which we now return. For the unstable mode (stencil (a), equations (10.29), (10.42)) and assuming kh_x is small (thus $a = 2|h_t/h_x|\tfrac{1}{2}kh_x = kh_t$):

$$\xi^2 = 1 \pm 2i|a| - a^2$$
$$= 1 \pm 2i\frac{h_t}{h_x}kh_x - h_t^2 k^2 \tag{10.47}$$

which agrees to first order in h_t, but differs at second order from the analytic result ξ_a.

For the conditionally stable ($|h_t/h_x| \lesssim 1$) explicit stencil (b) (equations (10.30), (10.43)) we again assume kh_x is small and keep terms only to $O(a^2)$. Then using ξ determined from the stability analysis

$$\xi^2 = \left(1 - \frac{a^2}{2} \pm ia\right)^2 + O(a^3)$$

$$= 1 - 2a^2 \pm 2ia + O(a^3). \tag{10.48}$$

Again using the small-k limit $a = kh_t$, we have

$$\xi^2 = 1 \pm 2ikh_t - 2k^2h_t^2 + O(kh_t)^3 \tag{10.49}$$

and this method agrees with the exact result to second order; it is a second-order method.

Finally, if we consider the implicit method (stencil (c), equations (10.31), (10.45)), it is easy to show that the discrete solution is only *first* order in time. The lesson to be learned is that only with symmetry can a three-point difference provide second-order finite differences.

10.8   Other stencils

We have been able to obtain second order accurate second derivatives using a three-point stencil. Reviewing that derivation shows that the symmetry in the description of the stencil was necessary to cancel the $O(h_x^3)$ terms, to yield a second-order approximation. However, this symmetry can only be accomplished at points *within* the computational domain. If on the other hand we want to evaluate a second derivative at a boundary point (say $x = x_{min} = 0$) we have

$$\downarrow \text{ evaluate second derivative here}$$

$$\begin{array}{cccc} \times & \times & \times & \cdots \\ 0 & h_x & 2h_x. & \end{array}$$

We find

$$f(h_x) = f(0) + h_x \left.\frac{\partial f}{\partial x}\right|_0 + \frac{(h_x)^2}{2} \left.\frac{\partial^2 f}{\partial x^2}\right|_0 + \frac{(h_x)^3}{3!} \left.\frac{\partial^3 f}{\partial x^3}\right|_0$$

$$+ \frac{(h_x)^4}{4!} \left.\frac{\partial^4 f}{\partial x^4}\right|_0 + O(h^5) \cdots \tag{10.50}$$

$$f(2h_x) = f(0) + 2h_x \left.\frac{\partial f}{\partial x}\right|_0 + \frac{(2h_x)^2}{2} \left.\frac{\partial^2 f}{\partial x^2}\right|_0 + \frac{(2h_x)^3}{3!} \left.\frac{\partial^3 f}{\partial x^3}\right|_0$$

$$+ \frac{(2h_x)^4}{4!} \left.\frac{\partial^4 f}{\partial x^4}\right|_0 + O(h^5) \cdots. \tag{10.51}$$

Solving for the second derivative $\frac{\partial^2 f}{\partial x^2}\big|_0$, we obtain

$$\frac{\partial^2 f}{\partial x^2}\bigg|_{x_0} = h_x^{-2}[f(0) - 2f(h_x) + f(2h_x)] - h\frac{\partial^3 f}{\partial x^3}\bigg|_0 + O(h^2). \quad (10.52)$$

Notice that this expression is the centred difference centred at point $l = 1$, obviously *not* at the point we want to evaluate the second derivative. And, as a consequence, this approximation to the second derivative is only first order in h_x (with coefficient $\frac{\partial^3 f}{\partial x^3}\big|_0$, which cancelled out of the previous *centred* expression).

We can improve the accuracy of the approximation by adding a fourth point to the stencil, with associated equation

$$\downarrow \text{ evaluate second derivative here}$$

$$\begin{array}{ccccc} \times & \times & \times & \times & \cdots \\ 0 & h_x & 2h_x & 3h_x & \end{array}$$

$$f(3h_x) = f_0 + 3h_x\frac{\partial f}{\partial x}\bigg|_0 + \frac{(3h_x)^2}{2}\frac{\partial^2 f}{\partial x^2}\bigg|_0 + \frac{(3h_x)^3}{3!}\frac{\partial^3 f}{\partial x^3}\bigg|_0 + \frac{(3h_x)^4}{4!}\frac{\partial^4 f}{\partial x^4}\bigg|_0 + O(h^5). \quad (10.53)$$

It is straightforward to verify that the following combination of coefficients (expressed for the one-sided derivative at a general point x_0) gives the correct result:

$$\frac{\partial^2 f}{\partial x^2}\bigg|_{x_0} = h_x^{-2}[2f(x_0) - 5f(x_0 + h_x) + 4f(x_0 + 2h_x) - f(x_0 + 3h_x)]$$

$$- \frac{11}{12}\frac{\partial^4 f}{\partial x^4}\bigg|_{x_0} h_x^2 + O(h^3). \quad (10.54)$$

By adding the extra stencil we have been able to get a second-order accurate estimate for $\frac{\partial^2 f}{\partial x^2}\big|_{x_0}$.

The difference between the differenced form and the analytical result is called the truncation error. It can be seen that even though we have managed to get a second-order approximation to the second derivative, we still have a truncation error in equation (10.54) equal to $-\frac{11}{12}\frac{\partial^4 f}{\partial x^4}h_x^2$, which differs from that of the central differenced stencil that we would naturally use in the middle of the domain: $-\frac{1}{12}\frac{\partial^4 f}{\partial x^4}h_x^2$; cf equation (10.29). Perhaps remarkably, this kind of discontinuity in the *coefficient* of the error causes difficulty in many applications. By adding yet another point and its associated equation, it is possible to make the coefficient of h_x^2 computed by a five-point stencil at the boundary equal to $-\frac{1}{12}\frac{\partial^4 f}{\partial x^4}$, as it is in the three-point centred difference case.

Even with this adjustment, the boundary expression for the second derivative has an error term of order h_x^3, which is zero by symmetry in the centred difference. It may be worthwhile to add yet another point (a six-point stencil), which will contain enough points to allow setting this third-order value to zero. This will be important in the following section. Note that we are required to extend the boundary stencil far inside the location of the points used by the adjacent interior symmetric difference, in order to obtain compatible error behaviour. Boundaries and boundary conditions are difficult issues in computation. In multiple dimensions boundary stencils become even more complicated.

10.9   Richardson extrapolation

Richardson [35] observed, as we have, that in smooth evolutions (no shocks) there is a smooth connection between the computational and the 'analytic' expressions

$$f_{\text{numerical}}(x) = f_{\text{analytic}}(x) + \sum_{i=1}^{\infty} C_i(x)h^i \tag{10.55}$$

where the sum is convergent, and the $C_i(x)$ are smooth bounded functions on the domain.

This form shows two very interesting features. First, although the numerical construction of a derivative involves division by h, which naively is expected to reduce the order of accuracy, in fact derivatives are as accurate as the function itself:

$$f'_{\text{numerical}} = f'_{\text{analytic}} + \sum_{i=1}^{\infty} C'_i(x)h^i \tag{10.56}$$

where the derivations $C'_i(x)$ are smooth and bounded.

Second, Richardson extrapolation can be used to improve the accuracy of the solution. Richardson extrapolation recognizes that for small h, the difference between the numerical and the analytic solutions (the truncation error) is dominated by the C_i with smallest index.

This can be used in two ways. The first is to actually validate the order of accuracy of the computation. For instance, one can compute

$$f^{(h)}_{\text{numerical}} = f_{\text{analytic}} + \tau^{(h)}. \tag{10.57}$$

(τ^h is truncation error for a discretization size of h.)

Compute

$$\frac{f^{(h)} - f^{\left(\frac{h}{2}\right)}}{f^{\left(\frac{h}{2}\right)} - f^{\left(\frac{h}{4}\right)}} = \frac{\tau^{(h)} - \tau^{\left(\frac{h}{2}\right)}}{\tau^{\left(\frac{h}{2}\right)} - \tau^{\left(\frac{h}{4}\right)}}. \tag{10.58}$$

We are thus comparing the computed result at three different resolutions, $h, \frac{h}{2}, \frac{h}{4}$. Assume the Richardson expansion gives

$$\tau^{(h)} = C_{i_{\min}} h^{i_{\min}} + O(h^{i_{\min}+1}) \tag{10.59}$$

where $i_{\min}$ is the smallest index of a non-vanishing C_i.

Hence,

$$\frac{f^{(h)} - f\left(\frac{h}{2}\right)}{f\left(\frac{h}{2}\right) - f\left(\frac{h}{4}\right)} = \frac{C_{i_{\min}}\left(h^{i_{\min}} - \left(\frac{h}{2}\right)^{i_{\min}}\right)}{C_{i_{\min}}\left(\left(\frac{h}{2}\right)^{i_{\min}} - \left(\frac{h}{4}\right)^{i_{\min}}\right)}(1 + O(h))$$

$$= 2^{i_{\min}}(1 + O(h)). \tag{10.60}$$

In the limit of small h, we expect a ratio of four for a second-order code, a ratio of eight for a third-order code etc. By comparing results of the code at three different refinements, we can actually verify the order of convergence of the code.

Obviously, the refinements need not be in a power of two ratio; in some cases that range of resolutions is very difficult or impossible to achieve; similar formulae can be derived for other ratios.

Once we have verified the dominant order of the code, we are ready to do Richardson extrapolation. With only two discretization levels, we write (again for the simple power of two case):

$$f^{(h)} = f_{\text{analytic}} + C_{i_{\min}} h^{i_{\min}} + O(h^{i_{\min}+1}) \tag{10.61}$$

$$f\left(\frac{h}{2}\right) = f_{\text{analytic}} + C_{i_{\min}}\left(\frac{h}{2}\right)^{i_{\min}} + O(h^{i_{\min}+1}). \tag{10.62}$$

Multiply the second equation by $2^{i_{\min}}$ and solve for f_{analytic}:

$$f^{(h)} - 2^{i_{\min}} f\left(\frac{h}{2}\right) = (1 - 2^{i_{\min}}) f_{\text{analytic}} + O(h^{i_{\min}+1})$$

$$f_{\text{analytic}} = \frac{2^{i_{\min}} f\left(\frac{h}{2}\right) - f^{(h)}}{2^{i_{\min}} - 1} + O(h^{i_{\min}+1}). \tag{10.63}$$

Thus by computing the solution at two different resolutions we have raised the order of accuracy of the computational solution by one. For instance, if it is a *second*-order computation ($i_{\min} = 2$)

$$f_{\text{analytic}} = \frac{4 f\left(\frac{h}{2}\right) - f^{(h)}}{3} + O(h^3). \tag{10.64}$$

If, as in the central differenced cases we have discussed, the third-order error actually vanishes, then by doing computations at two discretizations, we actually raise the accuracy to fourth order. This explains why one would take the trouble to use large stencils at boundaries in order to eliminate third-order terms, to achieve this extrapolatable behaviour. It is also worth noting that fourth-order methods, even interior stencils, are typically very complicated and difficult to code. It is likely that the Richardson extrapolation of second-order methods involves no more computation and, since it is based on much simpler second-order methods, is much easier to code. Choptuik [44] in particular has pointed out the utility of Richardson extrapolation in validating convergence, and in extracting high-accuracy results from computational simulations of gravitating systems.

10.10   The Binary Black Hole Grand Challenge

To be relevant to the detection of gravitational radiation in detectors like LIGO, VIRGO and GEO, computational relativity needs to provide accurate and reliable prediction of waveforms from astrophysical events. About five years ago the Binary Black Hole Alliance was initiated as a US National Science Foundation High Performance Computing and Communication Grand Challenge Project. Our plan was to do just that: to predict waveforms from interacting black holes.

The Alliance has made tremendous strides in the computational, theoretical, and infrastructural aspects of this problem. At present we have constructed all the pieces and a period of code integration, adjustment and refinement is beginning. We have posed data for generic evolution of binary black hole spacetimes, we have developed coordinate gauge schemes, we are beginning evolutions, we have extensively tested and demonstrated new ideas in computational relativity. In the process, we have developed, extended and refined concepts in theoretical relativity; computation requires you to get all the details right. We have pushed computers and computer technology, and we have developed new tools for visualizing and communicating these ideas.

The Alliance code is a finite-difference code. It consists of three modules; a central, *Cauchy* code, which uses the standard $3 + 1$ form (colloquially called the ADM form, though, importantly, the equations differ by a trace-reversal from those of the ADM paper [1]). It evolves from $3 + 1$ initial data for multiple black holes. It incorporates a matching interface, and an outer boundary module which provides stable boundary conditions for the Cauchy module, and propagates the radiation to infinity for extraction. It makes extensive use of computer science infrastructure to support code development and visualization. See figure 10.6.

The Cauchy module is three-dimensional in rectangular coordinates. It uses a causal differencing based on ideas by Thornburg [27, 28] and Unruh [29] and developed by a number of people [30, 36–40]. It was demonstrated for stationary (non-moving) Schwarzschild black holes by Anninos *et al* [31] and evolutions to $\sim 140\,M$ were given by Daues [33].

Our approach is to have the black holes move through the Cauchy rectangular grid the way a baseball travels through a Newtonian rectangular grid: the coordinate of the centre moves; as it moves some grid points which were outside the ball (hole) are temporarily inside the ball, but then reemerge. This approach is more general, is consistent with generic motions of the black hole and supersedes earlier efforts which kept the black hole surfaces at (essentially) fixed coordinate conditions. We expect this to be necessary for the situation with multiple black holes. The interior of the black hole (where the singularity resides) is excluded from the computation. As the black hole moves, some computational points are overrun and are dropped from the simulation while they are within the black hole; others 'reappear' behind the black hole. Figure 10.7 shows how this works. Notice that because of the tilt of the light cones inside (to the left of) the apparent horizon *a*, the innermost point on the spatial domain is not used in evolving to the next timestep. Hence *no* boundary condition is needed there, and we do not have to

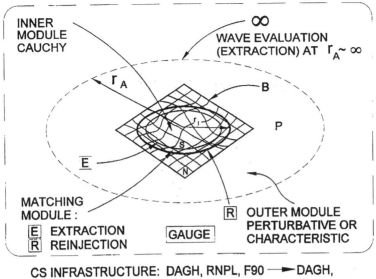

Figure 10.6.

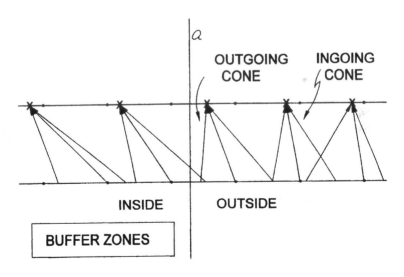

Figure 10.7.

integrate inward to the location of the singularity. The implementation in three dimensions is substantially more complicated than this one-dimensional example.

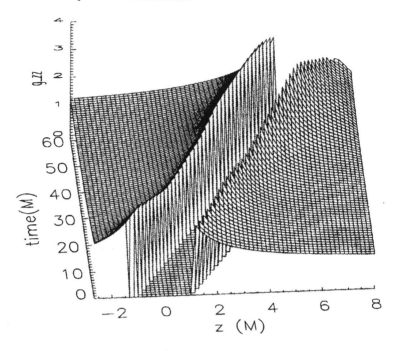

Figure 10.8.

This approach *works*: cf [41]. In that paper we show an evolution of a moving black hole, excising its interior and letting the hole move through a domain. Figure 10.8 gives the result g_{rr}, centred on the current location of the hole, for this simulation; the excised region is shown as the 'flat' area in the centre of the hole. The simulation shows the Schwarzschild hole moving about 1.5 diameters at a speed of 0.1 c. The data are initialized analytically as a boosted Schwarzschild black hole in Kerr–Schild coordinates. The 3 + 1 evolutions equations produce the subsequent evolution. This model is $33 \times 33 \times 65$ with domain $-(\frac{8}{3})$ M to $(\frac{8}{3})$ M in the x- and y-directions. The simulation uses a buffer region of five zones between the horizon and the inner boundary.

A similar simulation was carried out for a Kerr black hole, $a/M = 0.5$ (with spin axis in the x-direction), moving at speed $v = 0.5\ c$ in the y-direction. The hole moved more than one radius in this simulation [42].

Isolated non-moving black holes in this code run for hundreds of M, in some cases forever; see section 1.10 below. We are still exploring the very long-term stability of the solutions.

10.11   Maintenance of the constraints

The 3+1 equations are set using the constraints to pose initial data. The constraints for vacuum (e.g. black hole) spacetimes can be written:

$$\text{Hamiltonian constraint: } H = {}^3R + \tfrac{2}{3}K^2 - A^{ij}A_{ij} = 0 \tag{10.65}$$

where ${}^3R$ is the three-dimensional Ricci tensor, A^{ij} is the traceless extrinsic curvature tensor ($\equiv$ 'the momentum of the field') and K is the trace of the extrinsic curvature; and

$$\text{momentum constraint: } H^i = D_j A^{ij} - \tfrac{2}{3}K_{,j} = 0. \tag{10.66}$$

The *analytical* elementary proof that the constraints are maintained by the evolution uses the Bianchi identities:

$$G^{00}_{,0} + \cdots = 0 \tag{10.67}$$

and

$$G^{i0}_{,0} + \cdots = 0. \tag{10.68}$$

Here 0 refers to the time evolution. (Comma means partial derivative.) The ellipses in the equations (10.67) and (10.68) refer to spatial gradients of the $G^{\alpha\beta}$ and terms consisting of components $G^{\alpha\beta}$ times four-dimensional Christoffel symbols. Since time derivatives appear only as explicitly given in (10.67) and (10.68), if $G^{\alpha\beta} = 0$ at $t = 0$ and if $G^{ij} = 0$ (the spatial components) are maintained by solving these spatial equations as evolution equations, then $G^{00} = 0$ and $G^{0i} = 0$ are implied forever by these equations, since their time derivatives are zero.

Note that these equations are maintained by taking derivatives of the basic Einstein equation, and there initially was some worry, not clearly addressed in the discussion of computations at the time, that since extra derivatives mean an extra division by h_t or h_x, the order of accuracy would be lowered and the constraints would be maintained only to one order lower than the main computation, with unpredictable results.

In an important paper in 1991, Choptuik [44] pointed out that correctly differenced equations for the constraints involve only differencing the coefficients C_i as in equation (10.57); the C_i are smooth functions and their differentiation does not change the order of convergence. These results are verified frequently in the Alliance code: the Hamiltonian and non-linear constraints converge just as does the error in g_{xx}, say, i.e. at a second-order rate.

A related and at least equally important question concerns the stability of the codes to perturbations in the constraints. That is, suppose we introduce an error in the constraints as computational evolution inevitably does; what is the effect on the subsequent evolution?

Frittelli [43] showed that such disturbances obey a well posed characteristic problem, for the scheme we use, the $3 + 1$ $\dot{g}_{ij}$, $\dot{K}_{ij}$ form. In particular, in this formalism the spatial equations state the vanishing of the spatial components of

$^4R_{ij}$. Other formulations, e.g. the classical ADM [1], which is the trace reversed form and amounts to using the spatial components of $^4G_{ij}$, cannot be proven to have well behaved constraint behaviour. See also recent work by Frittelli and collaborators [56, 57] and by York and Anderson [58].

Since our $3 + 1$ $\dot{g}_{ij}$, $\dot{K}^{ab}$ form uses the spatial Ricci form of the equations, this gives credence that we are well maintaining the constraints, even in our free evolution. We do observe, however, that the Hamiltonian and momentum constraints become bad in the cases when the numerical solution eventually goes bad. Recently, Choquet-Bruhat and York and Anderson [58] have extended general relativity by coupling to the 36 components of a Riemann-like object. This object has two sets of antisymmetric pairs of spacetime indices, but *a priori* no other symmetries. When data are set initially imposing the Einstein initial value constraints and the symmetries of the Riemann-like object, Einstein's theory, plus a description of the Riemann tensor, ensues. The Einstein equations are of the spatial-Ricci-tensor form, and the perturbation of the constraints gives a well posed causal structure to their evolution. The Alliance group at Cornell University is investigating a computational implementation of this approach. A similar approach has been taken in [60].

Well-posedness controls the excursion of solutions by an initial data dependent bound which can grow as fast as an exponential in time. (The characteristic time in the exponent is independent of the data.) Although the solution can be converged to the analytic one by restricting the data, at late times there is no guarantee of smallness of error. A recent work [59] seeks to modify the form of Einstein's equations to a first-order hyperbolic set in which additional variables are used in a well posed system, which force errors in the constraints toward zero in the future. Examples in Maxwell theory and in linearized gravitation have been successfully treated this way [59].

10.12 Boundary conditions and wave extraction

To run black hole simulations reliably requires a 'good' boundary condition. Exactly what constitutes 'good' is difficult to state. By 'boundary condition' we mean 'outer boundary condition' because we have no need for boundary conditions at the inner excised location. If full knowledge of the boundary is available, then one already has the analytical solution to the problem. Hence boundary conditions are typically approximations, argued to little affect the physical behaviour of the solution. For many purposes one simply demands a *stable* boundary condition; one that allows the interior region to be computed for a very long time. This, combined with a proof of convergence (to expected analytic spacetimes, when an analytic solution is known), is effectively a demonstration of the correctness of the computations.

An example of a boundary condition that works satisfactorily in some situations is to take fixed Dirichlet outer conditions when computing (say) an isolated Schwarzschild black hole. For characteristic evolutions of such situations, one can take take such boundary conditions on an outer sphere, and set data on an initial spacelike (or ingoing, null) hypersurface. The inner boundary is located inside the apparent horizon, but excising the singularity. Computational techniques actually set an inner boundary a distance equal to a few grid spacings inside of the apparent horizon. For these cases the demonstrations are typically run at simulations up to 60^3, and cover out to about $r = 7\,M$. Both isolated Schwarzschild and isolated Kerr black holes have been run in these configurations.

Because the data are set analytically, but the evolution is via a discretized version of the equations, even for analytically stationary (and non-moving) data, there are immediate dynamics as the solution moves toward the discretized version of the solution. This is seen in both the characteristic and the Cauchy formulations. Figure 10.9 shows the evolution of the horizon mass as measured in a characteristic evolution of Kerr data. Note the extremely long time for this solution (examples have been carried out to $t \sim 60\,000\,M$). In *Cauchy* evolutions one finds that just fixing values at the outer boundary does not lead to very good evolution behaviour. Instead, we use *blended boundary conditions*. Outside an inner computational sphere we mix, by some simple (say linear) formula between evolved values and the analytic values. (See figure 10.10.) We find that this leads to much 'quieter' evolution. Figure 10.11 shows the l_2 norm of the relative Hamiltonian error for a Schwarzschild solution. (The Hamiltonian constraint involves a collection of terms of different sign; the relative Hamiltonian constraint is normalized by dividing by the sum of the absolute values of the constituent terms.) The rapid initial motion leads to rapid damped approach to a constant offset. Although the run was carried out to only a few hundred M, the time derivatives of dynamical quantities were at machine precision levels (parts in 10^{14}), a practical definition of a code that runs forever. The Cauchy module, however, is sensitive to the situation being formulated. At currently accessible resolutions, larger domains will run only out to a few hundred or before error norms become large.

Studies of the blended boundary condition began with an observation by Gomez of its behaviour for linearized gravitational radiation, where it allowed very long-term evolutions, and it has subsequently been transferred to the full Cauchy evolution problem.

One of the successful applications of the blending approach has been *perturbative matching* as an outer boundary condition, for weak and moderately strong waves (non-linear versions of Teukolsky waves). A number of matching techniques have been investigated. Perturbative matching solves a gauge invariant version of the perturbative problem out to very large radii. The inner, Cauchy evolution provides data for the propagation to infinity (where simple outgoing conditions can be applied and where waveforms are extracted) and the outer, perturbative evolution provides a boundary condition on the inner evolution. Because the perturbative simulation can be decomposed into angular harmonics, the one-dimensional radial equation can be evolved at very high resolution to very large outer radii.

The Alliance has studied (for both weak and moderate linear waves) matching via a perturbative Sommerfeld condition. For the extrinsic curvature, for instance,

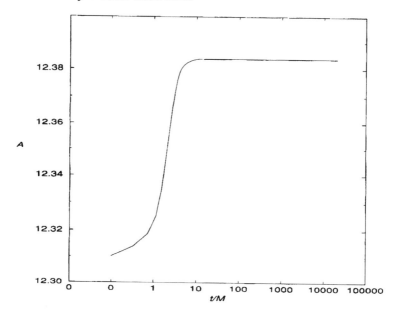

Figure 10.9.

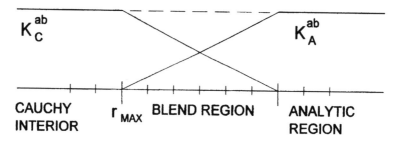

Figure 10.10.

we write

$$\frac{\partial}{\partial t}(K_{ij} - \kappa_{ij}) + \frac{\partial}{\partial r}(K_{ij} - \kappa_{ij}) + \frac{2}{r}(K_{ij} - \kappa_{ij}) = 0 \qquad (10.69)$$

which corrects the Sommerfeld boundary condition on K_{ij} by replacing the right-hand side (usually taken to be zero) by extra terms derived from κ_{ij}, the perturbation theory extrinsic curvature.

For longer (essentially infinite) evolution, a blended perturbative method is used instead. This is a direct blending of the perturbative computed and the Cauchy-computed values over a graded blending zone (typically 8–10 zones, linearly or quadratically graded). The behaviour of these extraction methods can

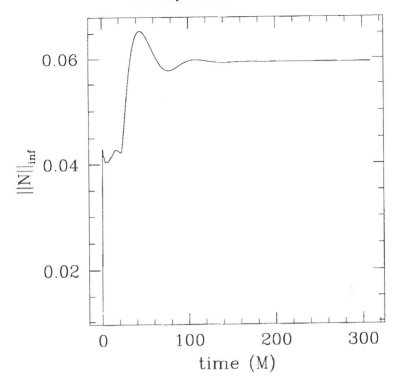

Figure 10.11.

be seen in these figures which show convergence of the extracted wave to the analytic expected value when the interior code is refined, and a comparison of long-term constant or declining error measures for the late time signal extraction. For black hole applications, the perturbative method will adjust the mass to the closest background fit. This approach to boundaries for computational relativity is fully described in [47], [46] and [45].

The ideal outer boundary condition is no condition at all. This means we really want to be able to evolve the outside of the boundary exactly. The perturbative matching approaches this, but does not accommodate the wave–wave non-linearity. A method which can in principle accomplish this is Cauchy-characteristic matching.

The characteristic evolution of a spacetime requires the construction of a reference frame in terms of characteristic hypersurfaces. Smooth hypersurfaces at null infinity become distorted by the gravitational lensing of masses encountered at finite locations. This focusing eventually leads to caustics, which destroy the possibility of using that characteristic surface inward from the caustic. Thus, in a general spacetime there is a minimum radius, outside the location of the mass,

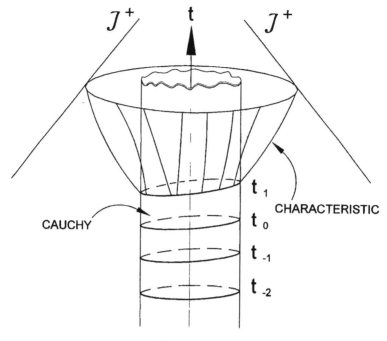

Figure 10.12.

where a consistent characteristic system can be established. In the algebraically *special* Kerr and Schwarzschild spacetimes the characteristics can be extended inward to inside the horizons. This is how the long-term black hole evolutions were accomplished via the Alliance characteristic module.

The Cauchy-characteristic matching matches the interior Cauchy evolution to outgoing characteristic surfaces. (See figure 10.12.) Because characteristic surfaces are phase surfaces of the wave fields, the fields change slowly $\propto \sum_{i=1}^{N} a_n r^{-n}$ as they approach $\mathcal{I}^+$, null radiative infinity. Hence the radial direction can be compactified (e.g. by $u = r^{-1}$), and the radial dependence is only polynomial in the new variables. Thus, a finite discretization covers the domain all the way to radiative infinity. The difficult requirement is to match the Cauchy to the characteristic evolution. If this is accomplished, one is solving the Einstein equations from the Cauchy interior to radiative to infinity.

For the case of non-linear scalar fields, this method has been demonstrated to be effective. The graphs show convergence to an 'analytic' result and show insensitivity to change of the extraction radius, for the non-linear scalar case. The computational effort for this exact boundary match is $\sim N^3$ where N^3 is the total number of interior points. Perturbative methods require a comparable, or larger, number of points. Simple boundary methods (e.g. Sommerfeld), on the other hand, are $O(N^2)$. However, other *exact* boundary methods are much more expensive ($O(N^4)$ or worse).

For the general relativity case, an added complication in the Cauchy-characteristic method is the fact that a coordinate transformation is required for the terms or quantities which are exchanged between the Cauchy and the characteristic frames. (A full presentation of the Cauchy-characteristic method is given by Winicour and collaborators [61–63].) This appears to increase sensitivity to discretization error and makes the match difficult to achieve at the currently accessible resolution. The best present demonstration of Cauchy-characteristic matching for the black hole case has achieved evolutions to only ~20 M for an isolated black hole. This area is under aggressive development.

10.13 Horizon locations

The essential point of black hole excision is to track the location of the horizons.

The apparent horizon is defined (within a particular $t = $ constant slice) as

$$D_i n_2^i + K^{ij} n_i n_j - K = 0. \tag{10.70}$$

This is an equation on the *unit* normal n^i to a closed 2-surface. (K^{ij} is the extrinsic curvature of the 3D space in the 4D spacetime and $K = K_i^i$.) By supposing that the surface is given by $f(r, \theta, \phi) = 0$ where we can write

$$f = r - h(\theta, \phi) = 0 \tag{10.71}$$

we have a way of parametrizing this equation.

Then

$$n_i = f_{,i} / \sqrt{f_{,i} g^{ij} f_{,j}} \tag{10.72}$$

and equation (10.70) becomes a very non-linear equation for f.

Three methods have been developed and implemented within the Alliance to solve this equation.

The first solves the resulting non-linear differential equation directly for $h(\theta, \phi)$. This is done on a grid of comparable resolution to that of the three-dimensional Cauchy volume, for instance a 33^2 discretization for a simulation in a 33^3 volume [48].

The second method uses a decomposition into orthogonal harmonics (e.g. spherical harmonics), and expands the function $h(\theta, \phi)$ in terms of these harmonics. The basic equation (10.70) defining the expansion is squared and the residual (the difference from zero) is integrated over the unit sphere. Coefficients in the expansion of $h(\theta, \phi)$ are then adjusted to minimize this integral [49–51].

The third method selects a trial surface and, using the sign of the expansion, flows points along the normal to reduce the value of the computed 4-expansion [52–54].

Although each of these methods has been implemented, none has yet been completely integrated into the Cauchy evolution module. This integration is currently being carried out.

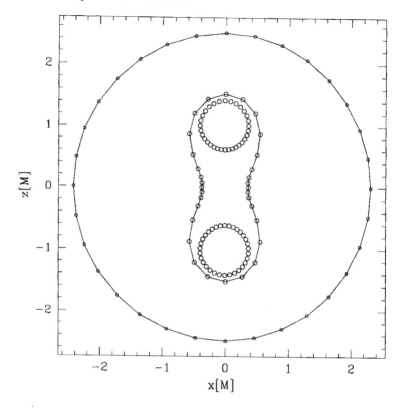

Figure 10.13.

One feature of an apparent horizon tracker must be the ability to locate individual black holes when they are well separated. In computational work we want to locate the outermost apparent horizon, because the inner boundary placement there will be the most economical for computation of the externally propagated signals. Hence a search mechanism must search from 'outside' the expected radius of an outer horizon. If the situation actually has no enveloping outer horizon, but only two distinct horizons, one for each black hole, then the horizon location must split the sphere into two, and locate the two horizons individually. Figure 10.13 shows this process being carried out by one of the Alliance solvers [54] at three different stages in the solve.

This will become even more important for the black hole merger problem, where individual black hole apparent horizons are swallowed by an enveloping apparent horizon. See figure 10.14, which gives early and late apparent configurations corresponding to the schematic black hole collision and merger shown in figure 10.15.

INITIALLY:

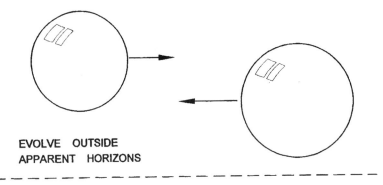

EVOLVE   OUTSIDE
APPARENT   HORIZONS

- -

LATER:

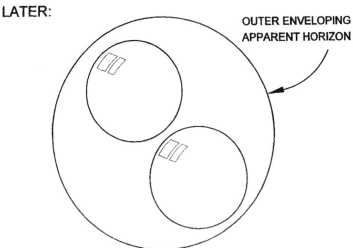

OUTER ENVELOPING
APPARENT HORIZON

Figure 10.14.

10.14   Moving black holes

The moving black hole is best posed in a $3 + 1$ form based on the (ingoing) Kerr–Schild system because the horizon is well described in that coordinate system, and moving black hole data can be easily obtained by boosting based on the Minkowski background. It is difficult to present the data visually because of the tensor nature of the variables. However, one can reasonably display the coordinate system describing a moving black hole.

Figure 10.16 represents the shift vector (line segments) and the lapse (cardioid surfaces) for a Schwarzschild black hole moving upward at speed $v = 0.5\,c$. The solid ellipse is the horizon surface, flattened in these coordinates due to Lorentz

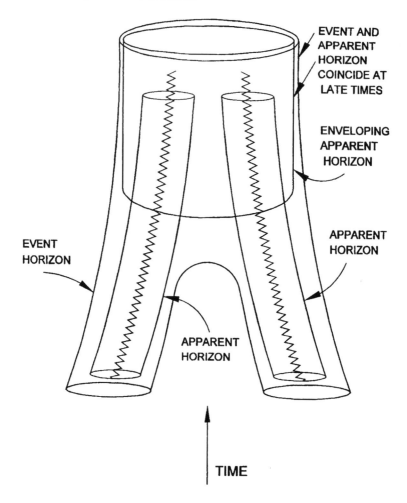

Figure 10.15.

contraction. The shift points everywhere away from the hole. This reflects the fact that the shift moves coordinate labels outward from the horizon to compensate for the fact that they are falling in toward the hole. For this moving hole the shift is smaller ahead of the hole (allowing coordinate points to fall inside the hole) and larger behind the hole where points emerge from the hole.

10.15   Binary black hole spacetimes

One of the Alliance theoretical developments has been a scheme to produce Kerr–Schild-like data for binary black hole spacetimes. The algorithm is a variant

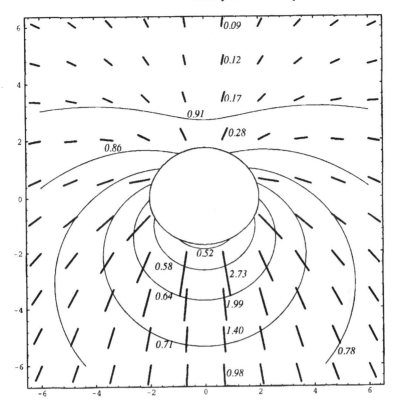

Figure 10.16.

of the conformal structure of York, except that one takes

$$\widehat{g}_{ij} = \delta_{ij} + {}_1H(r_1)_1l_{i\,1}l_j + {}_2H(r_2)_2l_{i\,2}l_j \tag{10.73}$$

as a conformal metric and

$$_0\widehat{K}^i{}_j = {}_1K^i{}_j + {}_2K^i{}_j \tag{10.74}$$

which defines $_0\widehat{E}^a{}_b = {}_0\widehat{K}^a{}_b - \frac{1}{3}\delta^a{}_bK$ as trial values. K is considered a given scalar and is not scaled. The 'hatted' values are taken as background values and one solves the coupled Hamiltonian constraint (for the ϕ, the conformal factor) and momentum constraint for a vector potential w^i.

The momentum constraint is solved in the standard way, working in the physical frame, related by the transformation

$$_0E^{ab} = \phi^{-10}\widehat{E}^{ab}. \tag{10.75}$$

We define

$$A^{cb} \equiv {}_0E^{cb} + (lw)^{cb} \tag{10.76}$$

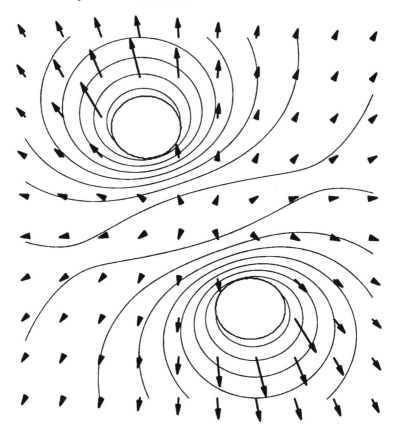

Figure 10.17.

where A^{cb} is the physical traceless extrinsic curvature, w^a is a vector to be solved for, and

$$(lw)^{cb} = D^c w^b + D^b w^c - \tfrac{1}{3} g^{bc} D_d w^d. \tag{10.77}$$

Thus the momentum constraint

$$D_b A^{cb} - \tfrac{2}{3} D^c K = 0 \tag{10.78}$$

becomes an elliptic equation for w^a. The Hamiltonian and momentum constraints are coupled because the conformal factor ϕ appears in each. One solves this coupled pair for ϕ and for w^a. This is a set of four coupled non-linear equations. Both finite element and multigrid solvers are being applied to this problem. Some recent results have shown the ability to solve this system with simple boundary conditions for ϕ and for w^a. The coordinates, at least for the initial configuration, can be simply adapted from the single black hole data. Figure 10.17 shows the

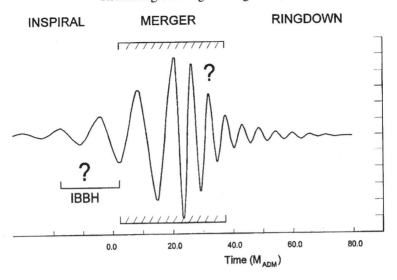

Figure 10.18.

lapse and shift assigned for interacting equal mass black holes in a non-head-on encounter. For instance, we have begun to study non-head-on binary black hole evolutions laid out with pairs of moving Kerr black holes. The evolution has proceeded only for about 4 M in time, where M is the mass of one of the black holes. We are currently debugging the implementation for this run to allow significantly longer runs [54]. We will then take up simulations involving black holes with astrophysically relevant initial momentum. An orbit of a test mass in Schwarzschild at $\sim$6 M separation has speed $\sim$0.4 c, and one orbit will require $\sim$100 M. This is consistent with the length of time that large Cauchy evolutions have already been carried out using the Alliance Cauchy module, which is one indication that we are ready to begin testing *physical* simulations.

10.16   Remaining challenges: integration, waveforms

The Alliance has developed its codes with a goal, which is to produce a flexible, general, simulation model. As I have outlined, this has required theoretical, infrastructural and computational efforts. We have functioning prototypes of each of the required components for the simulation of black hole mergers, though there are still rough edges that need to be smoothed in many cases (long-term stability for the Cauchy module, for instance). We have the means to set data and to set coordinate conditions for multiple black holes. The goal is to compute the strong-field merger epoch of gravitational radiation generation. The inspiral, and the long intermediate binary black hole epochs must be handled by other methods, or extensions of these methods. The ringdown can be described by perturbation theory, but the merger demands a full computation. See figure 10.18.

The task now is integration of the pieces we have developed. Early results for the Cauchy module have used, and will continue for a while to use, the blending boundary conditions. Horizon tracking is necessary and is being enabled in the Cauchy module. Matching for data extraction has been demonstrated via the gauge-invariant perturbative approach. The characteristic matching has also been demonstrated for brief evolution intervals. Hence we are poised to carry out the first binary-hole computations with wave extraction with this code. The demand for generality in our coding has meant a long development time for each of the components, but we expect the binary-hole performance of this code to rapidly come up to comparability with other codes, and hopefully expect improvements beyond that.

Acknowledgment

This work was supported by NSF grants ASC/PHY9318152 (the Binary Black Hole Grand Challenge), PHY93-10083, PHY9800722 and PHY9800725.

References

[1] Arnowitt R, Deser S and Misner C W 1962 *Gravitation, an Introduction to Current Research* ed L Witten (New York: Wiley)

[2] Misner C, Thorne K and Wheeler J 1973 *Gravitation* (San Francisco, CA: Freeman)

[3] Bishop N T, Gomez R, Lehner L, Maharaj M and Winicour J 1997 *Phys. Rev. D* **56** 6298

[4] Friedrich H 1996 *Class. Quantum Grav.* **13** 1451–69

[5] Friedrich H 1995 *J. Geom. Phys.* **17** 125–84

[6] Friedrich H 1998 Einstein's equation and conformal structure *Albert-Einstein Institüt Preprint* 019

[7] Friedrich H 1998 Gravitational fields near space-like and null infinity *Albert-Einstein Institüt Preprint* 022

[8] Bondi H, van der Burg M J G and Metzner A W K 1962 *Proc. R. Soc.* A **269** 21

[9] Sachs R K 1962 *Proc. R. Soc.* A **270** 103

[10] Penrose R 1963 *Phys. Rev. Lett.* **10** 66

[11] Isaacson R A, Welling J S and Winicour J 1983 *J. Math. Phys.* **24** 1824

[12] Winicour J 1983 *J. Math. Phys.* **24** 1193

[13] Winicour J 1984 *J. Math. Phys.* **25** 2506

[14] Friedrich H and Stewart J M 1983 *Proc. R. Soc.* A **385** 345

[15] Hübner P 1995 *Class. Quantum Grav.* **12** 791

[16] Hübner P 1996 *Phys. Rev.* **53** 701

[17] Bona C and Masso J 1992 *Phys. Rev. Lett.* **68** 1097

[18] Bona C, Stela J, Masso J and Seidel E 1995 A class of hyperbolic gauge conditions *General Relativity (MG7 Proc.)* ed R Ruffini and M Keiser (Singapore: World Scientific)

[19] Lichnerowicz A 1939 *Problèmes Globaux en Mécanique Relativiste* (Paris: Hermann)

[20] Choquet (Foures)-Bruhat Y 1956 *J. Rat. Mech. Anal.* **5** 951

[21] Choquet-Bruhat Y, York J W and Anderson A 1998 Curvature-based hyperbolic systems for general relativity gr-qc/9802027

[22] Choquet-Bruhat Y and York J W 1995 *C. R. Acad. Sci., Paris* **321** 1089

[23] Anderson A, Choquet-Bruhat Y and York J W 1996 *C. R. Acad. Sci., Paris* **323** 835

[24] Abrahams A, Anderson A, Choquet-Bruhat Y and York J W 1995 *Phys. Rev. Lett.* **75** 3377

[25] Miller W A 1997 *Class. Quantum Grav.* **14** L199–L204

[26] Regge T 1961 *Nuovo Cimento* **19** 558–71

[27] Thornburg J 1985 *Masters Thesis* University of British Columbia

[28] Thornburg J 1993 *PhD Dissertation* University of British Columbia

[29] Unruh W 1985 private communication to M Choptuik

[30] Seidel E and Suen W-M 1992 *Phys. Rev. Lett.* **69** 1845–8

[31] Anninos P, Daues G, Masso J, Seidel E and Suen W-M 1995 *Phys. Rev.* D **51** 5562–78

[32] Arnold D N, Mukherjee (Rutgers) A and Pouly L 1997 Adaptive finite elements and colliding black holes *Numerical Analysis 1997: Proc. 17th Bienn. Conf. on Numerical Analysis*

[33] Daues G 1996 *PhD Dissertation* Washington University

[34] Burnett S 1987 *Finite Element Analysis: from Concepts to Applications* (Reading, MA: Addison-Wesley)

[35] Richardson L F 1910 *Phil. Trans. R. Soc.* **210** 307

[36] Scheel M A, Shapiro S L and Teukolsky S A 1995 *Phys. Rev.* D **51** 4208

[37] Scheel M A, Shapiro S L and Teukolsky S A 1995 *Phys. Rev.* D **51** 4236

[38] Gomez R, Lehner L, Marsa R L and Winicour J 1998 *Phys. Rev.* D **57** 4778

[39] Marsa R 1994 *PhD Dissertation* University of Texas at Austin

[40] Marsa R L and Choptuik M W 1996 *Phys. Rev.* D **54** 4929–43

[41] The Binary Black Hole Grand Challenge Alliance (43 authors including R A Matzner) 1997 *Phys. Rev. Lett.* submitted

[42] Correll R 1998 *PhD Dissertation* The University of Texas at Austin

[43] Frittelli S 1997 *Phys. Rev.* D **55** 5992–6

[44] Choptuik M 1991 *Phys. Rev.* D **44** 3124

[45] Rezzolla L, Abrahams A M, Matzner R A, Rupright M E and Shapiro S L 1998 *Phys. Rev.* D **58** 3124

[46] Rupright M E, Abrahams A M and Rezzolla L 1998 *Phys. Rev.* D **58** 3124

[47] The Binary Black Hole Grand Challenge Alliance: A M Abrahams, L Rezzolla, M E Rupright, A Anderson, P Anninos, T W Baumgarte, N T Bishop, S R Brandt, J C Browne, K Camarda, M W Choptuik, G B Cook, C R Evans, L S Finn, G Fox, R Gomez, T Haupt, M F Huq, L E Kidder, S Klasky, P Laguna, W Landry, L Lehner, J Lenaghan, R L Marsa, J Masso, R A Matzner, S Mitra, P Papadopoulos, M Parashar, F Saied, P E Saylor, M A Scheel, E Seidel, S L Shapiro, D Shoemaker, L Smarr, B Szilagyi, S A Teukolsky, M H P M van Putten, P Walker, J Winicour and J W York Jr 1998 *Phys. Rev. Lett.* **80** 1812–15

[48] Huq M 1996 *PhD Dissertation* The University of Texas at Austin

[49] Anninos P, Bernstein D, Brandt S, Libson J, Masso J, Seidel E, Smarr L, Suen W-M and Walker P 1995 *Phys. Rev. Lett.* **74** 630

[50] Nakamura T, Kojima Y and Oohara K 1984 *Phys. Lett.* A **106** 235

[51] Oohara K, Nakamura T and Kojima Y 1984 *Phys. Lett.* A **107** 455

[52] Tod K P 1991 *Class. Quantum Grav.* **8** L115

[53] Huq M F, Klasky S A, Choptuik M W and Matzner R A 1998 *Phys. Rev.* to be submitted

[54] Shoemaker D *PhD Dissertation* (in progress) The University of Texas at Austin

[55] Choptuik M 1993 *Phys. Rev. Lett.* **70** 9

[56] Frittelli S and Lehner L 1998 *Phys. Rev.* submitted

[57] Frittelli S and Reula O 1996 *Phys. Rev. Lett.* **76** 4667–75

[58] Anderson A and York J Jr 1998 *Phys. Rev. Lett.* **81** 1154–7

[59] Brodbeck O, Frittelli S, Huebner P and Reula O *J. Mod. Phys.* submitted

[60] Choquet–Bruhat Y, York J Jr and Anderson A 1998 *Proc. 8th Marcel Grossmann Meeting on General Relativity* at press

[61] Gomez R, Marsa R L and Winicour J 1997 *Phys. Rev.* D **56** 6310

[62] Bishop N, Gomez R, Lehner L, Szilagyi B, Winicour J and Isaacson R 1996 *Phys. Rev. Lett.* **76** 4303

[63] Winicour J 1998 Characteristic evolution and matching *Living Reviews in Relativity* (Potsdam: Max-Planck-Gessellschaft, Max-Planck-Institüt für GravitationsPhysik)
http://www.livingreviews.org/Articles/Volume1/1998-5winicour/

Index